The Complete Laboratory Manual for Electricity

The Complete Laboratory Manual for Electricity
Fourth Edition

Stephen L. Herman

Southern Technical College

CENGAGE
Learning

Australia • Brazil • Japan • Korea • Mexico • Singapore • Spain • United Kingdom • United States

The Complete Lab Manual for Electricity, Fourth Edition
Stephen L. Herman

Senior Vice President, GM Skills & Product Planning: Dawn Gerrain

Product Team Manager: James DeVoe

Senior Director Development: Marah Bellegarde

Senior Product Development Manager: Larry Main

Senior Content Developer: John Fisher

Product Assistant: Andrew Ouimet

Vice President, Marketing Services: Jennifer Baker

Senior Production Director: Wendy A. Troeger

Production Director: Andrew Crouth

Senior Content Project Manager: Kara A. DiCaterino

Senior Art Director: Bethany Casey

Technology Project Manager: Joe Pliss

Cover Images:
©TomasMikula/Shutterstock.com
©Zffoto/Shutterstock.com

For product information and technology assistance, contact us at
Cengage Learning Customer & Sales Support, 1-800-354-9706

For permission to use material from this text or product,
submit all requests online at **www.cengage.com/permissions**.
Further permissions questions can be e-mailed to
permissionrequest@cengage.com

Library of Congress Control Number: 2014937118

ISBN: 978-1-133-67382-8

Cengage Learning
20 Channel Center Street
Boston, MA 02210
USA

Cengage Learning is a leading provider of customized learning solutions with office locations around the globe, including Singapore, the United Kingdom, Australia, Mexico, Brazil, and Japan. Locate your local office at:
www.cengage.com/global

Cengage Learning products are represented in Canada by Nelson Education, Ltd.

To learn more about Cengage Learning, visit **www.cengage.com**

Purchase any of our products at your local college store or at our preferred online store **www.cengagebrain.com**

Notice to the Reader

Publisher does not warrant or guarantee any of the products described herein or perform any independent analysis in connection with any of the product information contained herein. Publisher does not assume, and expressly disclaims, any obligation to obtain and include information other than that provided to it by the manufacturer. The reader is expressly warned to consider and adopt all safety precautions that might be indicated by the activities described herein and to avoid all potential hazards. By following the instructions contained herein, the reader willingly assumes all risks in connection with such instructions. The publisher makes no representations or warranties of any kind, including but not limited to, the warranties of fitness for particular purpose or merchantability, nor are any such representations implied with respect to the material set forth herein, and the publisher takes no responsibility with respect to such material. The publisher shall not be liable for any special, consequential, or exemplary damages resulting, in whole or part, from the readers' use of, or reliance upon, this material.

Printed in the United States of America
Print Number: 02   Print Year: 2016

Contents

Preface

The Complete Laboratory Manual for Electricity, fourth edition, contains hands-on experiments that range from basic electricity through motor control circuits. The components used in the experiments in this manual are readily obtainable from a variety of sources and vendors. The manual assumes that the laboratory has access to 120 volts AC and 208 volts three-phase. Although the manual is written with the assumption of a 208-volt power source, most of the experiments can be performed with a 240-volt power supply. A material list is provided that lists the components necessary to perform all laboratory exercises. A suggested list of vendors is given in the Material List section.

Each unit begins with an explanation of the circuit to be connected in the laboratory. Examples of the calculations necessary to complete the exercise are given in an easy-to-follow, step-by-step procedure. If the power source is 240 volts instead of 208 volts, the student should simply substitute a value of 240 for 208 when doing calculations during the experiment. Students are expected to calculate electrical values and then connect a circuit to make measurements of electrical values.

The Complete Laboratory Manual for Electricity, fourth edition, provides the student with hands-on experience in constructing a multitude of circuits such as series, parallel, combination, RL series and parallel, RC series and parallel, and RLC series and parallel. Section 2 of this manual provides instruction on the basic types of switches that electricians must install whether working in a residential, commercial, or industrial application. Section 3 contains exercises in the basic types of alternating current loads such as resistive, inductive, and capacitive. Section 4 provides experiments with both single-phase and three-phase transformers.

Section 5 provides the student with hands-on experience connecting motor control circuits. This section begins with a simple start-stop push-button circuit and progresses to control circuit design.

The Complete Laboratory Manual for Electricity, fourth edition, is a must-have text for any curriculum dedicated to training electricians to work in a construction or industrial environment. Basic electricity, AC theory, transformers, and motor controls—this text has it all.

New for the Fourth Edition

- Updated graphics.
- The fourth edition revision is the most comprehensive update since the text was first published. In previous editions, incandescent lamps were used as resistive loads because of their availability and cost. In the fourth edition, incandescent lamps have been replaced with fixed high-wattage resistors. The main reason for the change is that Congress has decided to phase out the availability of incandescent lamps over the next few years. Although fixed resistors are more expensive than incandescent lamps, they have the advantage that their resistance value is basically constant over a wide range of temperatures. This permits the student to make Ohm's law calculations before power is applied to the circuit and then make measurements to verify their calculations. The laboratory procedures for many of the units that are associated with Ohm's law have been rewritten to permit the students to make calculation and then measure the results.

Instructor Site

An Instructor Companion website containing supplementary material is available. This site contains an Instructor Guide and an image gallery of text figures.

Contact Cengage Learning or your local sales representative to obtain an instructor account.

Accessing an Instructor

Companion Website from

SSO Front Door

1. Go to http://login.cengage.com and login using the instructor e-mail address and password.
2. Enter author, title, or ISBN in the **Add a title to your bookshelf** search.
3. Click **Add to my bookshelf** to add instructor resources.
4. At the Product page, click the **Instructor Companion site** link.

Acknowledgments

The author and Cengage Learning wish to acknowledge and thank the members of the review panel for their suggestions and comments during development of this edition. Thanks go to:

Mel Elliston
Texas State Technical College
Marshall, TX

Cris Folk
Madison Area Technical College
Watertown, WI 53098

Matt Hanson
Sandburg Community College
Galesburg, IL

Sheila Horan
New Mexico State University
Las Cruces, NM

Marvin Moak
Hinds Community College
Raymond, MS

Gwen Oster
Northwest Technical College
Bemidji, MN

Chris Pittman
Wilson Technical College
Wilson, NC

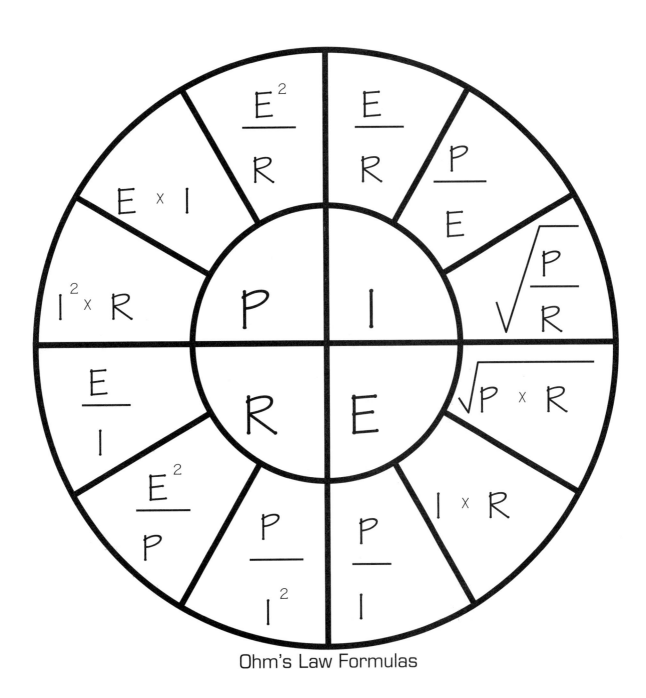

Ohm's Law Formulas

Material List

Quantity	Description
2	250-ohm 60 watt (minimum) resistors
6	150-ohm 100 watt (minimum) resistors
2	100-ohm 144 watt (minimum) resistors
10	Color-coded resistors with different values
3	0.5-kVA control transformers (two windings rated at 240 volts each and one winding rated at 120 volts)
1	7.5-μf AC capacitor with a voltage rating not less than 240 VAC
1	10-μf AC capacitor with a voltage rating not less than 240 VAC
1	25-μf AC capacitor with a voltage rating not less than 240 VAC
1	9-lead dual-voltage, three-phase motor (any horsepower)

CAUTION: The wattage rating given for the resistors is a minimum value. Resistors with a higher wattage rating can be used without a problem. The wattage rating indicates the amount of heat that the resistor can dissipate. It should also be noted that the resistors will often become very hot during experiments and they should not be place near objects that can burn or will be damaged by heat. Take care when handling high wattage components because they can cause severe burns.

2 ea.	3-way switches
1 ea.	4-way switch
	Two-conductor romex wire (number of feet is determined by individual laboratory conditions)
	Three-conductor romex wire (number of feet is determined by individual laboratory conditions)
1 ea.	Octagon metal box or PVC light fixture box
3 ea.	Metal or PVC switch boxes

Motor Controls

1 ea.	Control transformer to step your laboratory line voltage down to 120 VAC
3 ea.	Three-phase motor starter that contains at least 2 normally open and 1 normally closed auxiliary contact
3 ea.	Three-phase contactors (no overload relays) containing one normally open and one normally closed auxiliary contact
3 ea.	Three-phase motors 1/3 to 1/4 hp or simulated motor loads. (*Note*: Assuming a 208-volt three-phase 4-wire system, a simulated motor load can be constructed by connecting three lamp sockets to form a wye connection. These lamps will have a voltage drop of 120 volts each. If a 240-volt three-phase

system in is use, it may be necessary to connect two lamps in series for each phase. If two lamps are connected in series for each phase, these three sets of series lamps can then be connected wye or delta.)

1 ea.	Three-phase overload relay or three single-phase overload relays with the overload contacts connected in series
1 ea.	Reversing starter, or 2 three-phase contactors that contain 1 normally open and 1 normally closed contact, and 1 three-phase overload relay
4 ea.	Double-acting push-buttons
6 ea.	3-way toggle switches to simulate float switches, limit switches, etc.
4 ea.	Electronic timers (Dayton model 6A855 recommended)
3 ea.	11-pin control relays (120-volt coil)
3 ea.	8-pin control relays (120-volt coil)
4 ea.	11-pin tube sockets
3 ea.	8-pin tube sockets
3 ea.	Pilot light indicators
1 ea.	Three-phase power supply (This laboratory manual assumes the use of a 208-volt three-phase system. If an equivalent motor load is employed, the design may have to be modified to compensate for a higher voltage.)

Suppliers

Most of the parts listed can be obtained from Grainger Industrial Supply (www.grainger .com). The Dayton model 6A855 timer is recommended because of its availability and price. Also, it is a multifunction timer and can be used as both an on- and off-delay device. It will also work as a one-shot timer and a pulse timer. Although the Dayton timer is recommended, any 11-pin electronic timer with the same pin configuration can be used. One such timer is available from Magnecraft (model TDR SRXP-120, www.magnecraft.com). This timer as also available from Mouser Electronics (www.mouser.com). Other electronic timers can be employed, but if they have different pin configurations, the wiring connections shown in the text will have to be modified to accommodate the different timer. Mouser Electronics is also a supplier for the wire-wound resistors.

The 8- and 11-pin control relays and sockets can be purchased from Grainger, Mouser Electronics, or Newark Electronics (www.newark.com). The control transformer used in the controls sections can be purchased from Mouser Electronics or Sola/Hevi-duty (www.solaheviduty.com). Model E250JN is recommended because it has primary taps of 208/240/277 volts. The secondary winding is 120/24. It is also recommended that any control transformer used be fuse protected. Another control transformer that can be used is available from Grainger. It is rated at 150 VA and has a 208-volt primary and 120-volt secondary. The 0.5-kVA control transformers are available from Grainger or Newark Electronics. Transformers rated at 0.5 kVA are used because they permit the circuit to be load heavy enough to permit the use of clamp-on type ammeters.

Stackable banana plugs are available from both Grainger and Newark Electronics. The oil-filled capacitors listed are available from Grainger. Color-coded resistors can be obtained from Newark Electronics or Mouser Electronics.

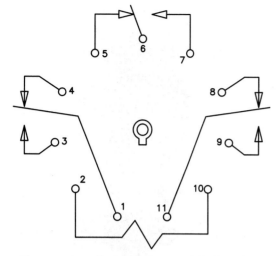

Connection diagram for an 11-pin relay.

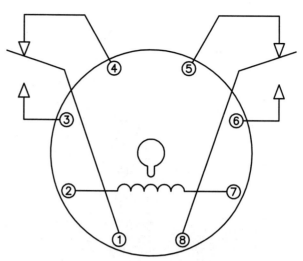

Connection diagram for an 8-pin relay.

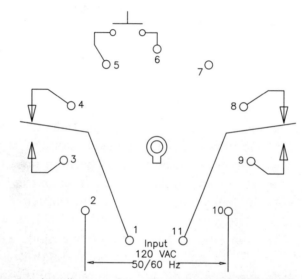

Connection diagram for a Dayton model 6A855 timer.

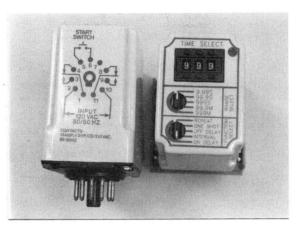

Dayton timer model 6A855. This timer mounts in an 11-pin tube socket and can be set to operate as a repeat timer, a one-shot timer, an interval timer, and an on-delay timer. The thumb-wheel switch sets the time value. Full range times can be set for 9.99 seconds to 999 minutes.

Three-phase contactor. This contactor contains one normally open auxiliary contact and three load contacts. The contactor differs from the motor starter in that the contactor does not contain an overload relay.

Control relays. These relays contain auxiliary contacts only and are intended to be used as part of the control circuit. They are capable of controlling low-current loads such as solenoid valves, pilot lights, and the like.

Eight-pin on-delay timing relay. This timer can be used as an on-delay timer only. Time setting is adjusted by the knob on top of the timer.

One single-phase and one three-phase overload relay. The three-phase overload relay contains three heaters but only one set of normally closed auxiliary contacts. If an overload should occur on any of the three lines, the contacts will open.

Three-phase motor starter with two normally open and one normally closed set of auxiliary contacts. Notice that a motor starter contains an overload relay as part of the unit.

Light sockets mounted on a metal plate. The sockets have been connected to form a wye connection. This can be used to simulate a three-phase motor load.

Eight- and 11-pin tube sockets. All wiring is done to the socket and the relay is then plugged into the socket.

Pneumatic off-delay timer. A microswitch has been added to the bottom to supply instantaneous contacts for the timer.

Three push-button station. The bottom push button is normally closed. The two top buttons are double action. Each contains both a normally open and normally closed set of contacts.

Eight- and 11-pin control relays contain two sets of double-acting contacts. Eleven-pin relays contain three sets of double-acting contacts.

Safety

Objectives

After studying this section, you should be able to:

- Discuss basic safety rules.
- Describe the effects of electric current on the body.

The purpose of this laboratory manual is to provide students of electricity with hands-on experience with electric circuits. Electricity is an extremely powerful force and should never be treated in a careless manner. This manual assumes a laboratory equipped with a 208/120 volt three-phase power system. Many of the experiments in this manual involve the use of full line voltage (208 or 120 volts). It is extremely important that you practice safety at all times. Please read and memorize the following safety rules:

- Never work on an energized circuit if it is possible to disconnect the power. When possible use a three-step check to make certain that the power is turned off. The three-step check is as follows:
  1. Test the meter on a known live circuit to make sure the meter is operating.
  2. Test the circuit that is to be de-energized with the meter.
  3. Test the meter on the known live circuit again to make certain that the meter is still operating.
- Install a warning tag at the point of disconnection to warn people not to restore power to the circuit.

General Safety Rules

Think

Of all the rules concerning safety, this one is probably the most important. No amount of safeguarding or "idiot proofing" a piece of equipment can protect a person as well as the person's taking time to think before acting. Many technicians have been killed by supposedly "dead" circuits. Do not depend on circuit breakers, fuses, or someone else to open a circuit. Test it yourself before you touch it. If you are working on high-voltage equipment, use insulated gloves and meter probes designed to be used on the voltage being tested. Your life is your own, so *think* before you touch something that can take it away.

Avoid Horseplay

Jokes and horseplay have a time and place, but the time or place is not when someone is working on an electric circuit or a piece of moving machinery. Do not be the cause of someone's being injured or killed and do not let someone else be the cause of your being injured or killed.

Do Not Work Alone

This is especially applicable when working in a hazardous location or on a live circuit. Have someone with you to turn off the power or give artificial respiration and/or cardiopulmonary resuscitation (CPR). One of the effects of severe electrical shock is that it causes breathing difficulties and can cause the heart to go into fibrillation.

Work with One Hand When Possible

The worst case of electrical shock is when the current path is from one hand to the other. This causes the current to pass directly through the heart. A person can survive a severe shock between the hand and one foot that would otherwise cause death if the current path was from one hand to the other. Working with one hand can sometimes be an unsafe practice by itself. The best procedure is to turn off the power. If it is not possible to disconnect the power, wear insulated gloves when handling "hot" circuits. Also wear shoes that have insulated soles and use rubber mats to cover energized conductors and components when possible.

Learn First Aid

Anyone working on electrical equipment should make an effort to learn first aid. This is especially true for anyone who must work with voltages above 50 volts. A knowledge of first aid, especially CPR, may save your life or someone else's.

Effects of Electric Current on the Body

Most people have heard that it's not the voltage that kills but the current. Although this is a true statement, do not be misled into thinking voltage cannot harm you. Voltage is the force that pushes the current through the circuit. Voltage can be compared to the pressure that pushes water through a pipe. The more pressure available, the greater the volume of water flowing through a pipe. Students often ask how much current will flow through the body at a particular voltage. There is no easy answer to this question. The amount of current that can flow at a particular voltage is determined by the resistance of the current path. Different people have different resistances. A body will have less resistance on a hot day when sweating because salt water is a very good conductor. What a person ate and drank for lunch can have an effect on a body's resistance. The length of the current path can affect the resistance. Is the current path between two hands or from one hand to one foot? All of these factors affect body resistance.

The chart in Figure SF-1 illustrates the effects of different amounts of current on the body. This chart is general; electricity affects most people in this way. Some people may have less tolerance to electricity and others may have a greater tolerance.

A current of 2 to 3 milliamperes will generally cause a slight tingling sensation. The tingling sensation will increase as current increases and becomes very noticeable at about 10 milliamperes. The tingling sensation is very painful at about 20 milliamperes. Currents between 20 and 30 milliamperes generally cause a person to seize the line and not be able to let go of the circuit. Currents between 30 and 40 milliamperes cause muscular paralysis, and currents between 40 and 60 milliamperes cause breathing difficulty. By the time the current increases to about 100 milliamperes breathing is extremely difficult. Currents from 100 to 200 milliamperes generally cause death because the heart goes into fibrillation. Fibrillation is a condition in which the heart begins to "quiver" and the pumping action stops. Currents above 200 milliamperes generally cause the heart to squeeze shut. When the current is removed, the heart will generally return to a normal pumping action. This is the principle of operation of a defibrillator. It is often said that 120 volts is the most

0.002–0.003 amp	Sensation (a slight tingling)
0.004–0.010 amp	Moderate sensation
0.010–0.020 amp	Very painful
0.020–0.030 amp	Unable to let go of the circuit
0.030–0.040 amp	Muscular paralysis
0.040–0.060 amp	Breathing difficulty
0.060–0.100 amp	Extreme breathing difficulty
0.100–0.200 amp	Death (fibrillation of the heart)

Figure SF-1 Effects of electric current on the body.

dangerous voltage to work with. The reason is that 120 volts generally causes a current flow between 100 and 200 milliamperes through the bodies of most people. Large amounts of current can cause severe electrical burns. Electrical burns are generally very serious because the burn occurs on the inside of the body. The exterior of the body may not look seriously burned, but the inside may be severely burned.

Review Questions

1. What is the most important rule of electrical safety?

   _____

   _____

   _____

2. Why should a person work with only one hand when possible?

   _____

   _____

   _____

3. What range of electric current generally causes death?

   _____

   _____

   _____

4. What is fibrillation of the heart?

   _____

   _____

   _____

5. What is the principle of operation of a defibrillator?

   _____

   _____

   _____

Measuring Instruments and Basic Electricity

Unit 1 Measuring Instruments

Objectives

After studying this unit, you should be able to

- Discuss the operation of a voltmeter.
- Describe differences between analog and digital voltmeters.
- Connect a voltmeter in a circuit.
- Discuss the operation of an ohmmeter.
- Measure resistance with an ohmmeter.
- Discuss differences between analog and digital ohmmeters.
- Discuss the operation of an ammeter.
- Describe the difference between in-line and clamp-on ammeters.
- Make a scale divider for clamp-on AC ammeters.

In order to conduct meaningful laboratory experiments, it is necessary to measure electrical quantities such as voltage, resistance, and current. In this unit basic measuring instruments such as voltmeters, ohmmeters, and ammeters will be presented. Their basic operation will be explained and examples of their use will be presented. It should be understood by those using electrical measuring instruments that they are not 100% accurate. Most meters have an accuracy of about +3%. This is the reason that two meters made by the same manufacturer and the same model may give different measurements. One meter may measure a voltage of 120 volts as 119.2 and the other measure the same source as 120.3. When making measurements with meters, always be aware that there may be some discrepancy between calculated and measured values. If the calculated and measured values are within about 3% of each other, they are generally considered to be the same.

Voltmeters

Voltmeters are one of the most useful instruments in the electrical field. They are used to measure the potential difference (voltage) between two points. Voltage is often thought of as "electrical pressure." It is like measuring the amount of pressure available to push liquid through a pipe. Voltmeters can be divided into two major types: analog and digital.

Analog Meters

Analog meters contain a pointer and scale. Most of them are part of a multimeter, which has the ability to measure several different electrical quantities (Figure 1-1). Multimeters are often called VOMs, which stands for volt-ohm-milliammeter. Since they are often part of a multimeter, it is generally necessary to choose the electrical quantity being measured. Most multimeters have the ability to measure both direct current (DC) and alternating current (AC). The meter shown in Figure 1-1, for example, can measure DC voltage (DCV), AC voltage (ACV), ohms (Rx1 and Rx10), and DC milliamps (DCmA). The position of the dial on the front of the meter determines the electrical quantity to be measured.

Analog multimeters can also be set to measure different ranges of voltage. The meter shown in Figure 1-1 has AC voltage ranges of 10, 50, 250, 500, and 1000 volts. These markings indicate the full-scale value of the meter when the indicator dial is set at one of them. Being able to set the meter for different full-scale ranges greatly increases the usefulness of the meter. If the meter were set for a value of 1000 volts, it would be very difficult to measure the voltage of a 24 volt system. If the meter were set to the 50 volt position, however, 24 volts could be measured accurately. When a voltage measurement is to be taken and the system voltage is not known, always start on the highest voltage range setting. A meter can never be hurt by connecting it to a voltage that is lower than the range setting. It is a simple thing to reset the meter to a lower range after an initial measurement has been made.

Helpful Hint

When a voltage measurement is to be taken and the system voltage is not known, always start on the highest voltage range setting.

Figure 1-1 Multimeter with an analog scale. (Courtesy of Triplett Test Equipment and Tools)

Another rule to remember when using a voltmeter is to make certain that the meter is set for the proper type of voltage. *Failure to do this can lead to serious injury.* If a DC voltage is to be measured and the meter has mistakenly been set for AC voltage, it will cause an inaccurate reading. The indicated voltage will be less than the actual voltage being measured.

Helpful Hint

When using a voltmeter, make certain that the meter is set for the proper type of voltage. Failure to do this can lead to serious injury.

If an AC voltage is to be measured and the meter has mistakenly been set to measure to DC voltage, the meter will indicate zero. Most analog-type meter movements are designed to operate on direct current. If an alternating current is applied to them, the pointer will try to move up scale during one-half cycle and down scale during the other. As a result, the pointer remains at zero. When the meter is set on the AC volt range, a rectifier inside the meter converts the AC voltage into DC, permitting the meter to operate.

Helpful Hint

If an AC voltage is to be measured and the meter has mistakenly been set to measure to DC voltage, the meter will indicate zero.

Analog Voltmeter Operation

Analog voltmeters operate by connecting resistance in series with the meter movement. Assume, for example, that a meter movement requires 100 μA (microamperes; 0.000100 amp) at 1 volt to make the pointer move full scale. This meter movement has the ability to measure 1 volt. If more voltage is applied to it, it can be damaged or destroyed. To permit the meter to measure higher voltages, resistance is connected in series with it (Figure 1-2). Assume that a resistor is to be added that will permit the meter to measure 15 volts full scale. In a series circuit, the sum of the voltage drops must equal the applied voltage. Since the total voltage to be measured is 15 volts and the meter has a voltage drop of 1 volt, the resistor must drop 14 volts (15 − 1 = 14). Another rule for series circuits is that the current

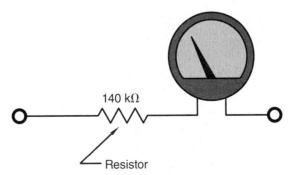

Figure 1-2 Resistance is connected in series with the meter movement to change the full-scale value.

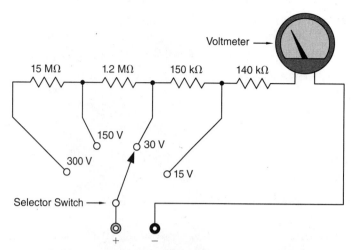

Figure 1-3 A selector switch permits the meter to be set for different full-scale values.

at any point in the circuit must be the same. Since the meter requires a current of 100 μA, the resistor must have the same current. The resistance value needed to permit the meter to measure 15 volts is

$$R = \frac{E}{I}$$

$$R = \frac{14}{0.000100}$$

$$R = 140,000 \ \Omega$$

A resistance of 140,000 Ω connected in series with the meter movement will convert the meter to a full-scale range of 15 volts.

Multirange settings are accomplished by connecting several resistors in series with the meter and permitting a range selection switch to change the amount of resistance connected in series with the meter. Assume that it is desirable to add full-scale range values of 30 volts, 150 volts, and 300 volts to the meter in Figure 1-2. The meter movement plus the 140,000 Ω resistor have a voltage drop of 15 volts at 100 μA. If the meter is to have a full-scale value of 30 volts, it will be necessary for the next resistor to have a value of 150,000 Ω (15/0.000100). The next resistor must have a voltage drop of 120 volts (150 − 30 = 120). The resistor value needed to produce a voltage drop of 120 volts at a current of 100 μA is 1,200,000 Ω (120/0.000100). The last resistor would have to have a voltage drop of 150 volts (300 − 150 = 150). The value needed for this resistor is 1,500,000 Ω (150/0.000100). A diagram of this type of multirange voltmeter is shown in Figure 1-3.

Digital Meters

Digital voltmeters display the voltage value with numeric figures instead of a pointer. Figure 1-4 shows a digital multimeter set to read AC voltage. Like analog voltmeters, digital voltmeters are generally part of a multimeter, also. It is necessary to set a selector switch to the electrical quantity to be measured, such as AC volts, DC volts, ohms, and so on, and sometimes it is necessary to set the maximum range value. Some digital meters contain an auto-ranging function making it unnecessary to select a particular range value.

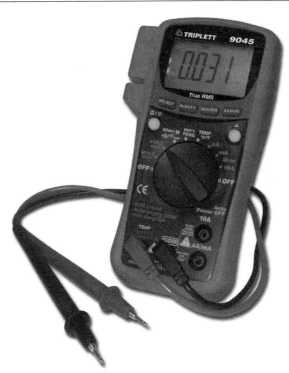

Figure 1-4 Digital multimeter. (Courtesy of Triplett Test Equipment and Tools)

Digital voltmeters are different from analog meters in more ways than appearance. As discussed previously, analog meters insert resistance in series with the meter movement to permit the meter to be used at different full-scale values. This means that the internal resistance of the meter is different for each voltage setting. Most analog meters are marked with the ohms per volt for the meter; 20,000 Ω/VDC and 10,000 Ω/VAC is typical. The resistance of the meter can be computed by multiplying the ohms per volt value by the full-scale value of the meter. Assume that a meter indicates a resistance of 10,000 Ω/VAC for a meter and that the meter is set to indicate a full-scale value of 250 VAC. The resistance of the meter is 2,500,000 Ω (10,000 × 250). Note that the resistance of the meter will change each time the voltage range is changed.

Digital meters maintain the same amount of input resistance for all ranges. Typical input resistance for a digital voltmeter is 10 million ohms. The constant input resistance is accomplished by using a voltage divider. Digital voltmeters use high-impedance electronic components to measure voltage. This means that it is only necessary to supply the circuit with extremely small amounts of current for the meter to operate. A basic construction for a digital meter is shown in Figure 1-5. Digital voltmeters can be employed to measure voltage in almost any circumstance, but they are especially necessary when measuring the voltage in low-power circuits such as electronic circuit boards.

Using the Voltmeter

Voltmeters are very high-resistance instruments. For this reason, they can be connected directly across the power source, as shown in Figure 1-6. When using a voltmeter, be certain that

1. The voltmeter is set to the proper quantity to be measured (DC or AC voltage).
2. The range setting is equal to or greater than the voltage to be measured. If there is any doubt, set the meter to the highest range and then reset the meter after taking an initial measurement.

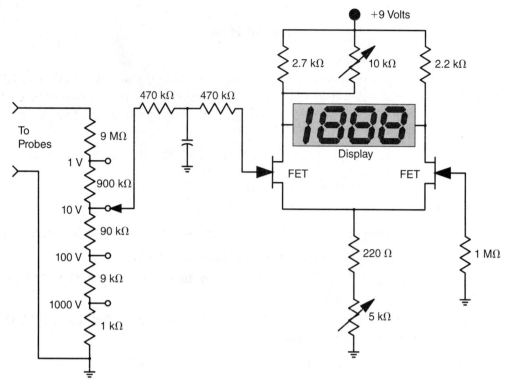

Figure 1-5 Digital voltmeter.

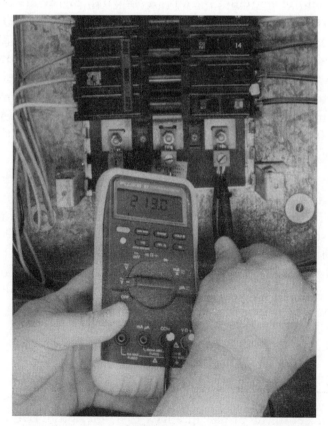

Figure 1-6 A voltmeter can be connected directly to a power source.

Helpful Hint

When using a voltmeter be certain that:
1. The voltmeter is set to the proper quantity to be measured (DC or AC voltage).
2. The range setting is equal to or greater than the voltage to be measured. If there is any doubt, set the meter to the highest range and then reset the meter after taking an initial measurement.

Safety Procedure

A set procedure is used when testing a circuit to ensure that the power is turned off.

1. Use the voltmeter to test a circuit that you know is energized.
2. Test the circuit in question with the voltmeter to make certain that the power is turned off.
3. Test the voltmeter again on a circuit that you know is energized to ensure that the voltmeter is still operating properly.

Safety is your responsibility. Your life is your own. Guard it well.

Voltmeters are also used to measure the voltage drop across components in a circuit. *Voltage drop* is the term used to describe the amount of voltage necessary to push current through a particular part of a circuit. The circuit shown in Figure 1-7 is a series circuit containing three resistors. Resistor R_1 has a resistance value of 80 Ω, R_2 has a value of 100 Ω, and R_3 has a resistance of 60 Ω. The circuit has an applied voltage of 120 volts. In a series circuit, the total resistance is the sum of the individual resistors. In this circuit the total resistance is 240 Ω (80 + 100 + 60). The total circuit current can be determined using Ohm's law (120/240 = 0.5 amp). Since the current is the same in a series circuit, 0.5 ampere

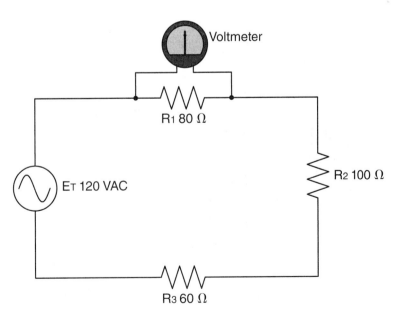

Figure 1-7   Voltmeter used to measure the voltage drop across a resistor.

flows through all resistors. Since the resistance and current are known, the amount of voltage drop across each resistor can be determined.

$$E_1 = R_1 \times I_1$$

$$E_1 = 80 \times 0.5$$

$$E_1 = 40 \text{ volts}$$

The voltmeter shown in Figure 1-7 would indicate a value of 40 volts. This voltage drop indicates that it takes 40 volts to push 0.5 ampere of current through 80 Ω of resistance. If the voltmeter were to be connected across resistor R_2, it would measure a voltage drop of 50 volts, indicating that it requires 50 volts to push 0.5 ampere through 100 Ω of resistance.

Ohmmeters

As with voltmeters, ohmmeters can be divided into two general categories: analog and digital. Analog ohmmeters use a pointer to indicate resistance value and are generally part of a multimeter. Ohmmeters must contain their own power source. They do not receive power from the circuit. The most important rule concerning the use of ohmmeters is that *ohmmeters should never be connected to a source of power!* Connecting an ohmmeter to a source of power will often result in destruction of the meter and sometimes injury to the person using the meter.

Helpful Hint

Ohmmeters should never be connected to a source of power!

A schematic for a basic analog ohmmeter is shown in Figure 1-8. It is assumed that the meter movement has a resistance of 1000 Ω and requires a current of 50 microamps (50 μA) to deflect the meter full scale. The power source will be a 3-volt battery. A fixed resistor with a value of 54 kΩ is connected in series with the meter movement, and a variable resistor with a value of 10 kΩ is connected in series with the meter and R1. These resistance values were chosen to ensure there would be enough resistance in the circuit to limit

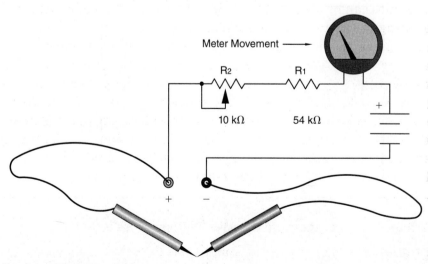

Figure 1-8 Basic analog ohmmeter.

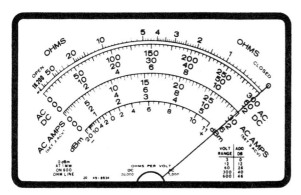

Figure 1-9 Adjusting the ohmmeter to zero.   **Figure 1-10** Reading the ohmmeter.

the current flow through the meter movement to 50 µA. If Ohm's law is used to compute the resistance needed (3 volts/0.000050 amp = 60,000 ohms), it will be seen that a value of 60 kΩ is needed. This circuit contains a total of 65,000 ohms (1,000 meter + 54,000 + 10,000). The circuit resistance can be changed to a value as low as 55,000 ohms by adjusting the variable resistor, however. The reason for this is to compensate for the voltage drop of the battery as it ages and becomes weaker.

When resistance is to be measured, the meter must first be zeroed. This is done with the ohms-adjust control, the variable resistor, located on the front of the meter. To zero the meter, connect the leads together (Figure 1-8), and turn the ohms-adjust knob until the meter indicates 0 at the far right end of the scale, as shown in Figure 1-9. When the leads are separated, the meter will again indicate infinity resistance at the left side of the meter scale. When the leads are connected across a resistance, the meter will again indicate up scale. Since resistance has been added to the circuit, less than 50 µA of current will flow and the meter will indicate some value other than zero. Figure 1-10 shows a meter indicating a resistance of 2.5 ohms, assuming the range setting is Rx1.

Helpful Hint

When resistance is to be measured, the meter must first be zeroed.

Ohmmeters can have different range settings such as Rx1, Rx100, Rx1000, or Rx10,000. These different scales can be obtained by adding different values of resistance in the meter circuit and resetting the meter to zero. An ohmmeter should always be readjusted to zero when the scale is changed. On the Rx1 setting, the resistance is measured straight off the resistance scale located at the top of the meter. If the range is set for Rx1,000, however, the reading must be multiplied by 1000. The ohmmeter reading shown in Figure 1-10 would be indicating a resistance of 2500 ohms if the range had been set for Rx1000. Notice that the ohmmeter scale is read backward from the other scales. Zero ohms is located on the far right side of the scale and maximum ohms is located at the far left side. It generally takes a little time and practice to read the ohmmeter properly. An analog ohmmeter being used to measure the resistance of an armature is shown in Figure 1-11.

Helpful Hint

The ohmmeter scale is read backward from the other scales.

Figure 1-11 Analog ohmmeter being used to measure armature resistance.

Digital Ohmmeters

Digital ohmmeters display the resistance in figures instead of using a meter movement. When using a digital ohmmeter, care must be taken to notice the scale indication on the meter. For example, most digital meters will display a "K" on the scale to indicate kilo-ohms or an "M" to indicate megohms (kilo means 1000 and mega means 1,000,000). If the meter is showing a resistance of 0.200 K, it means 0.200 × 1000 or 200 Ω. If the meter indicates 1.65 M, it means 1.65 × 1,000,000 or 1,650,000 Ω.

Appearance is not the only difference between analog and digital ohmmeters. Their operating principle is also different. Analog meters operate by measuring the amount of current change in the circuit when an unknown value of resistance is added. Digital ohmmeters measure resistance by measuring the amount of voltage drop across an unknown resistance. In the circuit shown in Figure 1-12, a constant current generator is used to supply a known amount of current to a resistor, Rx. It will be assumed that the amount of current supplied is 1 milliamp (0.001). The voltage dropped across the resistor is proportional to the resistance of the resistor and the amount of current flow. For example, assume the value of the unknown resistor is 4700 Ω. The voltmeter would indicate a drop of 4.7 volts when 1 ma of current flowed through the resistor. The scale factor of the ohmmeter can be changed by changing the amount of current flow through the resistor. Digital ohmmeters generally exhibit an accuracy of about ±1%. A digital ohmmeter being used to measure the resistance of an armature is shown in Figure 1-13.

Ammeters

The ammeter, unlike the voltmeter, is a very low-resistance device. Ammeters that are inserted into the circuit must be connected in series with the load to permit the load to limit the current flow, as shown in Figure 1-14. An ammeter has a typical impedance of

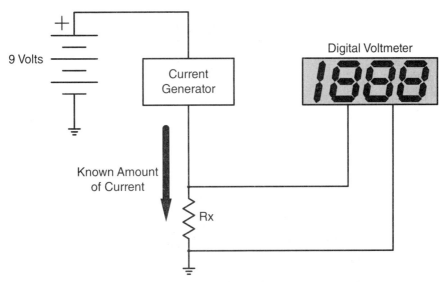

Figure 1-12 Digital ohmmeters operate by measuring the voltage drop across a resistor when a known amount of current flows through it.

less than 0.1 Ω. If this meter is connected in parallel with the power supply, the resistance of the ammeter is the only thing to limit the amount of current flow in the circuit. Assume that an ammeter with a resistance of 0.1 Ω is connected across a 240-volt AC line. The current flow in this circuit would be 2400 amps (240/0.1 = 2400). The blinding flash of light would be followed by the destruction of the ammeter. Ammeters connected directly into the circuit as shown in Figure 1-14 are referred to as in-line ammeters. Figure 1-15 shows an ammeter of this type.

Figure 1-13 Digital ohmmeter being used to measure resistance.

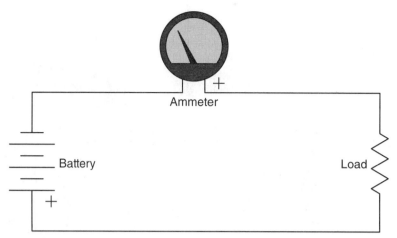

Figure 1-14 An ammeter connects in series with the load.

Figure 1-15 In-line ammeter.

DC Ammeters

Analog DC ammeters are constructed by connecting a common moving coil type of meter across a shunt. A shunt is a low-resistance device used to conduct most of the circuit current away from the meter movement. Since the meter movement is connected in parallel with the shunt, the voltage drop across the shunt is the voltage applied to the meter. Most ammeter shunts are manufactured to have a voltage drop of 50 mV (millivolts). If a 50 mV meter movement is connected across the shunt as shown in Figure 1-16, the pointer will move to the full-scale value when the rated current of the shunt is flowing. In the example shown, the ammeter shunt is rated to have a 50 mV drop when 10 amps of current is flowing in the circuit. Since the meter movement has a full-scale voltage of 50 mV, it will indicate the full-scale value when 10 amps of current is flowing through the shunt. An ammeter shunt is shown in Figure 1-17.

Ammeter shunts can be purchased to indicate different values. If the same 50 mV movement is connected across a shunt designed to drop 50 mV when 100 amps of current flows through it, the meter will now have a full-scale value of 100 amps.

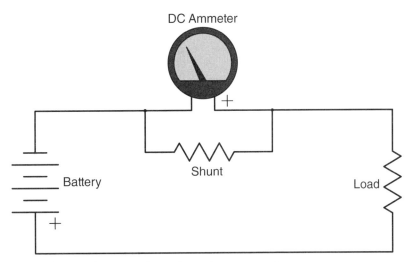

Figure 1-16 A shunt is used to set the value of an ammeter.

Multirange DC Ammeters

Many ammeters are designed to operate on more than one range. This is done by connecting the meter movement to different shunts. When this type of meter is used, care must be taken that the shunt is never disconnected from the meter. This would cause the meter movement to be inserted in series with the circuit and full-circuit current would flow through the meter. Two basic methods are used for connecting shunts to a meter movement. One method is to use a make-before-break switch. This type of switch is designed so that it will make contact with the next shunt before it breaks connection with the shunt it is connected to (Figure 1-18). This method does create a problem, however: contact resistance. Notice in Figure 1-18 that the rotary switch is in series with the shunt resistors. This causes the contact resistance to be added to the shunt resistance, which can cause inaccuracy in the meter reading.

Figure 1-17 Ammeter shunt.

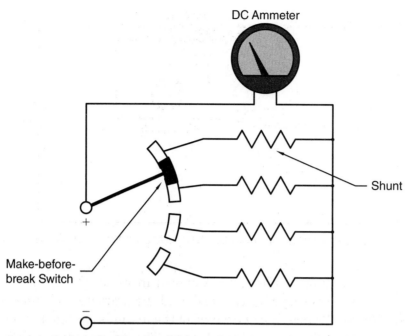

Figure 1-18 A make-before-break switch is used to change meter shunts.

The Ayrton Shunt

The second method is to use an Ayrton shunt (Figure 1-19). In this type of circuit, connection is made to different parts of the shunt and the meter movement is never disconnected from the shunt. Also, notice that the switch connections are made external to the shunt and meter. This prevents contact resistance from affecting the accuracy of the meter.

Alternating Current Ammeters

Shunts can be used with AC ammeters to increase their range but cannot be used to decrease their range. Most AC ammeters use a current transformer instead of shunts to

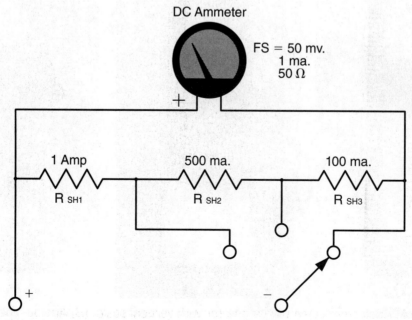

Figure 1-19 Ayrton shunt.

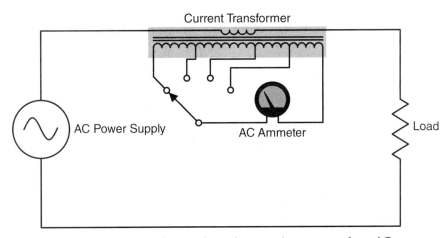

Figure 1-20 A current transformer is used to change the range of an AC ammeter.

change scale values. This type of ammeter is shown in Figure 1-20. The primary of the transformer is connected in series with the load, and the ammeter is connected to the secondary of the transformer. Notice that the range of the meter is changed by selecting different taps on the secondary of the current transformer. The different taps on the transformer provide different turns-ratios between the primary and secondary of the transformer.

Clamp-On Ammeters

Many electricians use the clamp-on type of AC ammeter (Figure 1-21 A, B, and C). To use this type of meter, the jaw of the meter is clamped around one of the conductors supplying power to the load, as seen in Figure 1-22. The meter is clamped around only one of the lines. If the meter is clamped around more than one line, the magnetic fields of the wires cancel each other and the meter indicates zero.

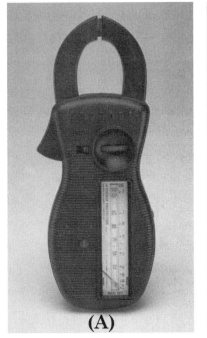
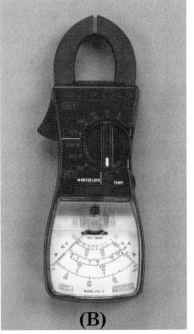
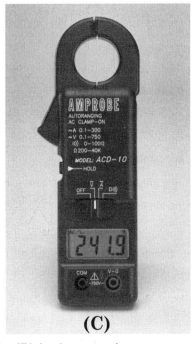

(A)   **(B)**   **(C)**

Figure 1-21 (A) Analog-type clamp-on ammeter with vertical scale. (B) Analog-type clamp-on ammeter with flat scale. (C) Clamp-on ammeter with digital scale. (Advanced Test Products)

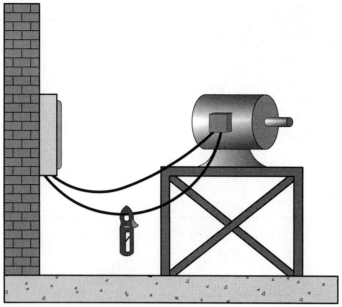

Figure 1-22 The clamp-on ammeter connects around only one conductor.

This type of meter also uses a current transformer to operate the meter. The jaw of the meter is part of the core material of the transformer. When the meter is connected around the current-carrying wire, the changing magnetic field produced by the AC current induces a voltage into the current transformer. The strength of the magnetic field and its frequency determine the amount of voltage induced in the current transformer. Since 60 Hz is a standard frequency throughout the United States and Canada, the amount of induced voltage is proportional to the strength of the magnetic field.

The clamp-on type of ammeter can have different range settings by changing the turns-ratio of the secondary of the transformer just as the in-line ammeter does. The primary of the transformer is the conductor around which the movable jaw is connected. If the ammeter is connected around one wire, the primary has one turn of wire as compared with the turns of the secondary. The turns-ratio can be changed by changing the turns of wire of the primary just as it can by changing the turns of the secondary. If two turns of wire are wrapped around the jaw of the ammeter, as shown in Figure 1-23, the primary winding now contains two turns instead of one, and the turns-ratio of the transformer is changed. The ammeter will now indicate double the amount of current in the circuit. The reading on the scale of the meter will have to be divided by 2 to get the correct reading. The ability to change the turns-ratio of a clamp-on

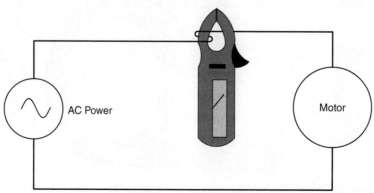

Figure 1-23 Looping the conductor around the jaw of the ammeter changes the ratio.

Figure 1-24 Scale divider used with clamp-on ammeter. The ammeter indicates a value of 3.25 amperes. If the scale divider has a ratio of 10:1, the actual current is 0.325 amp.

ammeter can be very useful for measuring low currents. Changing the turns-ratio is not limited to wrapping two turns of wire around the jaw of the ammeter. Any number of turns can be wrapped around the jaw of the ammeter and the reading will be divided by that number.

Ammeter Scale Divider

A very useful and simple-to-construct device that can be used to permit clamp-on-type ammeters to accurately measure low values of current is an ammeter scale divider (Figure 1-24). The scale divider is made by winding ten turns of wire around a nonconductive core material such as plastic. A piece of PVC pipe works well. The ten turns of wire are wrapped with tape to hold them together and alligator clips are attached to each end of the wire. The scale divider is connected in series with the load and the ammeter jaw is connected around the scale divider (Figure 1-25). Since the scale divider contains ten turns of wire, the ammeter

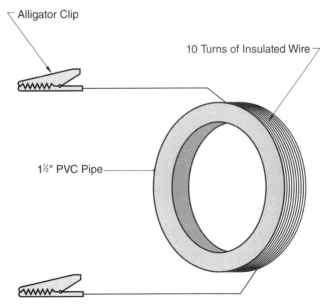

Figure 1-25 Construction of ammeter scale divider.

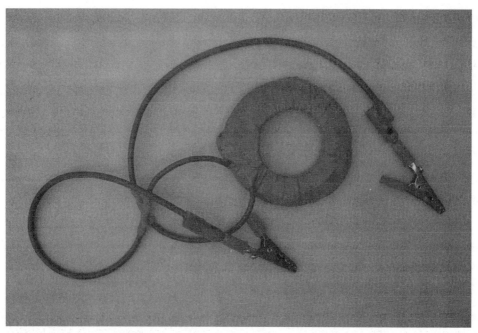

Figure 1-26 Ammeter scale divider.

scale can be divided by 10. If the ammeter has a full-range value of 6 amps, the ammeter will have a full range of 0.6 amp when the scale divider is used. The scale divider is a very useful tool when measuring low current values with a clamp-on type of ammeter. It is recommended that a scale divider be used when measuring low current values in the experiments presented in this manual. A typical scale divider is shown in Figure 1-26.

Review Questions

1. What are the two major types of voltmeters?

   _____

2. How can the full-scale voltage range of a voltmeter be increased?

   _____

3. A DC voltmeter has a resistance rating of 20,000 ohms per volt. The voltmeter is set for a value of 150 volts full scale. What is the resistance of the meter?

   _____

4. To measure voltage drop, should a voltmeter be connected across (in parallel) with the load or in series with the load?

   _____

5. A digital voltmeter has a resistance of 10 MΩ (megohms). If the meter is used to measure the voltage of a 480 volt panel, how much current will flow through the meter?

   _____

6. When using an analog ohmmeter, what should be done each time the meter is used to make a resistance measurement or the range setting is changed?

   _____

   _____

7. What is the most important rule concerning the use of ohmmeters?

   _____

8. Briefly explain the method used by digital ohmmeters to measure resistance.

   _____

   _____

9. When measuring the current in a circuit, should an in-line ammeter be connected across (in parallel) with the load or in series with the load?

   _____

10. Do ammeters have high internal resistance or low internal resistance?

    _____

11. A clamp-on–type AC ammeter is to be used to measure the current flow in a circuit. Should the meter be clamped around one conductor or more than one conductor to get the most accurate measurement?

    _____

12. A scale divider with ten turns of wire is connected in series with the load. The jaw of the clamp-on ammeter is connected around the scale divider. The ammeter is set for a full-scale range of 30 amperes. The ammeter indicates a current value of 12 amperes. What is the actual amount of current flow in the circuit?

    _____

Unit 2 Ohm's Law

Objectives

After completing this unit you should be able to
- Measure resistance with an ohmmeter.
- Measure current with an ammeter.
- Measure voltage with a voltmeter.
- Make Ohm's law calculations.

Ohm's law is the basic law concerning electric circuits. This unit will deal with the calculation and measurement of voltage, current, and resistance.

Materials Required

1 100 ohm resistor with a minimum watt rating of 175

1 150 ohm resistor with a minimum watt rating of 100

1 250 ohm resistor with a minimum watt rating of 100

1 AC voltmeter

1 AC ammeter (in-line or clamp-on)

1 Ohmmeter

Connection wires

120 volt AC power supply

Most conductors exhibit some amount of resistance change when they are heated. Most metals will increase their resistance with an increase of temperature; semi-conductor materials will decrease their resistance with an increase of temperature. This laboratory manual uses high-wattage wire-wound resistors for resistive loads. Caution must be exercised when using these components because they will become very hot and can cause severe burns. Also, the resistors should not be placed on or near materials that can burn. The resistive element of a wire-wound resistor is generally made of nichrome. Nichrome is an alloy that has a resistance of 675 ohms per mil-foot at 20 °C. Copper has a resistance of 10.4 ohms per mil-foot at 20 °C. Also, nichrome has a temperature coefficient of 0.0002 ohms per °C. This simply means that it exhibits very little change in resistance with a large change in temperature.

These high-wattage components are used in this manual to permit the circuit current to be great enough that it can be measured with a clamp-type ammeter. It may be necessary or convenient to use the scale divider discussed in Unit 1 to obtain more accurate current measurements.

LABORATORY EXERCISE

Name _____ Date _____

1. Use an ohmmeter to measure the resistance of the 100 Ω resistor. To make the measurement, zero the ohmeter (if necessary) and connect one probe to one end of the resistor and the other probe to the opposite end of the resistor, Figure 2-1.

   R = _____ Ω

2. Use an AC voltmeter to test the terminal voltage of the AC power supply to be used for this experiment. Make certain that the voltmeter is set to indicate AC volts and that the range setting is greater than 120 volts, Figure 2-2. Turn on the power and measure the voltage of the power supply.

   _____ Volts

3. **Make certain that the power is turned off.** Using the 100-ohm resistor, connect the circuit shown in Figure 2-3. A clamp-on-type AC ammeter may be substituted for the in-line ammeter shown in the drawing.

4. Using the Ohm's law formula shown, compute the amount of current that should flow in the circuit. Use 120 volts for the value of E.

$$I = \frac{E}{R}$$

I = _____ amps

Figure 2-1 Measuring the resistance of a resistor.

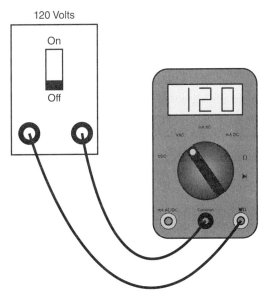

Figure 2-2 Testing the power supply with a voltmeter.

5. Turn on the AC power supply and measure the current flow through the 100 Ω resistor.

   I = _____ amps

6. Connect the AC voltmeter across the 100 Ω resistor as shown in Figure 2-4. Is the voltage the same as that measured at the power supply?

   _____

7. Use the measured values of voltage and current to compute the resistance of the resistor.

$$R = \frac{E}{I}$$

   R = _____ Ω

   Did the resistance increase or decrease? _____

   Amount of increase or decrease _____ Ω

8. **Turn off the power supply.**

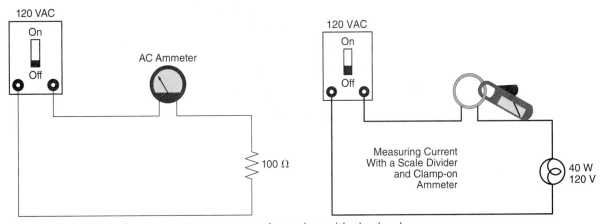

Figure 2-3 An in-line ammeter connects in series with the load.

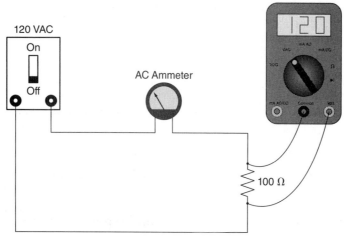

Figure 2-4   Measuring the voltage drop across the resistor.

9. Use an ohmmeter to measure the resistance of a 150-Ω resistor.

R = _____ Ω

10. Remove the 100-Ω resistor from the circuit and replace it with the 150-Ω resistor.

11. Using the formula shown, compute the amount of current that should flow in this circuit.

$$I = \frac{E}{R}$$

I = _____ amps

12. Turn on the power and measure the current flow in the circuit and the voltage drop across the resistor. If a clamp-type ammeter is used, it is recommended to connect the 10:1 scale divider in the circuit as shown in Figure 2-5. This can increase the accuracy of the measurement. If the ammeter is set on a 15-amp range, for example, the meter would have a full-scale range of 1.5 amps.

I = _____ amps

E = _____ volts

13. **Turn off the power supply.**

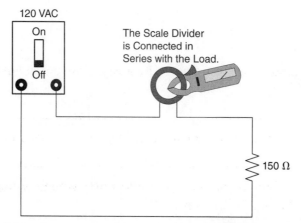

Figure 2-5   The scale divider is connected in series with the load.

14. Using Ohm's law, compute the resistance in the circuit.

$$R = \frac{E}{I}$$

R = _____ Ω

15. Compare this value with the measured resistance in step 9. What is the difference in resistance?

R = _____ Ω

16. Use an ohmmeter to measure the resistance of a 250-Ω resistor at room temperature.

R = _____ Ω

17. Remove the 150 Ω-resistor from the circuit and replace it with the 250-Ω resistor.

18. Using the formula shown, compute the amount of current that should flow in this circuit.

$$I = \frac{E}{R}$$

I = _____ amps

19. Turn on the power and measure the current flow in the circuit and the voltage drop across the lamp.

I = _____ amps

E = _____ volts

20. **Turn off the power supply.**

21. Using Ohm's law, compute the resistance in the circuit.

$$R = \frac{E}{I}$$

R = _____ Ω

22. Compare this value with the measured resistance in step 16. What is the difference in resistance?

R = _____ Ω

23. Disconnect the circuit and return the components to their proper place.

Review Questions

1. A 50 Ω resistor is connected to a 240 volt source. How much current will flow through the resistor?

_____

2. How much power (watts) is being consumed by the resistor in question 1?

_____

3. An electric heating element is rated at 1,500 watts when connected to 240 volts. How much current would flow in this circuit?

_____

4. What is the resistance of the heating element in question 3?

   _____

5. Assume that the voltage in   question 3 is reduced to 120 volts. How much current would flow in the circuit?

   _____

6. If the voltage connected to a 1,500 watt heating element in question 3 were to be reduced to 120 volts, how much power would the heating element actually consume?

   _____

7. Assume that a 100 watt lamp connected to a 120 volt circuit burns out and is replaced with a 150 watt lamp. How much more current will the 150 watt lamp draw?

   _____

8. A resistive heating element has a current draw of 2.3 amperes and a resistance of 12 Ω. How much power is the heating element consuming?

   _____

9. A heating element has a resistance of 52 Ω and a current draw of 4 amperes. How much voltage is connected to the heating element?

   _____

10. How much power is being consumed by the heating element in question 9?

   _____

Unit 3   Series Circuits

Objectives

After studying this unit, you should be able to
- List the three rules concerning series circuits.
- Determine the total resistance of a series circuit.
- Determine the voltage drops across individual components of a series circuit.
- Connect a series circuit.
- Measure values of voltage and current in a series circuit.

Series circuits are characterized by the fact that they contain only one path for current flow. There are three rules concerning series circuits that, when used with Ohm's law, permit values of current, voltage, and resistance to be determined.

Assume that an electron leaves the negative terminal of the battery in Figure 3-1 and must travel to the positive terminal. Notice that the only path the electron can travel is through each resistor. Since there is only one path for current flow, the current must be the same at any point in the circuit. Regardless of where an ammeter is connected in the circuit, it will indicate the same value. The first rule concerning series circuits states that the current must be the same at any point in the circuit.

Helpful Hint

The first rule concerning series circuits states that the current must be the same at any point in the circuit.

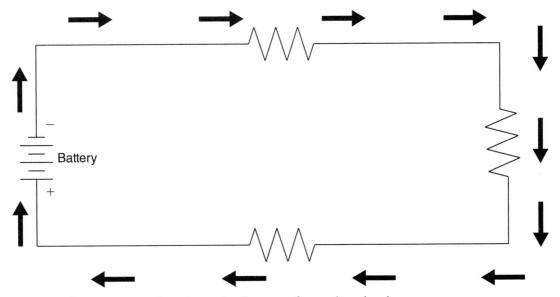

Figure 3-1   Current must flow through all parts of a series circuit.

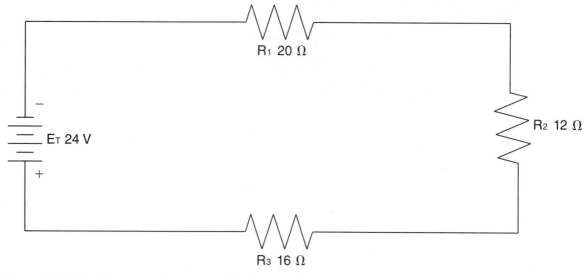

Figure 3-2 Resistance values are added to the circuit.

In Figure 3-2, values are added to the resistors shown in Figure 3-1. Resistor R_1 has a value of 20 Ω, R_2 has a value of 12 Ω, and resistor R_3 has a value of 16 Ω. Since current must pass through each of these resistors, each hinders the flow of current. The total amount of hindrance to current flow is the combined ohmic value of each resistor. The second rule of series circuits states that the total resistance is the sum of the individual resistances. The total circuit resistance can be determined by adding the values of all the resistors (20 + 12 + 16 = 48 Ω). The circuit shown in Figure 3-2 indicates a total voltage (E_T) of 24 volts for this circuit. Now that the total circuit resistance is known and the applied voltage is known, the circuit current can be determined using Ohm's law.

$$I = \frac{E}{R}$$

$$I = \frac{24}{48}$$

$$I = 0.5\,amp$$

Helpful Hint

The second rule of series circuits states that the total resistance is the sum of the individual resistances.

Since the current is the same at any point in a series circuit, 0.5 amp flows through each resistor. Now that the amount of current flowing through a resistor is known, the amount of voltage drop across each resistor can be computed using Ohm's law.

$$E_1 = I \times R_1$$
$$E_1 = 0.5 \times 20\,\Omega$$
$$E_1 = 10\,volts$$

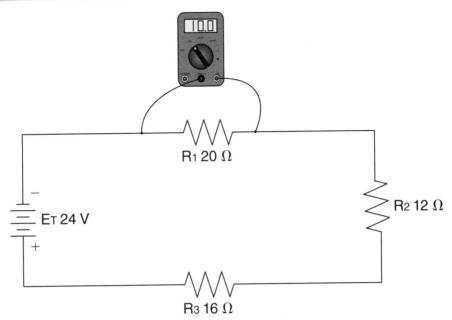

Figure 3-3 Measuring the voltage drop across resistor R_1

If a voltmeter were to be connected across resistor R_1, it would indicate a value of 10 volts (Figure 3-3). This 10 volts represents the amount of electrical pressure necessary to push 0.5 amp through 20 ohms of resistance. The amount of voltage drop across resistors R_2 and R_3 can be computed using Ohm's law also.

$$E_2 = I \times R_2$$

$$E_2 = 0.5 \times 12$$

$$E_2 = 6 \text{ volts}$$

$$E_3 = I \times R_3$$

$$E_3 = 0.5 \times 16$$

$$E_3 = 8 \text{ volts}$$

Notice that if all the voltage (pressure) drops are added, they will equal the applied voltage of the circuit ($10 + 6 + 8 = 24$ volts). The third rule of series circuits states that the total voltage is equal to the sum of the voltage drops around the circuit.

Another rule concerning circuits is that watts (power) will add in any type of circuit. This rule is true not only for series circuits but also for parallel and combination circuits. The total wattage of any type circuit will be the sum of all the individual power-consuming devices in the circuit.

Helpful Hint

The third rule of series circuits states that the total voltage is equal to the sum of the voltage drops around the circuit. Another rule concerning circuits is that watts (power) will add in any type of circuit.

LABORATORY EXERCISE

Name _____ Date _____

Materials Required

2 100 ohm resistors

1 150 ohm resistor

1 250 ohm resistor

1 AC ammeter (In-line or clamp-on. If a clamp-on type is used, the use of a scale divider is recommended.)

1 AC voltmeter

120 volt AC power supply

208 volt AC power supply

Connecting wires

1 Ohmmeter

1. Connect the circuit shown in Figure 3-4.

2. Compute the total resistance of the circuit using the following formula:

$$R_T = R_1 + R_2 + R_3 + R_N$$

Note: R_N simply means the number of resistors in the circuit, whether there are 3 or 23.

$R_T = $ _____ Ω

3. Disconnect the circuit from the power supply and measure the total resistance of the circuit with an ohmmeter as shown in Figure 3-5. Compare the measured value with the computed value.

$R_T = $ _____ Ω

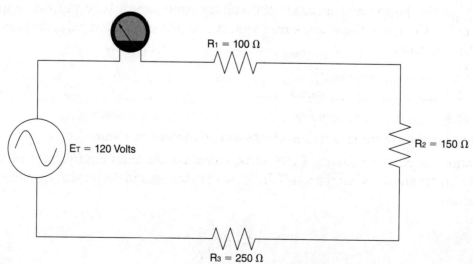

Figure 3-4 Connecting the first series circuit.

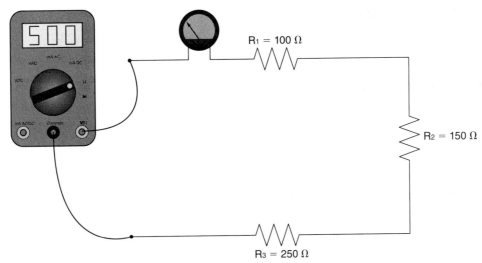

Figure 3-5 An ohmmeter is used to measure the total resistance of the circuit.

4. Assuming that the circuit is connected to a 120 volt source, calculate the total circuit current using the following formula:

$$I_T = \frac{E_T}{R_T}$$

$I_T =$ _____ A

5. Reconnect the circuit to the 120 volt power source. Turn on the power and measure the total circuit current. Compare this value with the computed value. **Turn off the power.**

$I_T =$ _____ A

6. Now that the total circuit current is known, compute the voltage drop across each of the resistors using the following formula:

$$E = I \times R$$

100 Ω = _____ volts

150 Ω = _____ volts

250 Ω = _____ volts

7. Turn on the power and measure the voltage drop across each resistor with an AC voltmeter. Compare these measured values with the calculated value in step 6. **Turn off the power.**

100 Ω = _____ volts

150 Ω = _____ volts

250 Ω = _____ volts

8. Reconnect the circuit to a source of 208 volts as shown in Figure 3-6.

9. Assuming a circuit voltage of 208 volts, calculate the total circuit current using the following formula: (Note: round off the answer to the second decimal place or hundredths of an amp.)

$$I_T = \frac{E_T}{R_T}$$

$I_T =$ _____ A

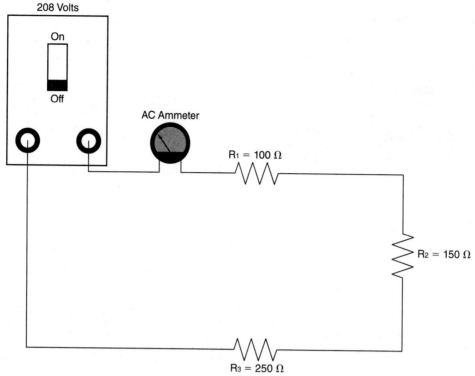

Figure 3-6   The circuit is connected to a 208-volt source.

10. Turn on the power and measure the total circuit current. Compare the measured value with the computed value. **Turn off the power.**

    I_T = _____ A

11. Calculate the voltage drop across each component using the following formula:

    $$E = I \times R$$

    100 Ω = _____ volts

    150 Ω = _____ volts

    250 Ω = _____ volts

12. Turn on the power and measure the voltage drop across each resistor. Compare the measured values with the computed values. **Turn off the power.**

    100 Ω = _____ volts

    150 Ω = _____ volts

    250 Ω = _____ volts

13. Turn on the power and measure the actual voltage applied to the circuit with a voltmeter. **Turn off the power.**

14. Add the measured values of voltage in step 12. Compare the sum to the measured value in step 13. Is the sum of the voltage drops across each resistor approximately the same as the circuit voltage?

    _____

15. Reconnect the circuit as shown in Figure 3-7.

16. Compute the total resistance of the circuit using this formula:

    $$R_T = R_1 + R_2 + R_3 + R_N$$

    R_T = _____ Ω

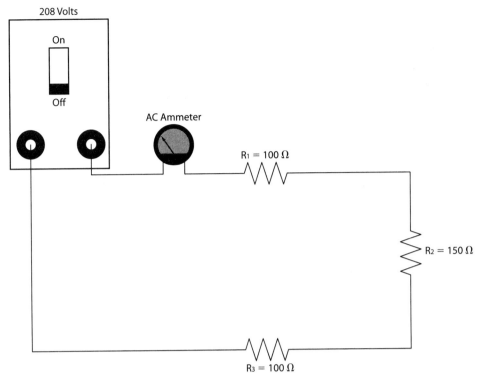

Figure 3-7 Second series circuit.

17. Disconnect the circuit from the power supply and measure the total resistance of the circuit with an ohmmeter. Compare the measured value with the computed value.

R_T = _____ Ω

18. Reconnect the circuit to a source of 208 volts as shown in Figure 3-7.

19. Assuming a circuit voltage of 208 volts, calculate the total circuit current using the following formula: (Note: Round off the answer to the second decimal place or hundredths of an amp.)

$$I_T = \frac{E_T}{R_T}$$

I_T = _____ A

20. Turn on the power and measure the total circuit current. Compare the measured value with the computed value. **Turn off the power.**

I_T = _____ A

21. Calculate the voltage drop across each component using the following formula:

$$E = I \times R$$

100 Ω = _____ volts

150 Ω = _____ volts

100 Ω = _____ volts

22. Turn on the power and measure the voltage drop across each resistor. Compare the measured values with the computed values. **Turn off the power.**

100 Ω = _____ volts

150 Ω = _____ volts

100 Ω = _____ volts

23. Turn on the power and measure the actual voltage applied to the circuit with a voltmeter. **Turn off the power.**

24. Add the measured values of voltage in step 22. Compare the sum to the measured value in step 23. Is the sum of the voltage drops across each resistor approximately the same as the circuit voltage?

_____

25. Disconnect the circuit and return the components to their proper place.

Review Questions

1. State the three rules for series circuits.

_____

_____

_____

_____

_____

_____

2. A series circuit has a total resistance of 1220 Ω. Resistor R_1 has a resistance of 220 Ω, R_2 has a resistance of 200 Ω, and R_3 has a resistance of 470 Ω. What is the value of resistor R_4?

_____

3. A circuit has three resistors connected in series. Resistor R_1 has a value of 16 Ω, R_2 has a value of 36 Ω, and R_3 has a value of 8 Ω. If the circuit is connected to a 12 volt battery, how much voltage is dropped across resistor R_1?

_____

4. A circuit contains two resistors connected in series. Resistor R_1 has a value of 1200 Ω and resistor R_2 has a value of 1600 Ω. Resistor R_1 has a voltage drop of 24 volts. What is the total voltage applied to the circuit?

_____

5. Three resistors are connected in series to a 120 volt power source. Resistor R_2 has a value of 200 Ω and a voltage drop of 50 volts. What is the total resistance of the circuit?

_____

6. Three resistors are connected in series to a 48 volt power source. Resistor R_2 has a voltage drop of 18 volts and resistor R_3 has a voltage drop of 16 volts. How much voltage is dropped across resistor R_1?

_____

7. A 150 watt incandescent lamp is connected to 120 volts. How much current will flow in this circuit?

_____

8. A series circuit contains four resistors. The circuit is connected to a 24 volt power source. The circuit consumes a total power of 360 watts. Resistor R_1 consumes 40 watts, R_2 consumes 65 watts, and R_3 consumes 85 watts. What is the resistance of resistor R_4?

_____

Unit 4 Parallel Circuits

Objectives

After studying this unit, you should be able to
  - List the three rules concerning parallel circuits.
  - Determine the total resistance of a parallel circuit.
  - Determine the current through individual components of a parallel circuit.
  - Connect a parallel circuit.
  - Measure values of voltage and current in a parallel circuit.

Parallel circuits are characterized by the fact that they have more than one path for current flow. There are three rules concerning parallel circuits that, when used in conjunction with Ohm's law, permit values of voltage, current, and resistance to be determined for almost any parallel circuit. As with series circuits, the total power consumption of the circuit is the sum of the power consumption of each component of the circuit.

A parallel circuit containing three resistors is shown in Figure 4-1. Assume that current leaves the negative terminal of the battery and must return to the positive terminal. Part of the circuit current will flow through resistor R_1 and return to the battery. The remainder of the current will flow to resistors R_2 and R_3. Part of the current will again split and flow through resistor R_2, permitting the remainder of the current to flow to resistor R_3. Notice in this circuit that there are three separate paths for the current to flow. The first rule for parallel circuits states that the total current is the sum of the currents through each branch of the circuit. Assume that the circuit shown in Figure 4-1 is connected to a 120 volt power source and that a current of 0.5 amp flows through resistor R_1, 1.5 amps flows through resistor R_2, and 2 amps flows through resistor R_3 (Figure 4-2).

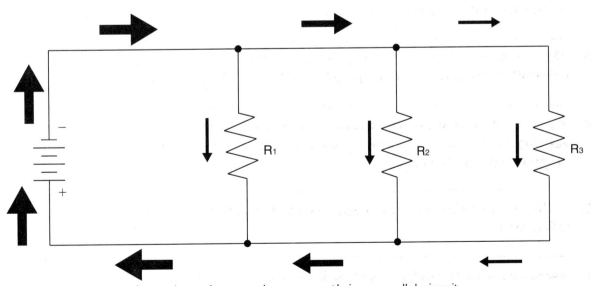

Figure 4-1 Current flows through more than one path in a parallel circuit.

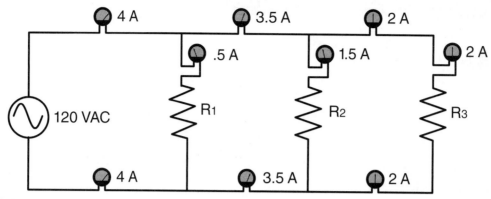

Figure 4-2   The total circuit current is equal to the sum of the currents through each branch.

Helpful Hint

The first rule for parallel circuits states that the total current is the sum of the currents through each branch of the circuit.

Notice in Figure 4-2 that each branch is connected directly to the power source. If a voltmeter were connected across each branch, it would be seen that the source voltage is applied across each branch. The second rule for parallel circuits states that the voltage is the same across all branches of a parallel circuit. Since the voltage across each branch is known and the amount of current flowing through each is known, the resistance of each branch can be determined using Ohm's law.

Helpful Hint

The second rule for parallel circuits states that the voltage is the same across all branches of a parallel circuit.

$$R_1 = \frac{E_1}{I_1}$$

$$R_1 = \frac{120}{0.5}$$

$$R_1 = 240 \ \Omega$$

$$R_2 = \frac{E_2}{I_2}$$

$$R_2 = \frac{120}{1.5}$$

$$R_2 = 80 \ \Omega$$

$$R_3 = \frac{E_3}{I_3}$$

$$R_3 = \frac{120}{2}$$

$$R_3 = 60 \ \Omega$$

It is also possible to determine the total resistance of the circuit using Ohm's law.

$$R_T = \frac{E_T}{I_T}$$

$$R_T = \frac{120}{4}$$

$$R_T = 30 \ \Omega$$

The circuit resistance values are shown in Figure 4-3. Also, notice that the total circuit resistance is less than any single resistor in the circuit. To understand this, recall that resistance is hindrance to the flow of current. Each time another branch is connected to the power source, another path for current flow is created. This reduces the hindrance to current flow for the circuit. The total resistance of a parallel circuit will always be less than the resistance of any single branch. There are three formulas used to calculate the total resistance of a parallel circuit when only resistance values are known. The first formula is called the "product over sum" formula. This formula can calculate the total resistance of only two parallel resistors at a time. The product over sum formula is

$$R_T = \frac{R_1 \times R_2}{R_1 + R_2}$$

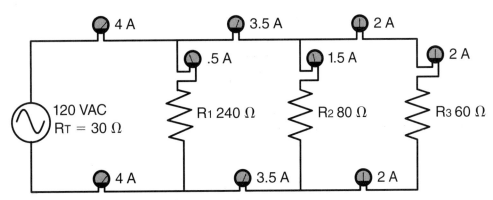

Figure 4-3 Determining circuit resistance values.

Helpful Hint

The total resistance of a parallel circuit will always be less than the resistance of any single branch.

To determine the total resistance of the circuit shown in Figure 4-3, the parallel resistance of the first two resistors will be determined first.

$$R_T = \frac{240 \times 80}{240 + 80}$$

$$R_T = \frac{19200}{320}$$

$$R_T = 60 \ \Omega$$

The total resistance of the first two resistors is 60 ohms. These two resistors are connected in parallel with another 60 Ω resistor, however. To find the parallel resistance of that connection, use the parallel resistance of the first two resistors as R_1 and the value of the next resistor as R_2 and repeat the calculation.

$$R_T = \frac{60 \times 60}{60 + 60}$$

$$R_T = \frac{3,600}{120}$$

$$R_T = 30 \ \Omega$$

This process is repeated for each resistor until all the resistors have been substituted in the formula.

The second formula used to calculate parallel resistance is called the "reciprocal formula." The reciprocal of any number can be determined by dividing that number into 1. The third rule for parallel circuits states that the reciprocal of the total resistance is equal to the sum of the reciprocals of each branch. The basic reciprocal formula is shown below.

$$\frac{1}{R_T} = \frac{1}{R_1} + \frac{1}{R_2} + \frac{1}{R_3} + \frac{1}{R_N}$$

Helpful Hint

The third rule for parallel circuits states that the reciprocal of the total resistance is equal to the sum of the reciprocals of each branch.

This formula can be modified to solve for RT instead of the reciprocal of RT.

$$R_T = \frac{1}{\dfrac{1}{R_1} + \dfrac{1}{R_2} + \dfrac{1}{R_3} + \dfrac{1}{R_N}}$$

This is the most-used formula for solving parallel resistance because almost all scientific calculators have a reciprocal key (1/X). Any number shown on the display of a calculator is on the X axis. When the 1/X key is pressed, the calculator will divide any number shown on the display into 1. To find the total resistance of the circuit shown in Figure 4-3 using the reciprocal formula, press the following keys on a scientific calculator:

$$240 \ 1/X + 80 \ 1/X + 60 \ 1/X =$$

The sum of the reciprocals of each branch is now on the calculator display. To find the total resistance, press the 1/X key again and this will give the reciprocal of the sum of the reciprocals.

Example: Find the total resistance of the following resistors connected in parallel:

1200 Ω, 2200 Ω, 1600 Ω, 3000 Ω, and 2400 Ω

Press the following calculator keys:

$$1200 \ 1/X + 2200 \ 1/X + 1600 \ 1/X + 3000 \ 1/X + 2400 \ 1/X = 1/X$$

The answer should be 375.5 Ω.

The third formula for calculating parallel resistance is special and can be used only when all resistor values are the same. The formula is

$$R_T = \frac{R}{N}$$

N stands for the number of resistors connected in parallel. Assume that four 100 Ω resistors are connected in parallel. To determine their total resistance, divide the resistance of one resistor by the total number of resistors.

$$R_T = \frac{100}{4}$$

$$R_T = 25 \ \Omega$$

LABORATORY EXERCISE

Name _____ Date _____

Materials Required

2 100 ohm resistors

1 150 ohm resistor

1 250 ohm resistor

1 AC ammeter (In-line or clamp-on. If a clamp-on type is used, the use of a scale divider is recommended.)

1 AC voltmeter

1 Ohmmeter

120 volt AC power supply

Connecting wires

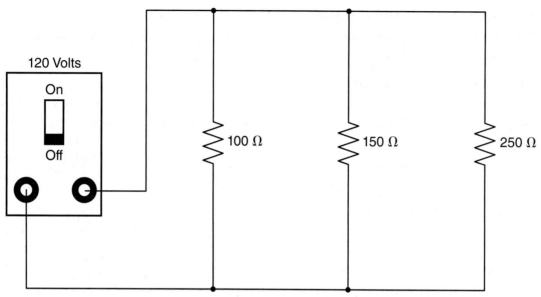

Figure 4-4 Connecting the first parallel circuit.

1. Connect the circuit shown in Figure 4-4.

2. Calculate the total circuit resistance using the following formula: (Note: Round off the answer to the second decimal place or hundredth of an ohm.)

$$R_T = \frac{1}{\dfrac{1}{R_1} + \dfrac{1}{R_2} + \dfrac{1}{R_3}}$$

   R_T = _____ Ω

3. Disconnect the circuit from the power supply and measure the total circuit resistance with an ohmmeter. Compare this value with the computed value. Are the values within 5 percent of each other?

   _____

4. Turn on the 120 VAC power and check the output voltage with an AC voltmeter. **Turn off the power.**

   _____ volts

5. Reconnect the circuit to the power source. Turn on the power and measure the voltage across each resistor using the AC voltmeter.

   (100 Ω) _____ volts

   (150 Ω) _____ volts

   (250 Ω) _____ volts

6. **Turn off the power supply.**

7. Calculate the current flow through each branch of the circuit using the following formula:

$$I = \frac{E}{R}$$

   (100 Ω) _____ A

   (150 Ω) _____ A

   (250 Ω) _____ A

8. Connect an ammeter in series with the 100 ohm resistor.

9. Turn on the power supply and measure the current flow through the resistor.

   $I_{(100\Omega)}$ _____ amps

10. **Turn off the power supply.**

11. Remove the AC ammeter from the branch containing the 100 ohm resistor and connect it in series with the 150 ohm resistor. Be sure to reconnect the 100 ohm resistor back into the circuit.

12. Turn on the power supply and measure the current flow through the 150 ohm resistor.

   $I_{(150\Omega)}$ _____ amps

13. **Turn off the power supply.**

14. Remove the AC ammeter from the branch containing the 150 ohm resistor and connect it in series with the 250 ohm resistor. Reconnect the 150 ohm resistor back into the circuit.

15. Turn on the power and measure the current flow through the 250 ohm resistor.

   $I_{(250\Omega)}$ _____ amps

16. **Turn off the power supply.**

17. Calculate the total circuit current. ($I_T = I_{(100\Omega)} + I_{(150\Omega)} + I_{(250\Omega)}$)

   $I_T =$ _____ amps

18. Disconnect the AC ammeter from the branch containing the 250 ohm resistor and reconnect it in series with one of the conductors connected to the power supply. This will permit the total circuit current to be measured. Reconnect the 250 ohm resistor back into the circuit.

19. Turn on the power supply and measure the total current of the circuit. Compare this value with the computed value.

   $I_T =$ _____ amps

20. **Turn off the power supply.**

21. Replace the 250 ohm resistor with a second 100 ohm resistor as shown in Figure 4-5.

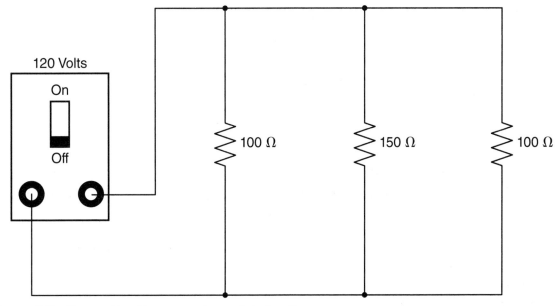

Figure 4-5 Connecting the second parallel circuit.

22. Calculate the total circuit resistance using this formula. (Note: Round off the answer to the second decimal place or hundredth of an ohm.)

$$R_T = \cfrac{1}{\cfrac{1}{R_1} + \cfrac{1}{R_2} + \cfrac{1}{R_3}}$$

R_T = _____ Ω

23. Disconnect the circuit from the power supply and measure the total circuit reistance with an ohmmeter. Compare this value with the computed value. Are the values within 5 percent of each other?

_____

24. Turn on the 120 VAC power and check the output voltage with an AC voltmeter.

_____ volts

25. Measure the voltage across each resistor using the AC voltmeter.

(100 Ω) _____ volts

(150 Ω) _____ volts

(250 Ω) _____ volts

26. **Turn off the power supply.**

27. Calculate the current flow through each branch of the circuit using the following formula:

$$I = \frac{E}{R}$$

(100 Ω) _____ A

(150 Ω) _____ A

(250 Ω) _____ A

28. Connect an ammeter in series with the first 100 ohm resistor.

29. Turn on the power supply and measure the current flow through the resistor.

$I_{(100\Omega)}$ _____ amps

30. **Turn off the power supply.**

31. Remove the AC ammeter from the branch containing the first 100 ohm resistor and connect it in series with the 150 ohm resistor. Be sure to reconnect the 100 ohm resistor back into the circuit.

32. Turn on the power supply and measure the current flow through the 150 ohm resistor.

$I_{(150\Omega)}$ _____ amps

33. **Turn off the power supply.**

34. Remove the AC ammeter from the branch containing the 150 ohm resistor and connect it in series with the second 100 ohm resistor. Reconnect the 150 ohm resistor back into the circuit.

35. Turn on the power and measure the current flow through the second 100 ohm resistor.

$I_{(100\Omega)}$ _____ amps

36. **Turn off the power supply.**

37. Calculate the total circuit current. $(I_T = I_{(100\Omega)} + I_{(150\Omega)} + I_{(100\Omega)})$

I_T = _____ amps

38. Disconnect the AC ammeter from the branch containing the second 100 ohm resistor and reconnect it in series with one of the conductors connected to the power supply. This will permit the total circiuit current to be measured. Reconnect the second 100 ohm resistor back into the circuit.

39. Turn on the power supply and measure the total current of the circuit. Compare this value with the computed value.

   I_T = _____ amps

40. **Turn off the power supply.** Disconnect the circuit and return the components to their proper place.

Review Questions

1. State the three rules for parallel circuits.

   _____

   _____

   _____

   _____

   _____

   _____

2. A circuit has four resistors connected in parallel. The resistor values are as follows: R_1 = 150 kΩ, R_2 = 220 kΩ, R_3 = 180 kΩ, and R_4 = 330 kΩ. What is the total resistance of this circuit?

   _____

3. Refer to the resistor values in question 2. Assume that resistor R_2 has a current of 0.945 mA flowing through it. How much current is flowing through resistor R_4?

   _____

4. A circuit contains three resistors connected in parallel and has a total power consumption of 14 watts. Resistor R_1 has a power consumption of 6.2 watts, R_2 consumes 4.8 watts. Resistor R_3 has a value of 192 Ω. What is the voltage applied to the circuit?

   _____

5. A circuit contains four resistors connected in parallel. Resistor R_1 has a current flow of 0.25 amp, R_2 has a current flow of 0.3 amp, R_3 has a current flow of 0.45 amp, and R_4 has a current flow of 0.7 amp. What is the total current flow in this circuit?

   _____

6. A circuit has the following resistors connected in parallel: 18 Ω, 24 Ω, 12 Ω, and 10 Ω. What is the total resistance of this circuit?

   _____

7. A parallel circuit contains three resistors. The total resistance of the circuit is 8 Ω. Resistor R_1 has a value of 24 Ω, and R_2 has a value of 18 Ω. What is the value of resistor R_3?

   _____

Unit 5 Combination Circuits

Objectives

After studying this unit, you should be able to

- Determine series and parallel paths through an electric circuit.
- Use measured values of voltage, current, and resistance in a combination circuit to determine unknown electrical quantities.
- Use measuring instruments to determine electrical quantities in a combination circuit.
- Solve combination circuit problems using Ohm's law.

Combination circuits contain both series and parallel paths in the same circuit. In order to solve unknown values, it is imperative to be able to identify which components are in series and which are in parallel. To determine which components are in series and which are in parallel, trace the current path through the circuit. In the circuit shown in Figure 5-1, resistors R_2 and R_3 are connected in parallel. Resistor R_1 is connected in series with R_2 and R_3. Assume that current leaves the negative terminal of the battery and must return to the positive terminal. All the current in the circuit must flow through resistor R_1 because there is no other path by which the current can travel from the negative to the positive terminal. Resistor R_1 is, therefore, in series with the rest of the circuit because the definition of a series circuit is a circuit that has only one path for current flow.

After leaving resistor R_1, the current path can divide. Part of the current will flow through resistor R_2 and part will flow through R_3. The amount of current that flows through each is determined by their resistance values. Since the current has more than one path, resistors R_2 and R_3 are connected in parallel with each other. After leaving resistors R_2 and R_3, the current proceeds to the positive battery terminal.

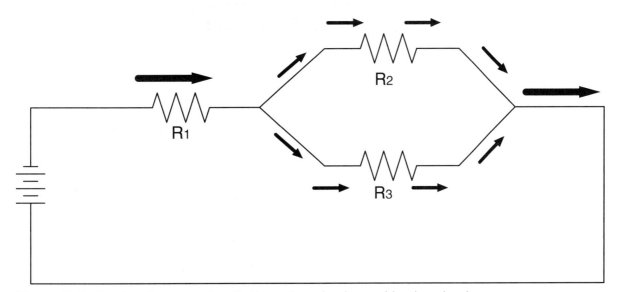

Figure 5-1 Tracing the current path through a simple combination circuit.

Example Circuit #1

Values of resistance and voltage have been added to the circuit in Figure 5-2. Resistor R_1 has a value of 36 Ω, R_2 has a value of 40 Ω, and R_3 has a value of 60 Ω. The battery has a terminal voltage of 30 volts. The following electrical values will be determined for this circuit:

R_T - Total resistance of the circuit

I_T - Total circuit current

I_1 - Current flow through resistor R_1

E_1 - Voltage drop across resistor R_1

E_2 - Voltage drop across resistor R_2

E_3 - Voltage drop across resistor R_3

I_2 - Current flow through resistor R_2

I_3 - Current flow through resistor R_3

When solving values for a combination circuit, it is generally helpful to reduce the circuit to a simple series or parallel circuit and then work back through the circuit in a step-by-step procedure. This is accomplished by combining series or parallel components to form a single resistance value. In the circuit shown in Figure 5-2, resistors R_2 and R_3 are connected in parallel. These two resistors can be combined into one resistor value by determining their total resistance value.

$$R_T = \frac{1}{\dfrac{1}{R_1} + \dfrac{1}{R_2} + \dfrac{1}{R_3} + \dfrac{1}{R_N}}$$

$$R_T = \frac{1}{\dfrac{1}{R_1} + \dfrac{1}{R_2}}$$

$$R_T = \frac{1}{40 + 60}$$

$$R_T = 24 \ \Omega$$

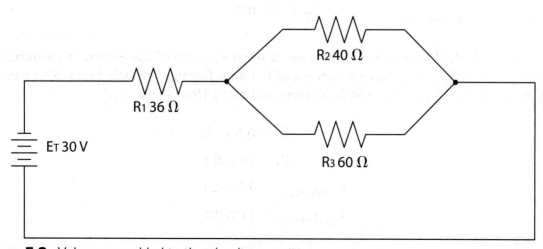

Figure 5-2 Values are added to the circuit.

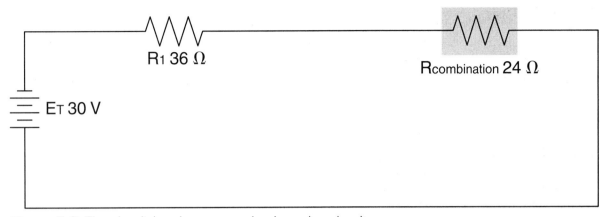

Figure 5-3 The circuit has become a simple series circuit.

The total resistance for resistors R_2 and R_3 will form resistor $R_{combination}$. The circuit has been redrawn in Figure 5-3. Notice that resistors R_2 and R_3 have been replaced by $R_{combination}$. The circuit is now a simple series circuit containing two resistors, R_1 and $R_{combination}$. One of the rules for series circuits states that the total resistance is equal to the sum of the individual resistances. The total circuit resistance can be determined by adding the two resistance values together.

$$R_T = R_1 + R_{combination}$$
$$R_T = 36 + 24$$
$$R_T = 60 \ \Omega$$

Now that the total resistance is known, the total current can be determined using Ohm's law.

$$I = \frac{E}{R}$$
$$I_T = \frac{30}{60}$$
$$I_T = 0.5 \ \text{amp}$$

In a series circuit, the current is the same through all parts of the circuit. Therefore, the resistors R_1 and $R_{combination}$ have a current of 0.5 amp flowing through them. The voltage drop across each resistor can now be determined using Ohm's law.

$$E_1 = 0.5 \times 36$$
$$E_1 = 18 \ \text{volts}$$
$$E_{combination} = 0.5 \times 24$$
$$E_{combination} = 12 \ \text{volts}$$

These added circuit values are shown in Figure 5-4.

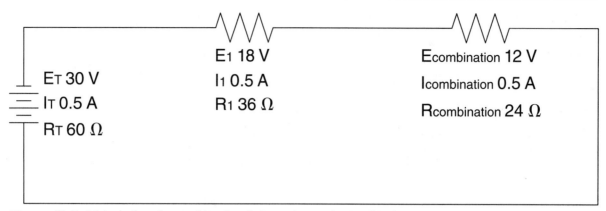

Figure 5-4   Values for the series circuit have been determined.

Resistor $R_{combination}$ is in reality resistors R_2 and R_3. The values that apply to $R_{combination}$, therefore, apply to resistors R_2 and R_3. If a voltmeter were to be connected across the parallel circuit containing R_2 and R_3, it would indicate the same voltage drop as that across $R_{combination}$, as shown in Figure 5-5. One of the rules of parallel circuits is that the voltage must be the same across all branches of the circuit. Therefore, 12 volts is dropped across resistors R_2 and R_3.

Now that the voltage across resistors R_2 and R_3 is known, the current flow through each can be determined using Ohm's law.

$$I_2 = \frac{12}{40}$$

$$I_2 = 0.3 \text{ amp}$$

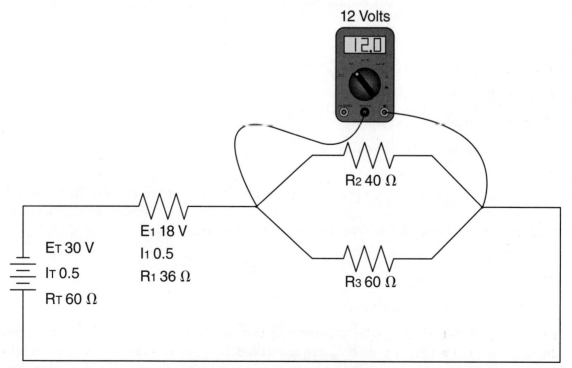

Figure 5-5   The same voltage is dropped across the two parallel resistors as is dropped across the combination resistor.

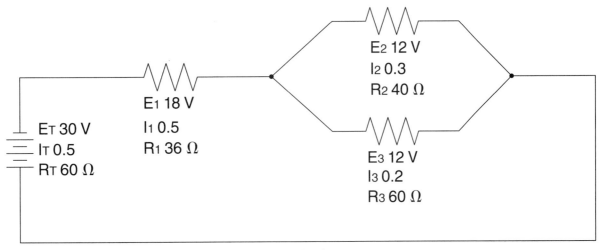

Figure 5-6 All circuit values have been determined.

$$I_3 = \frac{12}{60}$$

$$I_3 = 0.2 \text{ amp}$$

The circuit with all values is shown in Figure 5-6.

Example Circuit #2

The second example circuit is shown in Figure 5-7. The unknown values to be determined in this circuit are:

R_T - Total resistance of the circuit

I_T - Total current in the circuit

I_1 - Current flow through resistor R_1

I_2 - Current flow through resistor R_2

E_1 - Voltage drop across resistor R_1

E_2 - Voltage drop across resistor R_2

I_3 - Current flow through resistor R_3

E_3 - Voltage drop across resistor R_3

E_4 - Voltage drop across resistor R_4

E_5 - Voltage drop across resistor R_5

I_4 - Current flow through resistor R_4

I_5 - Current flow through resistor R_5

The first step in determining the unknown values for this circuit is to trace the current paths to determine which components are connected in series and parallel with each other. Assume that electrons leave the negative battery terminal and return to the positive terminal. Electrons can flow from the battery to the branch containing resistors R_1 and R_2.

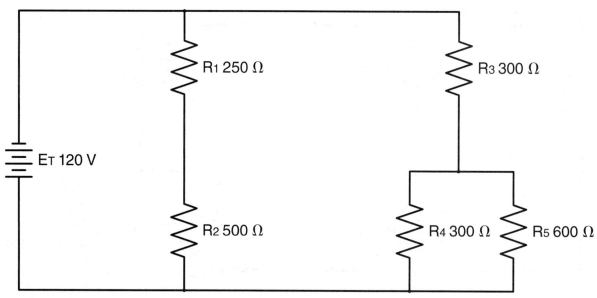

Figure 5-7 Example circuit 2.

Current can then flow through these two resistors and return to the positive battery terminal, as shown in Figure 5-8. Notice that there is only one path through the branch containing these two resistors. The same current must flow through both. Since the same current must flow through both resistors, they are connected in series with each other.

A second current path exists through the branch containing resistors R_3, R_4, and R_5, as shown in Figure 5-9. All of the current of that branch must flow through resistor R_3, but the current then divides through resistors R_4 and R_5. Resistors R_4 and R_5 are connected in parallel with each other because there is more than one path for current flow, but resistor R_3 is connected in series with R_4 and R_5. The branch containing resistors R_1 and R_2 is connected in parallel with the branch containing resistors R_3, R_4, and R_5.

The next step is to find the total resistance of the circuit. This can be accomplished by combining series and parallel connected resistors to form one single resistor. The procedure

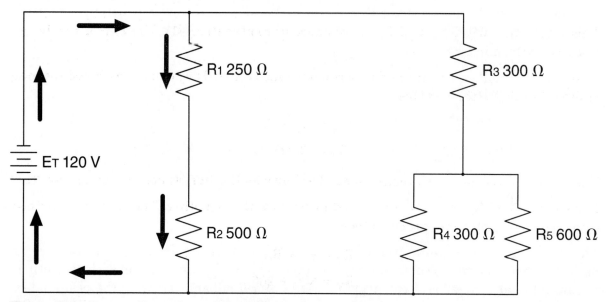

Figure 5-8 Current flows through resistors 1 and 2.

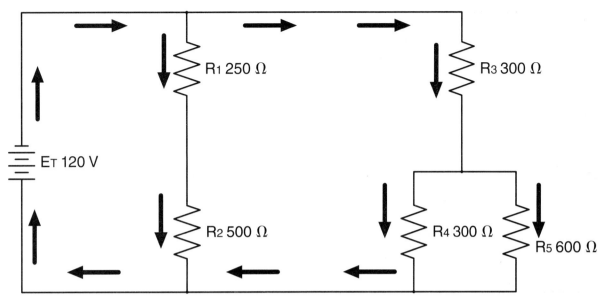

Figure 5-9 A current path also exists through resistors 3, 4, and 5.

is continued until there is a simple series or parallel circuit. The first two resistors to be combined are R_4 and R_5. Since these two resistors are connected in parallel, their total resistance can be determined using the following formula:

$$R_T = \frac{1}{\frac{1}{R_4} + \frac{1}{R_5}}$$

$$R_T = \frac{1}{\frac{1}{300} + \frac{1}{600}}$$

$$R_T = 200 \ \Omega$$

This total value will be called R_{C1} (resistance of combination #1). The circuit can be re-drawn as shown in Figure 5-10.

Resistors R_1 and R_2 are connected in series with each other. They can be combined into one resistor by adding their values.

$$R_T = R_1 + R_2$$

$$R_T = 750 \ \Omega$$

The total value of these two resistors will be shown as R_{C2} (resistance of combination #2).

Resistors R_3 and R_{C1} are also connected in series with each other. They can be combined into one resistor by adding their values.

$$R_T = R_3 + R_{C1}$$

$$R_T = 300 + 200$$

$$R_T = 500 \ \Omega$$

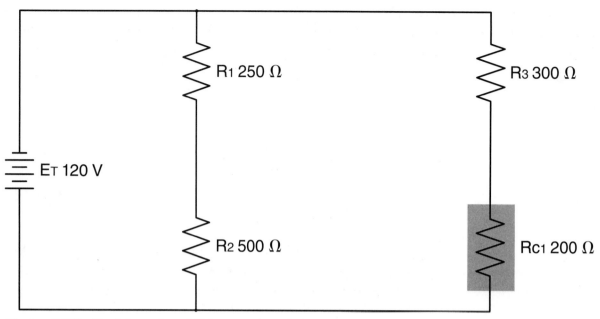

Figure 5-10   Resistors 4 and 5 form one resistor.

The total values of these two resistors will be shown as R_{C3} (resistance of combination #3). The circuit can now be redrawn as a simple parallel circuit, as shown in Figure 5-11.

The total resistance of the circuit can now be determined.

$$R_T = \frac{1}{\frac{1}{750} + \frac{1}{500}}$$

$$R_T = 300 \ \Omega$$

Now that the total resistance is known, the total circuit current can be computed using Ohm's law.

$$I_T = \frac{120}{300}$$

$$I_T = 0.4 \ \text{amp}$$

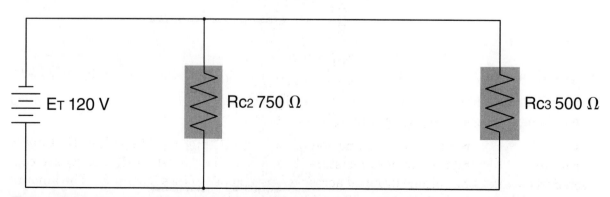

Figure 5-11   The circuit has been reduced to a simple parallel circuit.

In a parallel circuit, the voltage must be the same across all branches. Therefore, a voltage of 120 volts is applied across resistors R_{C2} and R_{C3}. The current flow through these branches can be determined using Ohm's law.

$$I_{C2} = \frac{120}{750}$$

$$I_{C2} = 0.16 \text{ amp}$$

$$I_{C3} = \frac{120}{500}$$

$$I_{C3} = 0.24 \text{ amp}$$

Note that if the currents flowing through resistors R_{C2} and R_{C3} are added together, they will equal the total circuit current ($0.16 + 0.24 = 0.4$). One of the rules for parallel circuits states that the total current of a parallel circuit is equal to the sum of the currents through the branches.

Resistor R_{C2} is, in reality, a combination of resistors R_1 and R_2. The values that apply to R_{C2}, therefore, apply to resistors R_1 and R_2. Since R_1 and R_2 are connected in series, the current flowing through R_{C2} flows through both of them. The voltage drop across R_1 and R_2 can now be determined using Ohm's law.

$$E_1 = 0.16 \times 250$$

$$E_1 = 40 \text{ volts}$$

$$E_2 = 0.16 \times 500$$

$$E_2 = 80 \text{ volts}$$

Note that if the voltage drops across resistors R_1 and R_2 are added together, they will equal the voltage applied across the branch ($40 + 80 = 120$). Recall that one of the rules concerning series circuits states that the sum of the voltage drops must equal the applied voltage.

Resistor R_{C3} is, in reality, the combination of resistors R_3 and R_{C1}. Since R_3 and R_{C1} are connected in series, they will have the same current flowing through them that flows through R_{C3}. The voltage drop across these two resistors can be determined with Ohm's law.

$$E_3 = 0.24 \times 300$$

$$E_3 = 72 \text{ volts}$$

$$E_{C1} = 0.24 \times 200$$

$$E_{C1} = 48 \text{ volts}$$

The values for resistors R_1, R_2, R_3, and R_{C1} are shown in Figure 5-12.

Resistor R_{C1} is, in reality, the combination of resistors R_4 and R_5. The values that apply to resistor R_{C1}, therefore, apply to resistors R_4 and R_5. Since resistors R_4 and R_5 are connected in parallel, the voltage dropped across R_{C1} is dropped across R_4 and R_5. The amount of current flowing through resistors R_4 and R_5 can now be computed.

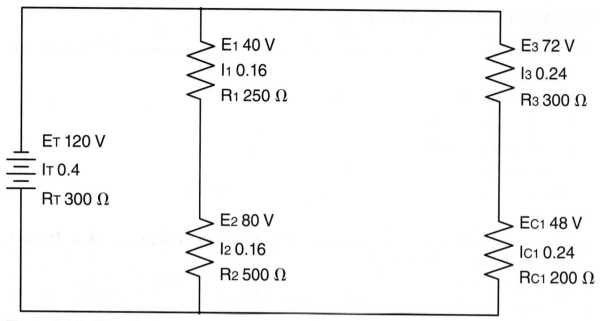

Figure 5-12   The unknown values for R_1, R_2, R_3, and R_{C1} have been determined.

$$I_4 = \frac{48}{300}$$

$$I_4 = 0.16 \text{ amp}$$

$$I_5 = \frac{48}{600}$$

$$I_5 = 0.08 \text{ amp}$$

The circuit with all values is shown in Figure 5-13.

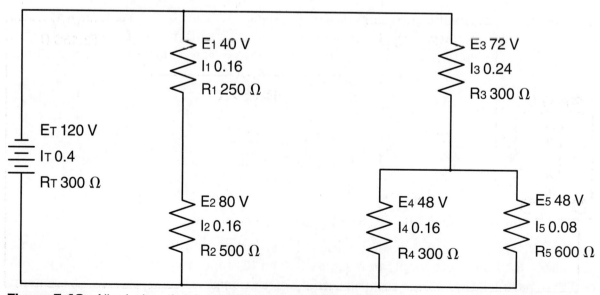

Figure 5-13   All missing circuit values have been determined.

LABORATORY EXERCISE

Name _____ Date _____

Materials Required

1 208 volt AC power supply

1 120 volt AC power supply

2 100 ohm resistors

2 150 ohm resistors

2 250 ohm resistors

1 AC ammeter (in-line or clamp-on may be used. If a clamp-on type is used, a 10:1 scale divider is recommended.)

1 AC voltmeter

1 Ohmmeter

Connecting wires

1. Connect the circuit shown in Figure 5-14. **Make sure that the AC power remains turned off until you are told to turn it on.**

2. Determine the combined resistance of resistors R_2 and R_3. These resistors are connected in series with each other. Therefore, the total resistance is the sum of the two values.

$$R_{C\,2\&3} = R_2 + R_3$$

$R_{C\,2\&3}$ _____ Ω

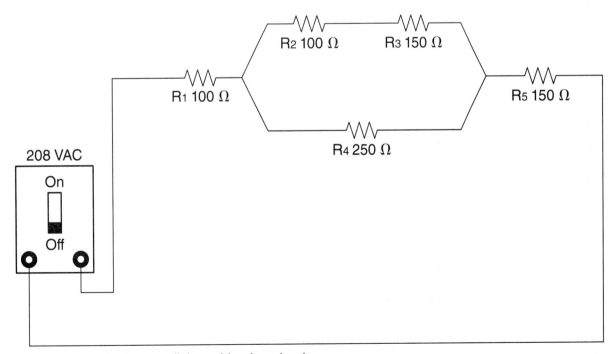

Figure 5-14 Series-parallel combination circuit.

3. Resistors R_2 and R_3 are connected in parallel with resistor R_4. The combined resistance of these three resistors can be determined using the following formula:

$$R_{C\,2,3,4} = \frac{1}{\dfrac{1}{R_{C\,2,3}} + \dfrac{1}{R_4}}$$

Alternatively, because the combined resistance of resistors R_2 and R_3 is the same as the value of resistor R_4, the total resistance can be determined using the following formula

$$R_{C\,2,3,4} = \frac{R}{N}$$

where R is the value of the resistors and N is the number of resistors.

$R_{C\,2,3,4} =$ _____ Ω

4. The circuit has now become a simple series circuit with resistors R_1, $R_{C\,2,3,4}$, and R_5 connected in series. Determine the total resistance of the circuit using the following formula:

$$R_T = R_1 + R_{C\,2,3} + R_5$$

$R_T =$ _____ Ω

5. Disconnect the circuit from the power supply and measure the total resistance of the circuit with an ohmmeter. Compare this value with the computed value. Are the values within 5 percent of each other? After making the measurement, reconnect the circuit to the power source but do not turn the power on.

_____ Ω

6. Assuming a voltage of 208 volts, compute the total circuit current using the following formula: (Note: Round off the answer to the second decimal place or hundredth of an amp.)

$$I_T = \frac{E_T}{R_T}$$

$I_T =$ _____ A

7. Resistors R_1 and R_5 are connected in series with the combined resistors $R_{C\,2,3,4}$. Therefore, the total circuit current flows through resistors R_1 and R_5. Determine the voltage drop across these two resistors using Ohm's law.

$$E = I \times R$$

$E_1 =$ _____ volts

$E_5 =$ _____ volts

8. The voltage drop across combined resistors 2, 3, and 4 can be determined using Ohm's law. The total circuit current flows through this combination. The voltage drop across this combination is equal to the total circuit current and the combined resistance of the three resistors. Determine the voltage drop across the combination of resistors.

$E_{C\,2,3,4} =$ _____ volts

9. Resistor R_4 is a single resistor in the combination of resistors 2, 3, and 4. Therefore, the voltage dropped across the combination is dropped across resistor R_4. Determine the amount of current flow through resistor R_4.

$$I = \frac{E}{R}$$

$I_4 =$ _____ A

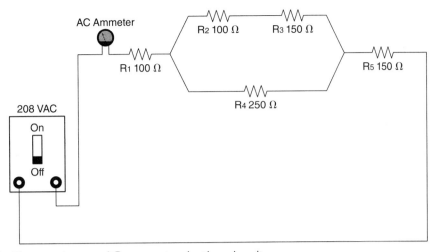

Figure 5-15 Connecting an AC ammeter in the circuit.

10. The voltage dropped across the combined resistors 2, 3, and 4 is also dropped across resistors R_2 and R_3. The current flow through these two resistors can be calculated using the voltage drop across them and the total resistance of the two resistors. Determine the current flow through these resistors using Ohm's law.

    $I_{2\&3}$ = _____ A

11. The voltage drop across resistors R_2 and R_3 can be determined using Ohm's law. Calculate the voltage drop across resistors R_2 and R_3.

    E_2 = _____ volts

    E_3 = _____ volts

12. Connect an ammeter in series with the circuit as shown in Figure 5-15. Turn on the power and measure the total circuit current. **Turn off the power**. Compare the measured value with the value computed in step 6. Are these two values within 5 percent of each other?

    _____

13. Connect an AC voltmeter across resistor R_1 as shown in Figure 5-16. Turn on the power and measure the voltage drop across the resistor. **Turn off the power**. Compare the measured value with the value computed in step 7. Are the two values within 5 percent of each other?

    _____

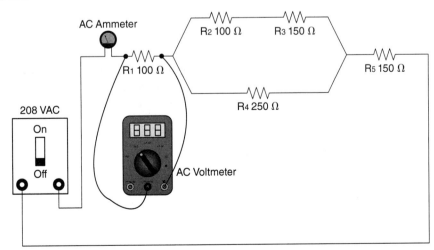

Figure 5-16 A voltmeter measures the voltage drop across resistor R_1.

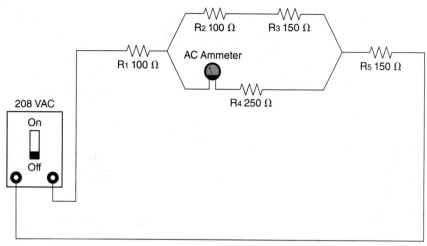

Figure 5-17   The ammeter measures the current flow through resistor R$_4$.

14. Reconnect the AC voltmeter across resistor R$_5$. Turn on the power and measure the voltage drop across the resistor. **Turn off the power.** Compare the measured value with the value computed in step 7. Are the two values within 5 percent of each other?

   _____

15. Reconnect the AC ammeter in the circuit to measure the current flow through resistor R$_4$ as shown in Figure 5-17. Turn on the power and measure the current. **Turn off the power.** Compare the measured value with the value computed in step 9. Are the two values within 5 percent of each other?

   _____

16. Connect the AC voltmeter across resistor R$_4$. Turn on the power and measure the voltage drop across the resistor. **Turn off the power.** Compare the measured value with the value computed in step 8. Are the two values within 5 percent of each other?

   _____

17. Reconnect the circuit as shown in Figure 5-18. The AC ammeter has been connected in series with resistors R$_2$ and R$_3$. Turn on the power and measure the current flow through resistors R$_2$ and R$_3$. **Turn off the power.** Compare the measured value with the value computed in step 10. Are the two values within 5 percent of each other?

   _____

18. Connect an AC voltmeter across resistor R$_2$. Turn on the power and measure the voltage drop across the resistor. **Turn off the power.** Compare the measured value with the value computed in step 11. Are the two values within 5 percent of each other?

   _____

19. Connect an AC voltmeter across resistor R$_3$. Turn on the power and measure the voltage drop across the resistor. **Turn off the power.** Compare the measured value with the value computed in step 11. Are the two values within 5 percent of each other?

   _____

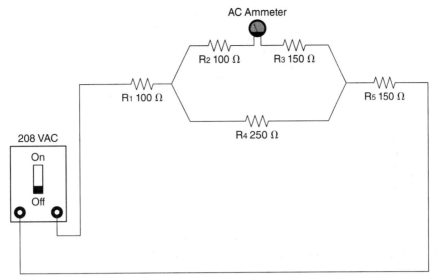

Figure 5-18 The ammeter measures the current flow through resistors R_2 and R_3.

20. Connect the circuit shown in Figure 5-19. Do not turn on the power until you are instructed to do so.

21. Calculate the combined resistance of resistors R_4 and R_5. These two resistors are connected in parallel with each other. The total resistance can be determined using the following formula:

$$R_{4,5} = \frac{1}{\dfrac{1}{R_4} + \dfrac{1}{R_5}}$$

$R_{4,5}$ = _____ Ω

22. The combined resistors R_4 and R_5 are connected in series with resistor R_3. Calculate the resistance of the combined resistors $R_{4,5}$ and R_3 with the following formula:

$$R_{3,4,5} = R_3 + R_{4,5}$$

$R_{3,4,5}$ = _____ Ω

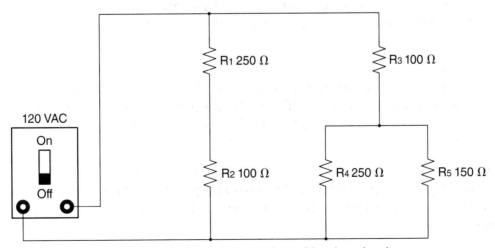

Figure 5-19 Connection schematic for the second combination circuit.

23. Resistors R_1 and R_2 are connected in series. Determine the total combined resistance of resistors R_1 and R_2 using the following formula:

$$R_{1,2} = R_1 + R_2$$

$R_{1,2} = $ _____ Ω

24. The combination or resistors R_1 and R_2 is connected in parallel with the combination of resistors R_3, R_4, and R_5. Calculate the total circuit resistance using the following formula: (Note: Round off the answer to the second decimal place or hundredth of an ohm.)

$$R_T = \frac{1}{\dfrac{1}{R_{1,2}} + \dfrac{1}{R_{3,4,5}}}$$

$R_T = $ _____ Ω

25. Disconnect the circuit from the power supply and measure the total resistance of the circuit with an ohmmeter. Compare the measured value with the value calculated in step 24. Are the values within 5 percent of each other?

_____

26. Assuming a voltage of 120 volts, calculate the total circuit current using the following formula: (Note: Round the answer off to the second decimal place or hundredth of an amp.)

$$I_T = \frac{E}{R_T}$$

$I_T = $ _____ A

27. In the circuit shown in Figure 5-19, resistors R_1 and R_2 form one branch of a parallel circuit. Therefore, 120 volts are connected across the two series resistors. Because the resistors are connected in series, they will each have the same current. Calculate the current through resistors R_1 and R_2 using the following formula: (Note: Round the answer off to the second decimal place or hundredth of an amp.)

$$I_{1,2} = \frac{E}{R_{1,2}}$$

$I_{1,2} = $ _____ A

28. Because the current flow through resistors R_1 and R_2 is now known, the voltage drop across each resistor can be calculated using Ohm's law.

$$E = I \times R$$

$E_1 = $ _____ volts

$E_2 = $ _____ volts

29. In a series circuit, the sum of the voltage drops must equal the applied voltage. Therefore, the sum of the voltage drops across R_1 and R_2 should equal the voltage applied across the branch. Add the values of E_1 and E_2. Does the sum equal the 120 volts applied across the branch? (Note: There may be a slight difference caused by rounding off values.)

_____

_____

_____

30. The second branch of the circuit comprises resistor R_3 and the combination of resistors R_4 and R_5. Determine the current flow through the second branch using the following formula: (Note: Round the answer off to the second decimal place or hundredth of an amp.)

$$I_{3,4,5} = \frac{E}{R_{3,4,5}}$$

$I_{3,4,5} = $ _____ A

31. The voltage drop across resistor R_3 can be calculated using the following formula:

$$E = I \times R_3$$

$E_3 = $ _____ volts

32. The voltage drop across combination resistor $R_{4,5}$ can be calculated using the following formula: (Note: Round the answer off to the second decimal place or hundredth of a volt.)

$$E_{4,5} = I \times R_{4,5}$$

$E_{4,5} = $ _____ volts

33. Resistors R_3 and combination resistor $R_{4,5}$ are connected in series with each other. Therefore, the sum of the voltage drops should equal the applied voltage of 120 volts. Does the sum of E_3 and $E_{4,5}$ equal the applied voltage of 120 volts? (Note: A slight difference caused by rounding off values may occur.)

_____

_____

_____

34. Resistors R_4 and R_5 are connected in parallel with each other. Therefore, the voltage drop across combination resistor $R_{4,5}$ is dropped across each of the two resistors that comprise the combination. The current flow through each resistor can be determined with Ohm's law.

$$I_4 = \frac{E}{R_4}$$

$I_4 = $ _____ A

$$I_5 = \frac{E}{R_5}$$

$I_5 = $ _____ A

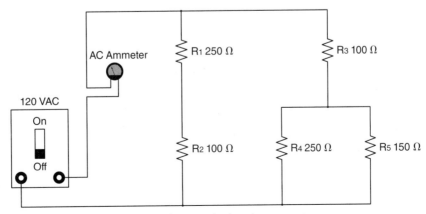

Figure 5-20   The ammeter measures the total circuit current.

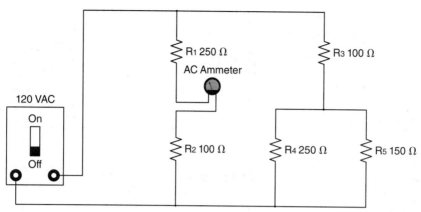

Figure 5-21 The ammeter measures the current through resistors R_1 and R_2.

35. Reconnect the circuit to the power supply and install an AC ammeter in the circuit as shown in Figure 5-20. Turn on the power and measure the total circuit current. **Turn off the power**. Compare the measured value with the value computed in step 26.

    $I_T = $ _____ A

    Are the values within 5 percent of each other? (Note: If there is a significant difference in the measured and calculated values, measure the input voltage with an AC voltmeter. The calculations were made with the assumption that the input voltage is 208 volts. If the voltage is significantly greater, the current will be greater than the calculated value. Also, the measured voltage drops will be greater than calculated. If the input voltage is significantly less, the current and voltage drop measurements will also be less.)

    _____

36. Reconnect the circuit as shown in Figure 5-21. Turn on the power and measure the current flow through resistors R_1 and R_2. **Turn off the power**. Compare the measured value of current with the value computed of current in step 27.

    $I_{1,2} = $ _____ A

    Are the measured and calculated values within 5 percent of each other?

    _____

37. Turn on the power and measure the voltage drop across resistors R_1 and R_2 with an AC voltmeter. **Turn off the power**. Compare the measured values with the values computed in step 28.

    $E_1 = $ _____ volts

    $E_2 = $ _____ volts

    Are the measured and computed values within 5 percent of each other?

    _____

38. Reconnect the circuit as shown in Figure 5-22. Turn on the power and measure the current flow through resistor R_3 and combination resistors R_4 and R_5. **Turn off the power**. Compare the measured value with the value calculated in step 30.

    $I_{3,(4,5)}$ _____ A

    Are the measured and calculated values within 5 percent of each other?

    _____

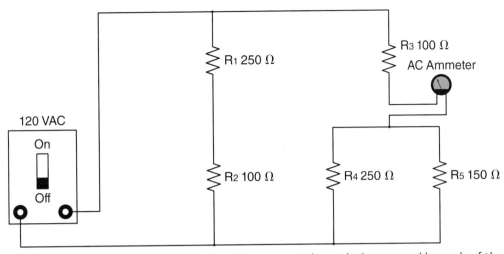

Figure 5-22 The ammeter meter measures the current through the second branch of the circuit.

39. Turn on the power and measure the voltage drop across resistor R_3. **Turn off the power**. Compare the measured value with the value computed in step

    $E_3 =$ _____ volts

    Are the measured and calculated values within 5 percent of each other?

    _____

40. Resistors R_4 and R_5 are connected in parallel with each other and form the combination resistance $R_{4,5}$. The voltage drop across these two resistors will be the same. Turn on the power and measure the voltage drop across these resistors with an AC voltmeter as shown in Figure 5-23. **Turn off the power**. Compare the measured value with the value calculated in step 32.

    $E_{4,5} =$ _____ volts

    Are the measured value and calculated value within 5 percent of each other?

    _____

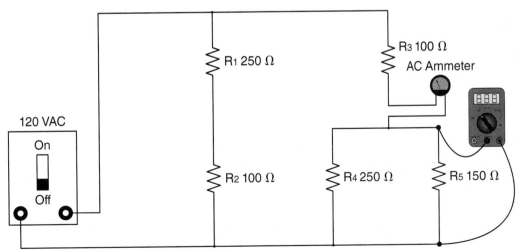

Figure 5-23 The voltmeter measures the voltage drop across resistors R_4 and R_5.

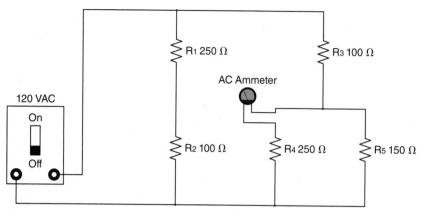

Figure 5-24   The ammeter measures the current flow through resistor R₄.

41. Reconnect the circuit as shown in Figure 5-24. Turn on the power and measure the current through resistor R_4. **Turn off the power.** Compare the measured value with the value calculated in step 34.

$I_4 = $ _____ A

Are the measured value and calculated value within 5 percent of each other?

_____

42. Reconnect the circuit as shown in Figure 5-25. Turn on the power and measure the current flow through resistor R_5. **Turn off the power.** Compare the measured value with the value calculated in step 34.

$I_5 = $ _____ A

Are the measured value and calculated value within 5 percent of each other?

_____

43. Disconnect the circuit and return the components to their proper places.

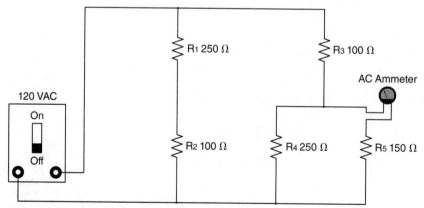

Figure 5-25   The ammeter measures the current flow through resistor R₅.

Review Questions

1. State the three rules for series circuits.

_____

_____

_____

2. State the three rules for parallel circuits.

To answer the following questions, refer to the circuit shown in Figure 5-26.

3. What is the total resistance of resistors R_2 and R_3?

4. What is the total resistance of resistors R_4 and R_5?

5. What is the total resistance of the parallel block containing resistors R_2, R_3, R_4, and R_5?

6. What is the total resistance of this circuit?

7. What is the total amount of current flow in this circuit?

8. How much current flows through resistors R_1 and R_6?

9. What is the voltage drop across resistor R_1?

10. How much voltage is dropped across resistor R_6?

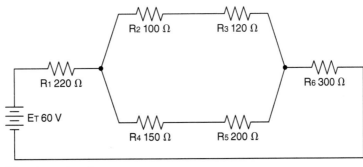

Figure 5-26 Combination circuit.

11. How much voltage is dropped across the parallel block containing resistors R_2, R_3, R_4, and R_5?

_____

12. The voltage drop across the parallel block containing resistors R_2, R_3, R_4, and R_5 is dropped across resistors R_2 and R_3. How much current is flowing through resistors R_2 and R_3?

_____

13. What is the voltage drop across resistor R_2?

_____

14. What is the voltage drop across resistor R_3?

_____

15. What is the sum of the voltage drops across R_2 and R_3?

_____

16. Is the sum of these two voltage drops approximately equal to the voltage across the parallel block containing resistors R_2, R_3, R_4, and R_5?

_____

17. How much current is flowing through resistors R_4 and R_5?

_____

18. What is the voltage drop across resistor R_4?

_____

19. What is the voltage drop across resistor R_5?

_____

20. Is the sum of the two voltage drops across R_4 and R_5 approximately equal to the voltage drop across the parallel block containing resistors R_2, R_3, R_4, and R_5?

_____

21. Add the amount of current flowing through R_2 and R_3 to the amount of current flowing through R_4 and R_5. Is the sum of these two currents approximately equal to the total circuit current?

_____

Unit 6 Resistor Color Code

Objectives

After studying this unit, you should be able to

- Determine the resistance value by the color bands on the resistor.
- Determine the tolerance range of a resistor.
- Determine the value of a 1 percent resistor with five bands of color.
- Measure resistance with an ohmmeter.

The values of a resistor can often be determined by the color code. Many resistors have bands of color that are used to determine the resistance value, tolerance, and in some cases reliability. The color bands represent numbers. Each color represents a different numerical value. The chart shown in Figure 6-1 lists the color and the number value assigned to that color. The resistor shown below the color chart illustrates how to determine the resistor's value. Resistors can have from three to five bands of color. Resistors that have a tolerance of 20 percent have only three color bands. Most resistors contain four bands of color. For resistors with tolerances that range from 10 percent to 2 percent, the first two color bands represent number values. The third color band is called the multiplier. *Multiply the first*

Color	Number Value	Resistor Tolerance
Black	0	
Brown	1	Brown +/-1%
Red	2	Red +/-2%
Orange	3	Gold +/-5%
Yellow	4	Silver +/-10%
Green	5	No color +/-20%
Blue	6	
Violet	7	Gold (0.1 Multiplier)
Gray	8	Silver (0.01 Multiplier)
White	9	

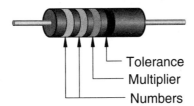

Tolerance
Multiplier
Numbers

Figure 6-1 Resistor color code chart.

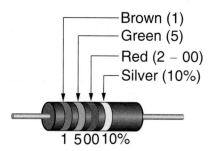

Figure 6-2 Determining resistor values using the color code.

two numbers by 10 the number of times indicated by the number value of the third band. The fourth band indicates the tolerance. For example, assume a resistor has color bands of brown, green, red, and silver, as shown in Figure 6-2. The first two bands represent the numbers 1 and 5 (brown is 1 and green is 5). The third band is red, which has a number value of 2. The number 15 should be multiplied by 10 two times. The value of the resistor is 1500 Ω. Another method that is simpler to understand is to add the number of zeros indicated by the multiplier band to the first two numbers. The multiplier band in this example is red, which has a numeric value of 2. Add two zeros to the first two numbers. The number 15 becomes 1500.

Tolerance

The fourth band is the tolerance band. The tolerance band in this example is silver, which means ±10%. This resistor should be 1500 Ω plus or minus 10 percent. To determine the value limits of this resistor, find 10 percent of 1500.

$$1500 \times 0.10 = 150$$

The value can range from 1500 + 10% or 1500 + 150 = 1650 Ω to 1500 − 10% or 1500 − 150 = 1350 Ω.

Five Band Resistors

Resistors that have a tolerance of ±1% and some military resistors contain five bands of color.

Example #1: The resistor shown in Figure 6-3 contains the following bands of color:

   First band = Brown

   Second band = Black

   Third band = Black

   Fourth band = Brown

   Fifth band = Brown

The brown fifth band indicates that this resistor has a tolerance of ±1%. To determine the value of a 1 percent resistor, the first three bands are numbers and the fourth band is the multiplier. In this example, the first band is brown, which has a number value of 1.

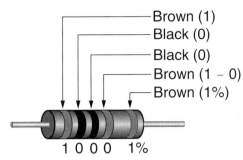

Brown (1)
Black (0)
Black (0)
Brown (1 – 0)
Brown (1%)

1 0 0 0   1%

Figure 6-3 Determining the value of a 1 percent resistor.

The next two bands are black, which represent a number value of 0. The fourth band is brown, which means add one zero to the first three numbers. The value of this resistor is 1000 ±1%.

Example #2: A five-band resistor has the following color bands:

First band = Red

Second band = Orange

Third band = Violet

Fourth band = Red

Fifth band = Brown

The first three bands represent number values. Red is 2, orange is 3, and violet is 7. The fourth band is the multiplier; in this case red represents 2. Add two zeros to the number 237. The value of the resistor is 23,700 ohms. The fifth band is brown, which indicates a tolerance of ±1%.

Military resistors often have five bands of color. The ohmic value and tolerance of these resistors are read in the same manner as a resistor with four bands of color. The fifth band can represent different things. A fifth band of orange or yellow is used to indicate reliability. It has been known for many years that if a resistor is going to fail, it will generally fail within so many hours. The military often pays companies to put resistors in a circuit and operate them for so many days. Resistors that do not fail within the test period are considered to be more reliable than untested resistors. Resistors with a fifth band of orange have a reliability good enough to be used in missile systems, and a resistor with a fifth band of yellow can be used in space flight equipment. A military resistor with a fifth band of white indicates the resistor has solderable leads.

Resistors with tolerance ratings ranging from 0.5% to 0.1% will generally have their values printed directly on the resistor.

Gold and Silver as Multipliers

The colors gold and silver are generally found in the fourth band of a resistor, but they can be used in the multiplier band also. When the color gold is used as the multiplier band, it means to *divide* the first two numbers by 10. If silver is used as the multiplier band, it means to *divide* the first two numbers by 100. For example, assume a resistor has color bands of orange, white, gold, and gold. The value of this resistor is 3.9 ohms with a tolerance of ±5%

(orange = 3; white = 9; gold means to divide 39 by 10, which equals 3.9; and gold in the fourth band means 5 percent tolerance).

Standard Resistance Values

Fixed resistors are generally produced in standard values. The higher the tolerance value, the fewer resistance values available. Standard resistor values are listed in the chart shown in Figure 6-4. In the column under 10%, there are only twelve values of resistors listed. These standard values, however, can be multiplied by factors of 10. Also, notice that one of the standard values listed is 33 ohms. There are also standard values in 10 percent resistors of 0.33, 3.3, 330, 3,300, 33,000, 330,000, and 3,300,000 ohms. The 5% column lists twenty four resistor values and the 1% column lists ninety six values. All of the values listed in the chart can be multiplied by factors of 10 to obtain other resistance values.

Power Rating

Resistors also have a power rating in watts, which should not be exceeded or damage will occur to the resistor. The amount of heat that must be dissipated by the resistor can be determined by the use of one of the following formulas:

$$P = \frac{E^2}{R}$$

$$P = I^2 R$$

$$P = EI$$

Example:

The resistor shown in Figure 6-5 has a value of 100 Ω and a power rating of 0.5 watt. If the resistor is connected to a 10 volt power supply, will it be damaged?

Solution:

Using the formula shown below, determine the amount of heat that will be dissipated by the resistor.

$$P = \frac{E^2}{R}$$

$$P = \frac{100}{100}$$

$$P = 1 \text{ watt}$$

Since the resistor has a power rating of 0.5 watt and the amount of heat that will be dissipated is 1 watt, the resistor will be damaged.

STANDARD RESISTANCE VALUES

.1%, .25% .5%	1%	.1%, .25% .5%	1%	.1%, .25% .5%	1%	.1%, .25% .5%	1%	.1%, .25% .5%	1%
10.0	10.0	17.2	-	29.4	29.4	50.5	-	86.6	86.6
10.1	-	17.4	17.4	29.8	-	51.1	51.1	87.6	-
10.2	10.2	17.6	-	30.1	30.1	51.7	-	88.7	88.7
10.4	-	17.8	17.8	30.5	-	52.3	52.3	89.8	-
10.5	10.5	18.0	-	30.9	30.9	53.0	-	90.9	90.9
10.6	-	18.2	18.2	31.2	-	53.6	53.6	92.0	-
10.7	10.7	18.4	-	31.6	31.6	54.2	-	93.1	93.1
10.9	-	18.7	18.7	32.0	-	54.9	54.9	94.2	-
11.0	11.0	18.9	-	32.4	32.4	55.6	-	95.3	95.3
11.1	-	19.1	19.1	32.8	-	56.2	56.2	96.5	-
11.3	11.3	19.3	-	33.2	33.2	56.9	-	97.6	97.6
11.4	-	19.6	19.6	33.6	-	57.6	57.6	98.8	-
11.5	11.5	19.8	-	34.0	34.0	58.3	-		
11.7	-	20.0	20.0	34.4	-	59.0	59.0		
11.8	11.8	20.3	-	34.8	34.8	59.7	-		
12.0	-	20.5	20.5	35.2	-	60.4	60.4		
12.1	12.1	20.8	-	35.7	35.7	61.2	-		
12.3	-	21.0	21.0	36.1	-	61.9	61.9		
12.4	12.4	21.3	-	36.5	36.5	62.6	-		
12.6	-	21.5	21.5	37.0	-	63.4	63.4		
12.7	12.7	21.8	-	37.4	37.4	64.2	-	2%, 5%	10%
12.9	-	22.1	22.1	37.9	-	64.9	64.9	10	10
13.0	13.0	22.3	-	38.3	38.3	65.7	-	11	-
13.2	-	22.6	22.6	38.8	-	66.5	66.5	12	12
13.3	13.3	22.9	-	39.2	39.2	67.3	-	13	-
13.5	-	23.2	23.2	39.7	-	68.1	68.1	15	15
13.7	13.7	23.4	-	40.2	40.2	69.0	-	16	-
13.8	-	23.7	23.7	40.7	-	69.8	69.8	18	18
14.0	14.0	24.0	-	41.2	41.2	70.6	-	20	-
14.2	-	24.3	24.3	41.7	-	71.5	71.5	22	22
14.3	14.3	24.6	-	42.2	42.2	72.3	-	24	-
14.5	-	24.9	24.9	42.7	-	73.2	73.2	27	27
14.7	14.7	25.2	-	43.2	43.2	74.1	-	30	-
14.9	-	25.5	25.5	43.7	-	75.0	75.0	33	33
15.0	15.0	25.8	-	44.2	44.2	75.9	-	36	-
15.2	-	26.1	26.1	44.8	-	76.8	76.8	39	39
15.4	15.4	26.4	-	45.3	45.3	77.7	-	43	-
15.6	-	26.7	26.7	45.9	-	78.7	78.7	47	47
15.8	15.8	27.1	-	46.4	46.4	79.6	-	51	-
16.0	-	27.4	27.4	47.0	-	80.6	80.6	56	56
16.2	16.2	27.7	-	47.5	47.5	81.6	-	62	-
16.4	-	28.0	28.0	48.1	-	82.5	82.5	68	68
16.5	16.5	28.4	-	48.7	48.7	83.5	-	75	-
16.7	-	28.7	28.7	49.3	-	84.5	84.5	82	82
16.9	16.9	29.1	-	49.9	49.9	85.6	-	91	-

Figure 6-4 Standard resistance values.

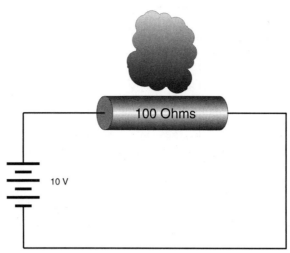

Figure 6-5 Exceeding the power rating causes damage to the resistor.

LABORATORY EXERCISE

Name _____ Date _____

Materials Required

10 color-coded resistors of various values

Ohmmeter

1. Using the table provided in Figure 6-6, list the color of each resistor band in the spaces provided. Then list the resistance value and tolerance according to the color bands. Determine the upper and lower limits of tolerance for each resistor. Next, measure the resistance with an ohmmeter, and, finally, indicate whether the resistor is within its tolerance.

    Example: A resistor has color bands of red, yellow, yellow, and gold. This resistor has been listed on the first line of the chart. After listing the colors and determining the value and tolerance, calculate the upper and lower limit. The example resistor has a marked value of 240,000 Ω with a tolerance of ±5%. The upper and lower limits are determined by the tolerance of the resistor.

    240,000 × 5% (0.05) = 12,000 Ω. Upper limit (240,000 + 12,000 = 252,000 Ω). Lower limit (240,000 − 12,000 = 228,000 Ω). The measured value is determined by measuring the resistance value with an ohmmeter. In this example it is assumed that the ohmmeter measured a resistance of 246,000 Ω. Since this value is within the upper and lower limits, the resistor is within tolerance.

Review Questions

1. A resistor has color bands of orange, orange, orange, and silver. What is the resistance value and tolerance of this resistor?

FIRST COLOR	SECOND COLOR	THIRD COLOR	FOURTH COLOR	MARKED VALUE	TOLERANCE VALUE	UPPER LIMIT	LOWER LIMIT	MEASURED VALUE	WITHIN TOLERANCE
Red	Yellow	Yellow	Gold	240,000	5%	252,000	228,000	246,000	Yes

Figure 6-6 Determining resistor value.

2. A resistor has color bands of brown, red, black, and gold. What is the resistance and tolerance of this resistor?

_____

3. What color bands should be found on a 5100 Ω resistor with a tolerance of ±2%?

_____

4. Is it possible to find a resistor with color bands that are orange, blue, brown, and silver?

_____

5. A resistor has the following color bands: red, yellow, orange, red, and brown. What is the resistance value and tolerance for this resistor?

_____

6. A resistor has the following color bands: green, blue, gold, and red. What is the resistance value and tolerance of this resistor?

_____

7. A 14,000 Ω resistor is needed in a circuit. Is it possible to obtain this resistor in a standard value?

_____

8. A 470 Ω half-watt resistor is connected across 12 volts. Will this resistor be damaged?

_____

9. A resistor has color code bands of orange, orange, red, and red. An ohmmeter is used to check the resistor's value and indicates a value of 3400 Ω. Is the resistor within its tolerance?

_____

10. A resistor has color bands of brown, black, black, red, and brown. An ohmmeter indicates that the resistor has a value of 9950 Ω. Is this resistor within its tolerance?

_____

Basic Switch Connections

Unit 7   Single-Pole Switches

Objectives

After studying this unit, you should be able to

- Discuss the operation of a single-pole switch.
- Identify a single-pole switch.
- Define *switch leg*.
- Employ two methods of connecting single-pole switches.
- Determine the amount of current flow in a neutral conductor.

One of the most common jobs of an electrician is to make basic switch connections. There are three main types of switches used when connecting lighting circuits: single-pole, 3-way, and 4-way. Single-pole switches are used when a device, such as a light or receptacle outlet, is to be controlled from one location. Lights are controlled by interrupting the current flow in one of the circuit conductors. In a common 240/120 volt residential or commercial service, a center-tapped transformer is employed to provide 240 or 120 volts (Figure 7-1). The transformer converts the power line voltage (primary) into 240 volts at the secondary. The secondary winding contains a center tap that is grounded. The grounded center tap is generally referred to as the *neutral*. A voltmeter connected across the entire secondary winding will indicate a value of 240 volts. If the voltage is measured from the center tap to either side of the secondary winding, a voltage of 120 volts will be indicated.

Current Relationships

When connecting loads to a center-tapped transformer, the center tap will carry the sum of the unbalanced loads between the other two conductors. In other words, the center tap connection will carry the difference between the other two conductors. Assume that a transformer of this type is connected to a load that produces 10 amperes in each leg (Figure 7-2). Since each of the ungrounded or hot conductors is carrying the same amount of current, the neutral or grounded conductor will carry no current.

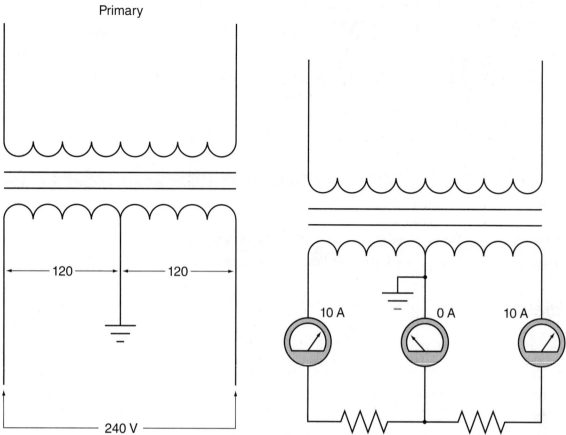

Figure 7-1 Typical 240/120 volt service.   **Figure 7-2** A common single-phase service.

Now assume that one of the ungrounded conductors has a current flow of 10 amperes, and the other has a current flow of 7 amperes. The neutral conductor will now carry a current of 3 amperes (10 − 7 = 3), as shown in Figure 7-3.

Single-Pole Switch Construction

A single-pole switch contains one movable and one stationary contact (Figure 7-4). This type of switch is designated as single-pole single-throw (SPST). The movable contact is called the switch pole. Since this switch is single-pole, it has only one movable contact. The switch pole will make connection with a stationary contact when switched or *thrown* in only one position. The switch is, therefore, called a single-throw. Single-pole switches can be easily identified by the following:

1. They contain only two terminal screws (Figure 7-5). Switches will often contain an extra green screw used for grounding. The bare copper grounding wire connects to the green screw.

2. The words OFF and ON are shown on the switch lever. The switch contacts will be open (OFF) or closed (ON) when the switch lever is thrown in one position.

Basic Switch Connection

Loads intended to operate on 120 volts are connected between the grounded neutral conductor and the ungrounded hot conductor. These circuits are referred to as *branch circuits*. Branch circuits are protected by fuses or circuit breakers installed at the panel box. A branch circuit that supplies power to one lamp is shown in Figure 7-6.

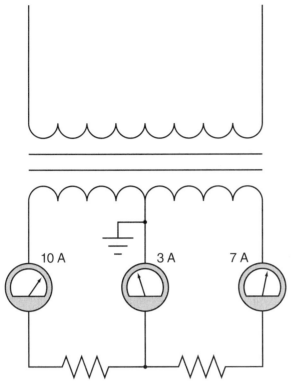

Figure 7-3 The neutral carries the difference between the two hot conductors.

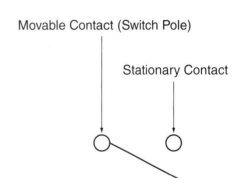

Figure 7-4 Basic construction of a single-pole switch.

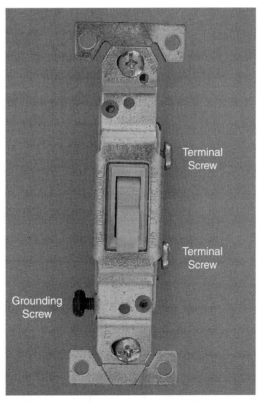

Figure 7-5 A single-pole switch contains only two terminal screws, and the words OFF and ON are shown on the switch lever.

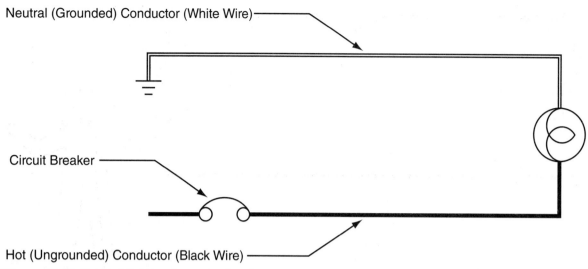

Figure 7-6 Typical lighting branch circuit.

The lamp is controlled by breaking connection between the circuit breaker and lamp (Figure 7-7). Notice that the switch is placed in the hot or ungrounded conductor. The light could be controlled by placing the switch in the neutral conductor, but the *National Electrical Code® (NEC®)* does not permit the neutral conductor to be broken. The only exception to this is if both the neutral (grounded) and hot (ungrounded) conductors are broken at the same time. The reason for this is safety. If the switch were to be placed in the neutral conductor, it would turn the lamp on or off, but power would still be connected to the lamp. If a person were to attempt to work on the lamp with the light switched off, he or she would still be working on a *hot* circuit. This could result in a severe electrical shock (Figure 7-8).

The circuit shown in Figure 7-7 is a schematic diagram of a single-pole switch connection. Schematic diagrams show components in their electrical sequence and are used to illustrate circuit logic. However, they do not indicate how or where the components are placed. Switches are generally located away from the light, not at the light. Switches are commonly installed beside a door that enters a room and the light is installed in the ceiling. When this is the case, the hot conductor must be extended to permit the switch to make or break the circuit (Figure 7-9).

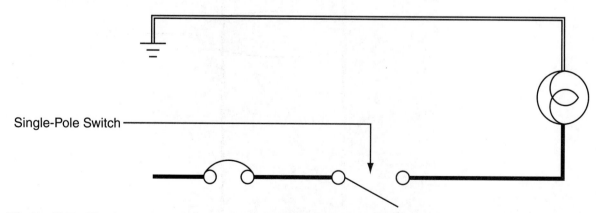

Figure 7-7 The lamp is controlled by breaking the connection between the circuit breaker and the lamp.

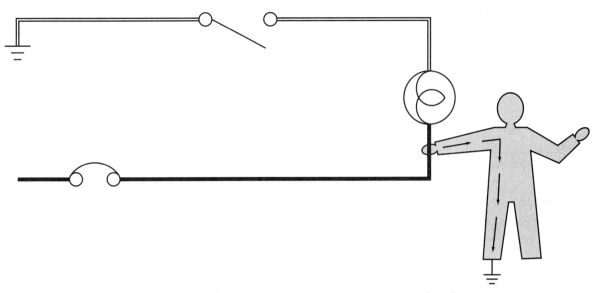

Figure 7-8 Placing the switch in the neutral conductor creates a safety hazard.

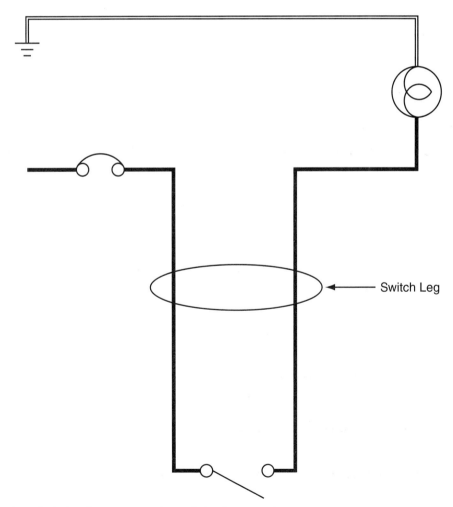

Figure 7-9 A switch leg is an extension of the hot conductor.

Wiring Consideration

When switches are installed in a home or business, several factors must be taken into account.

1. *Switch connections are made using two- and three-conductor cables.* Two-conductor cables actually contain three wires: black, white, and bare copper (Figure 7-10). The black and white wires are actual circuit conductors. The bare copper wire is not considered a circuit conductor because it is there to provide a low-resistance path to ground in the event of a grounded circuit. The bare copper wires connect together throughout the entire building and are connected to green grounding screws on switches and receptacle outlets. Three-conductor cables contain four wires: black, white, red, and bare copper (Figure 7-11). As with two-conductor cables, the bare copper wire is not considered a circuit conductor.

2. *All connections must be made inside a box.* Figure 7-9 shows that a switch leg is an extension of the hot or ungrounded conductor. In reality, this connection must be made inside a box, not in the middle of the conductor.

3. *The white wire is connected to neutral.* The *NEC* requires that the white wire be connected to neutral. For many years the *NEC* permitted white wires to be connected to the hot conductor when they were used as switch legs. The *NEC* now requires that the white wire be reidentified by marking it with colored tape or paint when it is used as a switch leg.

4. *When connection is made to a device such as a lamp or outlet receptacle, the wires must be identified.* This simply means that when connection is made to the lamp, the neutral wire will be white and the hot wire will be red or black. In this way, if an electrician is working on a device, he or she will know immediately which wire is neutral and which is hot.

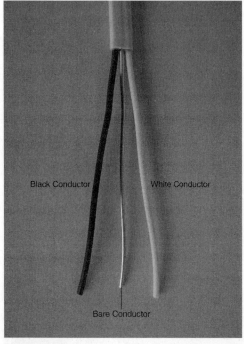

Figure 7-10 Two-conductor cables contain three wires: black, white, and bare copper.

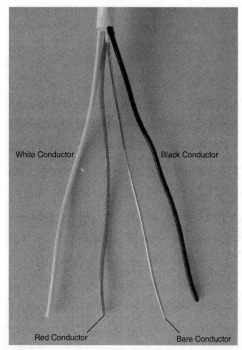

Figure 7-11 Three-conductor cables contain four wires: black, white, red, and bare copper.

Standard Connections for Single-Pole Switches

There are two standard connections used for single-pole switches. One involves bringing power to the switch and then running a two-conductor cable from the switch to the lamp (Figure 7-12). When making this connection, remember that the neutral conductor is never broken. All switching is done in the hot or ungrounded conductor. Therefore, the neutral conductors (white wires) will be connected together inside the switch box, and the hot conductors (black wires) will be connected to the switch. The black and white wires will then be connected to the lamp (Figure 7-13).

The second connection involves supplying power to the light and bringing a switch leg from the light to the switch (Figure 7-14). To make this connection, connect the neutral conductor (white wire in the power cable) directly to the lamp. The *NEC®* requires that the wires that connect to the lamp (device) be identified, so connect the black wire of the switch leg to the other side of the lamp. The white wire of the switch leg is used to carry power down to the switch. Therefore, the white wire of the switch leg connects to the black wire of

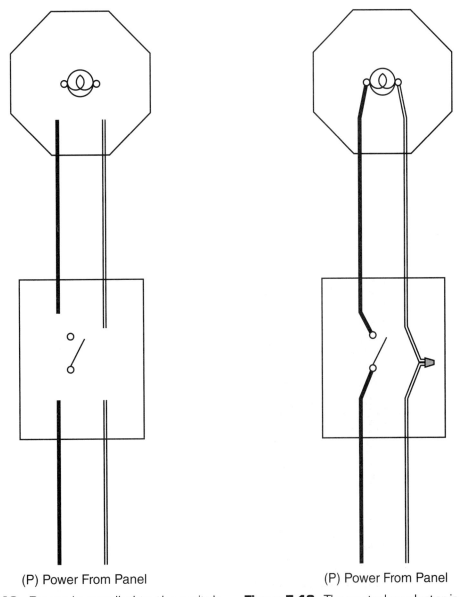

(P) Power From Panel   (P) Power From Panel

Figure 7-12 Power is supplied to the switch.   **Figure 7-13** The neutral conductor is not broken.

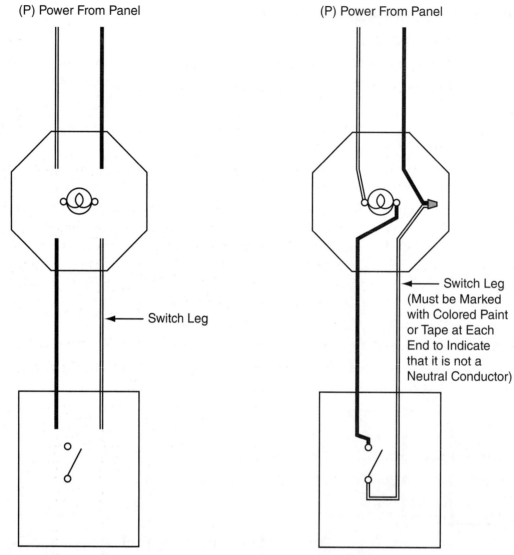

Figure 7-14 Power is brought to the light.

Figure 7-15 The white wire of the switch leg brings power to the switch.

the power cable (Figure 7-15). This is the most common method of making this connection. Be sure to reidentify the white switch leg conductor by marking it in both the light box and the switch box with colored tape or paint.

LABORATORY EXERCISE

Name _____ Date _____

Materials Required

120-volt AC power supply

1 single-pole switch

1 120-volt lamp

1 switch box mounted to a wall stud or on a board

1 standard octagon box mounted on a rafter or on a board

Two-conductor cable (length will be decided by the individual laboratory)

1. Mount the switch box and octagon box as indicated by your instructor.

2. Run a two-conductor cable between the octagon box and the switch box. Be sure to strip the cable at both the octagon box and switch box so that approximately 6 inches of individual wire are available to work with.

3. **Test and verify that the power is turned off.** Connect a two-conductor cable from the power source to the switch box.

4. Connect the circuit shown in Figure 7-16. (Note: The bare grounding wire connection is not shown due to space limitations. The bare grounding wires should connect together in the switch box, and one should be placed under the green screw on the switch if one exists. If the octagon or switch box is metal, a grounding clamp should be used to ground the bare copper wire to the box. If the boxes are made of plastic, fold the bare copper wires in the box so that they are out of the way.

5. Turn on the power and test the circuit by turning the switch on and off.

6. **Turn off the power** and disconnect the circuit.

7. Reroute the power wire so that it enters the octagon box used as the light box.

8. Run a two-conductor cable from the octagon box to the switch box.

9. Connect the circuit illustrated in Figure 7-17.

10. Turn on the power and test the circuit by turning the switch on and off.

11. **Turn off the power** and disconnect the circuit.

12. Return the components to their proper place.

Review Questions

1. How many movable contacts are contained in a single-pole single-throw switch?

   _____

2. State two characteristics that can be used to identify a single-pole switch.

   _____

3. What are the three main types of switches used for connecting lighting circuits?

   _____

4. A 240/120 volt residential service has a load of 12 amperes on one hot leg and 8 amperes on the other. How much current is flowing through the neutral conductor?

   _____

5. Define *switch leg*.

   _____

6. Is it permissible to control a light by placing the switch in the hot or ungrounded conductor?

   _____

7. Many switches contain an extra screw that is green in color. Which wire is connected to this green screw?

   _____

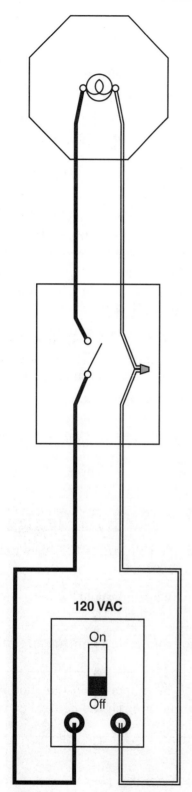

Figure 7-16 Laboratory circuit 1.

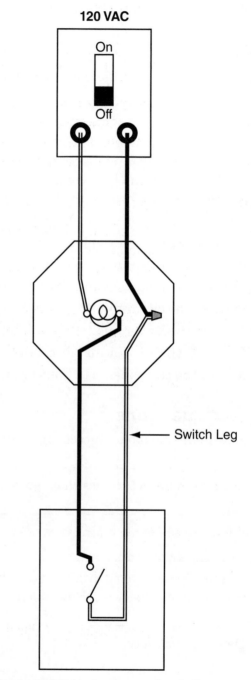

Figure 7-17 Laboratory circuit 2.

Unit 8 3-Way Switches

Objectives

After studying this unit, you should be able to
- Describe the construction of a 3-way switch.
- Determine which terminal is common and which terminals are for travelers.
- Draw a schematic diagram of a 3-way switch connection.
- Connect a 3-way switch circuit to control a lamp from two locations.

Three-way switches are used when it is desirable to control a light or outlet receptacle from two locations. These connections are very common in rooms that have more than one entrance or exit door, in long hallways, and for lights used to illuminate stairs. Making 3-way switch connections is one of the most common jobs for an electrician.

Switch Construction

A 3-way switch is a single-pole double-throw (SPDT) switch. Since the switch is a single-pole, it has only one movable contact (Figure 8-1). Double-throw indicates that the movable contact will make connection with a stationary contact when thrown in either direction. The switch, therefore, contains two stationary contacts. Three-way switches can be identified because they contain three terminal screws. The terminal screw that connects to the movable contact is called the *common* terminal and is generally a different color than the two terminal screws that connect to the stationary contacts (Figure 8-2). Some manufacturers

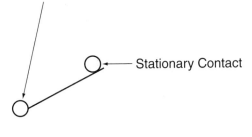

Figure 8-1 A 3-way switch contains one movable and two stationary contacts.

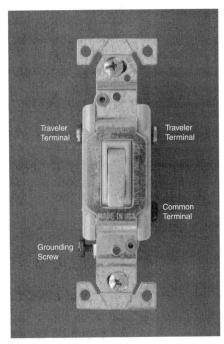

Figure 8-2 The common terminal screw is a different color than the two screws that connect to stationary contacts.

Figure 8-3 Many 3-way switches will have the word *common* written beside the common screen terminal.

also print the word *common* on the back of the switch beside the common screw terminal (Figure 8-3). The common terminal is so well identified because it is necessary to know which terminal is common when connecting a 3-way switch circuit.

Another way of identifying 3-way switches is that they do not have OFF or ON printed on the switch lever as do single-pole switches. Three-way switches can turn a light or receptacle outlet on or off when thrown in either direction.

3-Way Switch Logic

The circuit shown in Figure 8-4 illustrates the logic behind a 3-way switch connection. Notice that the switches are connected in the hot or ungrounded conductor only. The neutral conductor is connected directly to the lamp and is not broken by a switch. The conductors that connect the two switches together are called *travelers*. To understand how this connection works, trace the current path of the hot (ungrounded) conductor from the circuit breaker through the switches, the lamp, and back to neutral. In the circuit shown, a complete path exists from the circuit breaker through the lamp and back to neutral. Therefore, the lamp is turned on with the switches in the position shown in Figure 8-5.

If either of the two 3-way switches is toggled to a different position, the current path will be broken and a circuit will no longer be complete through the lamp, as shown in Figure 8-6. If either of the two switches is again toggled to a different position, the current path will be reestablished through the lamp, as shown in Figure 8-7. Regardless of which switch is toggled to a different position, the light will be alternately turned on or off.

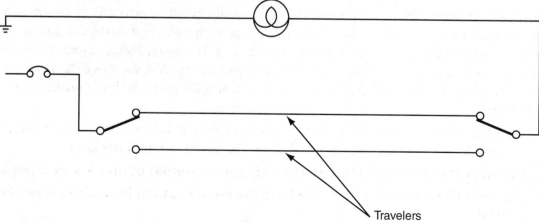

Travelers

Figure 8-4 A basic 3-way switch connection.

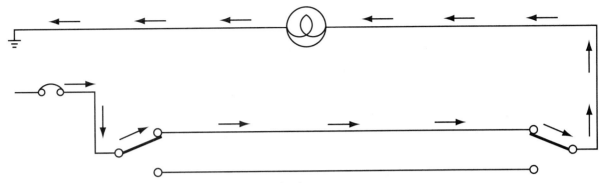

Figure 8-5 A current path exists through the lamp.

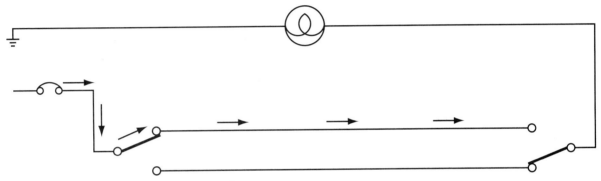

Figure 8-6 The current path is broken.

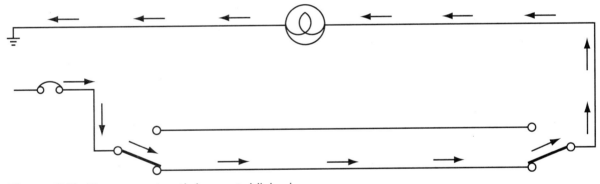

Figure 8-7 The current path is reestablished.

3-Way Switch Connections

Figure 8-5, Figure 8-6, and Figure 8-7 schematically illustrate the logic of a 3-way switch connection. In reality, connection is made with two- and three-conductor cables. When installing the wiring for 3-way switches, a three-conductor cable is connected between the two switches. There are several ways in which 3-way switches can be connected. Making the proper connections is not difficult, however, if the following four rules are followed:

1. **Connect the neutral to the light.** The neutral conductor is not to be broken by a switch. It must be a continuous path from the power panel to the lamp.

2. **Connect the hot conductor to the common terminal of one 3-way switch.**

3. **Connect the other side of the light to the common terminal of the other 3-way switch.**

4. **Connect the travelers.**

Example Connection 1

In the circuit shown in Figure 8-8, power is brought from the panel box to one 3-way switch. A three-conductor cable is connected between the two switches, and a two-conductor cable runs from the other 3-way switch to the lamp. To connect this circuit, follow these four rules for connecting 3-way switches.

1. *Connect the neutral to the light.* The neutral is the white wire of the power cable from the panel box. Neutral conductors should be color-coded white, so connect the neutral conductor to the white wire in the three-conductor cable. Then connect the white wire of the three-conductor cable to the white wire in the two-conductor cable that runs from the second 3-way switch to the light. Connect the lamp to the white wire (Figure 8-9). Note that the neutral conductor is continuous from the power wire to the light. It has not been broken by a switch at any point.

2. *Connect the hot conductor to the common terminal of one 3-way switch.* Since the power cable enters the box of one 3-way switch, the black wire of the power cable will be connected to the common terminal of that 3-way switch (Figure 8-10).

3. *Connect the other side of the light to the common terminal of the other 3-way switch.* The black wire of the two-conductor cable that runs between the switch box and the light is connected to the common terminal of the second 3-way switch (Figure 8-11).

4. *Connect the travelers.* The travelers are used to connect the two remaining terminals on each 3-way switch (Figure 8-12). The red and black conductors of the three-conductor cable are used to make this connection.

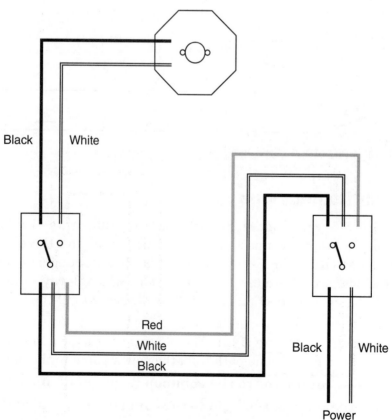

Figure 8-8 Example 1 of 3-way switch.

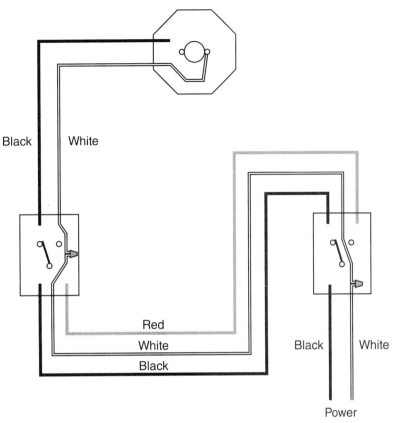

Figure 8-9 Rule 1: Connect the neutral to the light.

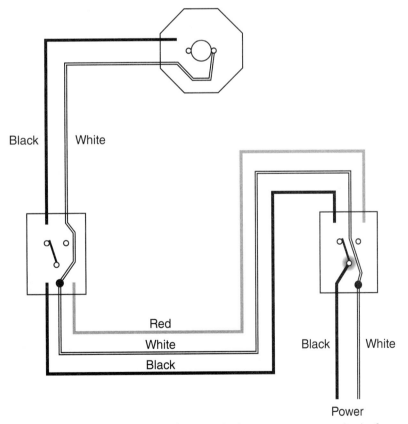

Figure 8-10 Rule 2: Connect the hot conductor to the common terminal of one 3-way switch.

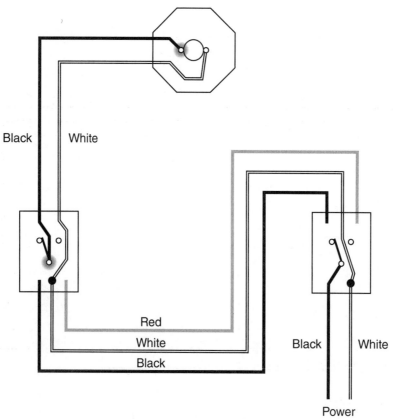

Figure 8-11 Rule 3: Connect the other side of the light to the common terminal of the second 3-way switch.

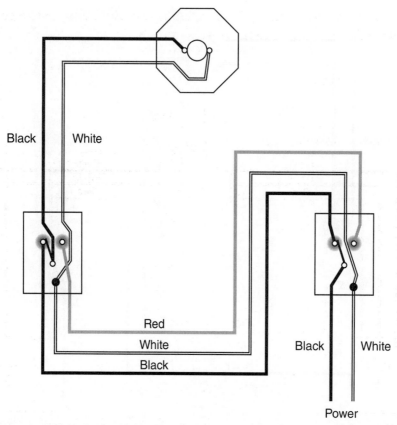

Figure 8-12 Rule 4: Connect the travelers.

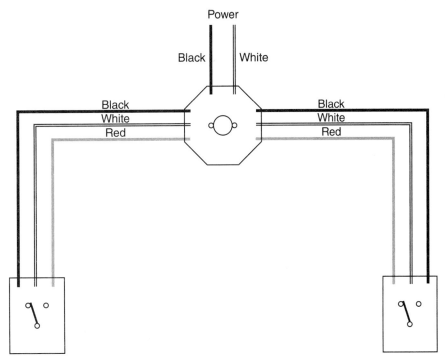

Figure 8-13 Example 2 of 3-way switch.

Example Connection 2

In the next example, power is brought to the light box, and a three-conductor cable is run from the light box to each 3-way switch box (Figure 8-13). The four rules for making 3-way switch connections will be followed.

1. *Connect the neutral to the light.* The neutral is the white wire in the power cable. Connect this wire directly to the light, as in Figure 8-14.

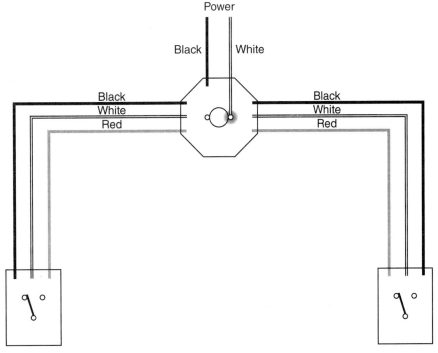

Figure 8-14 The white wire of the power cable is connected to the light.

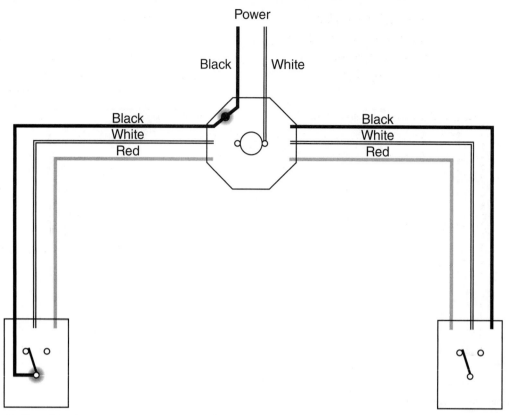

Figure 8-15   The hot conductor is connected to the common terminal of one 3-way switch.

2. *Connect the hot conductor to the common terminal of one 3-way switch.* The black wire of one three-conductor cable will be connected to the hot conductor in the power cable, shown in Figure 8-15. The other end of the black wire will be connected to the common terminal of the 3-way switch. Since the three-conductor cables are used as switch legs, the red or white wires could have been employed to carry the hot wire to the common terminal of the switch. Another helpful rule that can be followed when making 3-way switch connections is to *always* use a black wire to connect to the common terminal of a switch. A black wire can always be used to connect to the common terminal in any type of 3-way switch connection, but this is not true of the white or red wires. When this is done, the electrician will always know which of the wires connects to the common terminal when installing a 3-way switch.

3. *Connect the other side of the light to the common terminal of the other 3-way switch.* The black wire of the other three-conductor cable will be used to make this connection, as shown in Figure 8-16. Notice that the black wire is connected to the common terminal of the switch.

4. *Connect the travelers.* The red and white wires of the three-conductor cables are used as travelers (Figure 8-17). The *NEC*® requires that the white conductor be reidentified by marking it with colored tape or paint because it is a switch leg. The white switch legs should be marked in the light box and each of the switch boxes.

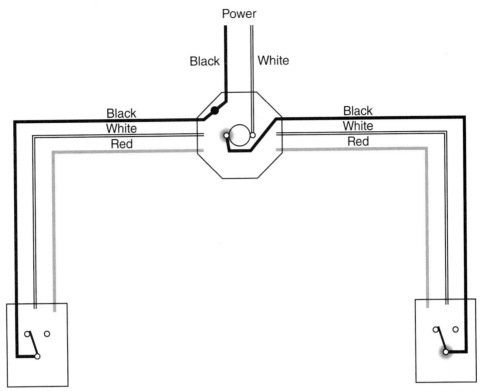

Figure 8-16 The other side of the light is connected to the common terminal of the second 3-way switch.

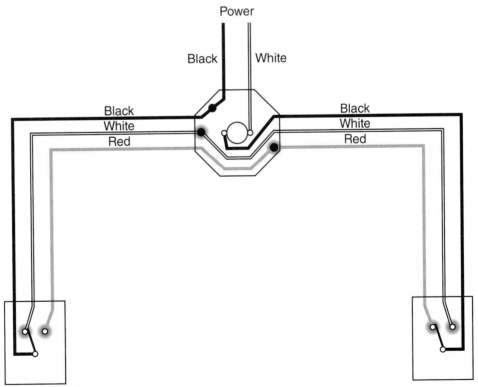

Figure 8-17 The travelers connect the two switches.

Example Connection 3

In this example the power wire is again brought to the light box. A two-conductor cable is run between the light box and one switch box, and a three-conductor cable is connected between the two switch boxes (Figure 8-18). To connect this circuit, the four rules for making 3-way switch connections will again be employed.

1. *Connect the neutral to the light.* The white wire of the power cable is connected directly to the light (Figure 8-19).

2. *Connect the hot conductor to the common terminal of one 3-way switch.* One of the requirements of the *NEC* is that the wires connected to the light must be identified. The neutral conductor must be white and the other wire must be a color other than white or green. Therefore, the black conductor that runs between the light box and the switch box should be connected to the light. That leaves the white wire of the switch leg to carry power to the common terminal of one of the 3-way switches. In the previous example, it was discussed that a black conductor can always be used as the wire that connects to the common terminal of a 3-way switch. To do that, connect the white switch leg in the light box to the hot power wire. Then connect the other end of the white switch leg to the black wire in the three-conductor cable that runs between the two switches. The black wire then connects to the common terminal of the 3-way switch (Figure 8-20).

3. *Connect the other side of the light to the common terminal of the other 3-way switch.* The black conductor of the switch leg that runs from the light box to the switch will be used (Figure 8-21).

4. *Connect the travelers.* The red and white wires of the three-conductor cable are used to connect the traveler terminals of the two switches (Figure 8-22). The white switch leg wire should be reidentified with colored tape or paint at each location.

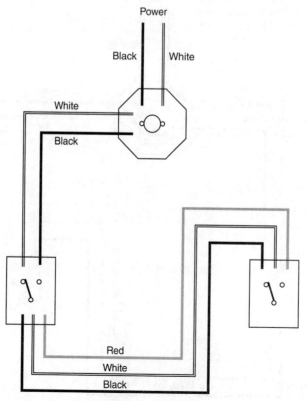

Figure 8-18 Example 3 of 3-way switch.

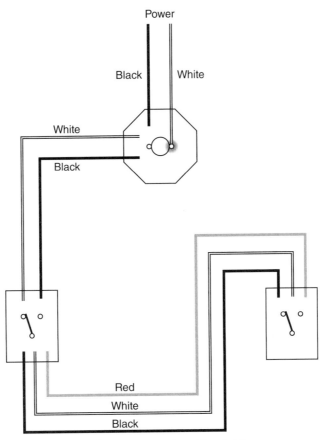

Figure 8-19 The neutral conductor is connected to the light.

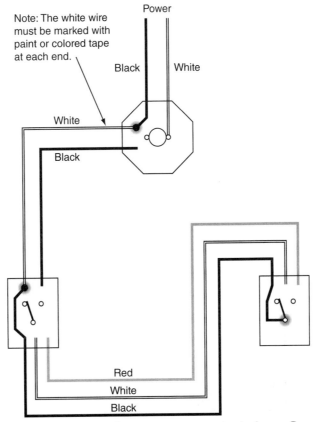

Figure 8-20 The hot wire connects to the common terminal of one 3-way switch.

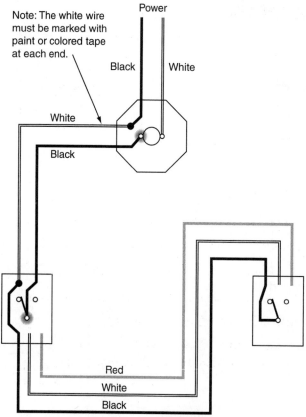

Figure 8-21 The other side of the light is connected to the common terminal of the second 3-way switch.

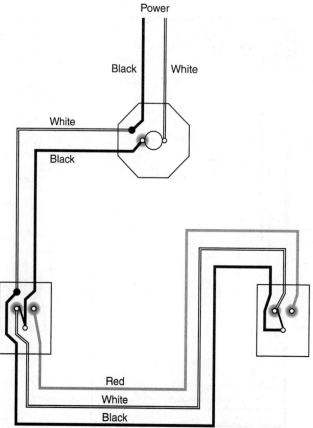

Figure 8-22 Connecting the travelers.

LABORATORY EXERCISE

(See Figure 8-23.)

Materials Required

120-volt AC power supply

1 120-volt lamp (any wattage)

2 3-way switches

2 switch boxes mounted to wall studs or on a board

1 standard octagon box mounted on a rafter or on a board

1 lamp socket that will mount to the octagon box

Two-conductor cable (length will be decided by the individual laboratory)

Three-conductor cable (length will be decided by the individual laboratory)

1. **Test and verify that the power is turned off.**

2. Using the materials listed, mount two switch boxes and one octagon box on a board or wall studs according to the provisions of the laboratory.

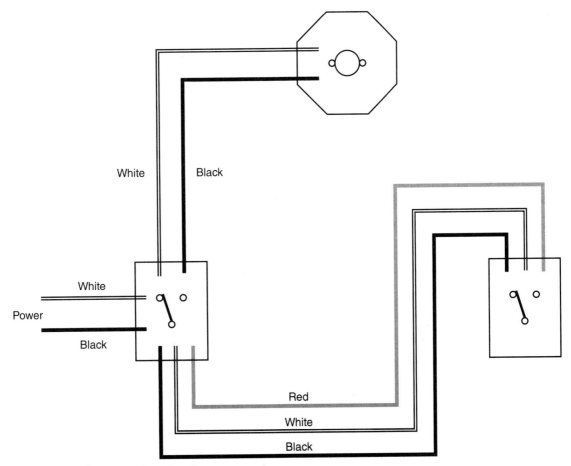

Figure 8-23 Connect the circuit.

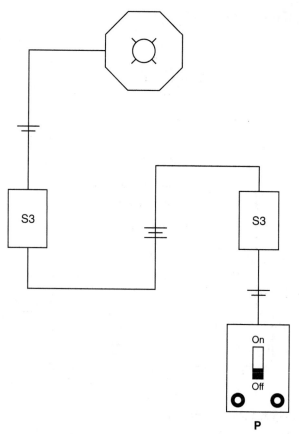

Figure 8-24 First circuit for connection.

3. Connect two- and three-conductor cables between the power supply and the boxes as shown in Figure 8-24. (Note: Blueprints generally indicate electrical cables as lines with hash marks. The same notation will be used in this laboratory exercise. A line with two hash marks indicates a two-conductor cable. A line with three hash marks indicates a three-conductor cable.)

4. Use the four rules for connecting 3-way switches to connect the circuit.

5. Turn on the power and test the circuit by alternately changing the position of each 3-way switch.

6. **Turn off the power** and disconnect the circuit.

7. Reposition the two- and three-conductor cables as shown in Figure 8-25.

8. Use the four rules for connecting 3-way switches to connect the circuit.

9. Turn on the power and test the circuit by alternately changing the position of each 3-way switch.

10. **Turn off the power** and disconnect the circuit.

11. Reposition the two- and three-conductor cables as shown in Figure 8-26.

12. Use the four rules for connecting 3-way switches to connect the circuit.

13. Turn on the power and test the circuit by alternately changing the position of each 3-way switch.

14. **Turn off the power** and disconnect the circuit. Return the components to their proper place.

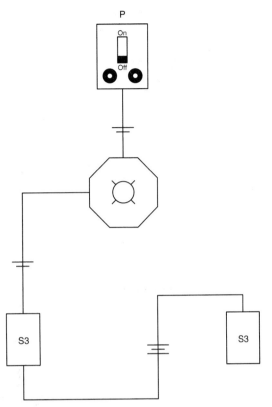

Figure 8-25 Second circuit for connection.

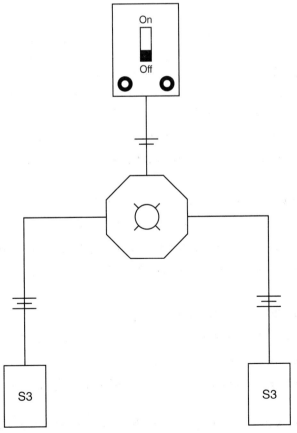

Figure 8-26 Third circuit for connection.

Review Questions

1. How many terminal screws are contained on a 3-way switch?

   _____

2. Name two methods commonly employed to identify the common terminal screw on a 3-way switch.

   _____

3. List four rules for connecting 3-way switches.

   _____

4. What color wire is it possible to always connect to the common terminal of a 3-way switch?

   _____

5. Refer to Figure 8-23. Connect the wires for proper switch operation. Connect the circuit so that a black conductor will supply the common terminal on each 3-way switch.

   _____

6. According to the *NEC*, when white wires are employed as switch legs, what should be done to reidentify the conductors?

   _____

7. What does *SPDT* stand for in reference to switches?

   _____

Unit 9   4-Way Switches

Objectives

After studying this unit, you should be able to

- Discuss the operation of a 4-way switch.
- Identify a 4-way switch.
- Draw a schematic illustrating the operation of a 4-way switch.
- Connect a 4-way switch in a circuit.

Four-way switches are used when it is desirable to control a light or outlet receptacle from more than two locations. Two 3-way switches are always used when a device is controlled from more than one location, but the number of switches above two will be 4-way switches. If a light was to be controlled from seven locations, for example, it would require two 3-way switches and five 4-way switches.

Switch Construction

Four-way switches are double-pole double-throw (DPDT) switches. This means that the switch contains two movable (pole) contacts, and each movable contact can make connection to two stationary contacts. A 4-way switch is constructed like a DPDT knife blade switch with the stationary contacts cross-connected (Figure 9-1). Although DPDT switches normally contain six connection terminals, because the stationary contacts are cross-connected, the 4-way switch requires only four terminal connections (Figure 9-2). When the switch lever

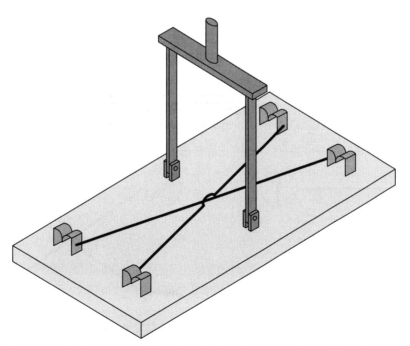

Figure 9-1 A 4-way switch is a double-pole double-throw switch with the stationary contacts cross-connected.

Figure 9-2 The 4-way switch contains four terminal connections.

is in one position, connection will be made straight across the screw terminals as shown in Figure 9-3A. When the switch lever is moved to the other position, the screw terminals are cross-connected as shown in Figure 9-3B.

Four-way switches should not be confused with double-pole single-throw (DPST) switches, which are often used to control the operation of 240 volt devices. DPST switches have four terminal connection screws also. A simple method of identifying the difference between the two switches is that DPST switches have OFF and ON printed on the switch lever and 4-way switches do not.

Basic Switch Logic

The logic for 4-way switches is basically the same as that of 3-way switches discussed in Unit 8. As mentioned previously, when a device is to be controlled from more than one location, two 3-way switches are required. In the circuit shown in Figure 9-4, a light is controlled

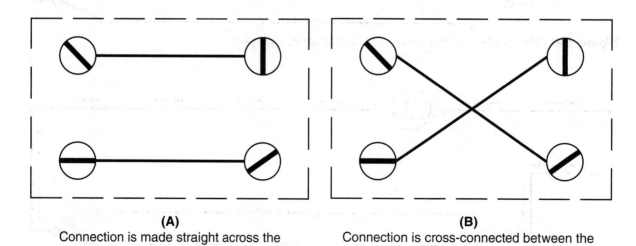

(A)	(B)
Connection is made straight across the terminal screws.	Connection is cross-connected between the terminal screws.

Figure 9-3 Switch positions of 4-way switches.

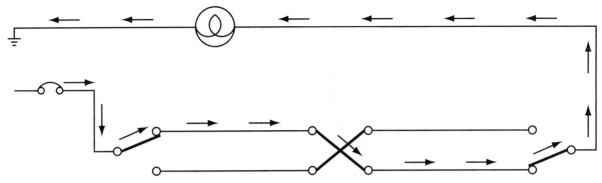

Figure 9-4 A complete circuit exists.

by three switches. Two of the switches are 3-way and the third is a 4-way. Note that the 4-way switch connects in the traveler wires between the two 3-way switches. In the example shown, a complete circuit exists through the lamp. If the switch lever of any of the three switches is changed, the circuit will be broken and the lamp will turn off (Figure 9-5). In this example the lever of the 4-way switch has been changed. There is no longer a complete circuit through the lamp.

Now assume that the lever of one of the 3-way switches is changed (Figure 9-6). A complete circuit again exists through the lamp. Regardless of which switch position is changed, the lamp will toggle from off to on or on to off. Any number of 4-way switches can be connected in the traveler circuit (Figure 9-7). Changing the position of any switch in the circuit will change the lamp from on to off or off to on.

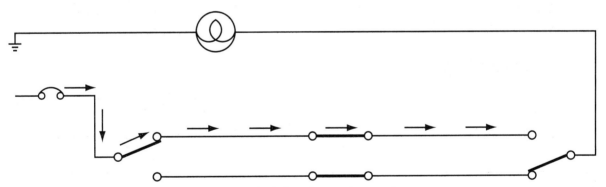

Figure 9-5 The position of the 4-way switch has been changed.

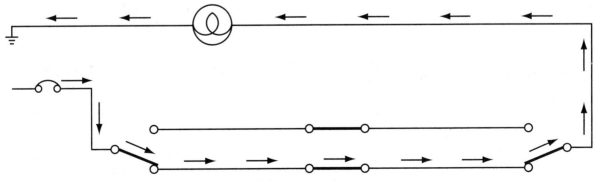

Figure 9-6 The position of a 3-way switch has been changed.

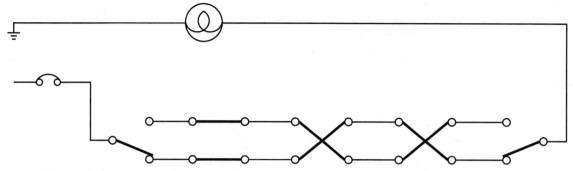

Figure 9-7   Any number of 4-way switches can be connected in the travelers.

Circuit Connections

The schematics shown in Figures 9-4 through 9-7 are used to explain the logic of 4-way switch circuits. In actual practice, connection is made with two- and three-conductor cables. In Unit 8, four rules were given for the connection of 3-way switches. These same four rules can be employed when connecting 4-way switches. The only exception is rule #4, which states "connect the travelers." When connecting 4-way switches, the switch must be connected in the travelers that connect the stationary contact terminals of the two 3-way switches together.

Example Connection 1

In the first example, power is brought to one of the 3-way switch boxes. A three-conductor cable runs from that switch box to the switch box containing the 4-way switch. The

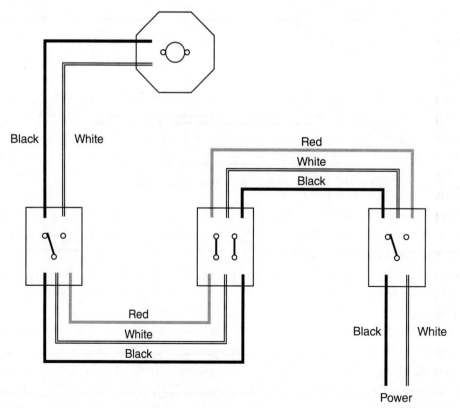

Figure 9-8   Example 1 of 4-way switch.

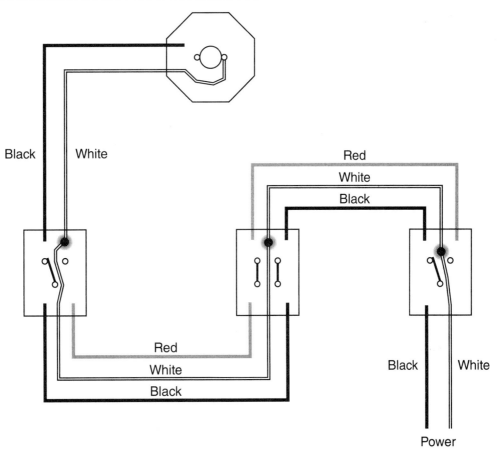

Figure 9-9 The neutral is connected to the lamp.

three-conductor cable proceeds to the second 3-way switch box, and a two-conductor cable runs from that box to the light (Figure 9-8). To connect the circuit, follow the four rules for connecting 3-way switches.

1. *Connect the neutral to the light.* The neutral is the white wire of the two-conductor cable that enters the switch box. It will connect to the white conductor of the three-conductor cable. The two white wires in the 4-way switch box will connect together, and the white wire in the second 3-way switch box will connect to the white wire that runs between the switch box and the lamp box. Then the white wire will connect to one side of the lamp (Figure 9-9).

2. *Connect the hot conductor to the common terminal of one 3-way switch.* Since the black wire of the power cable is hot, it will be connected to the common terminal of the 3-way switch (Figure 9-10).

3. *Connect the other side of the light to the common terminal of the other 3-way switch.* The black wire of the switch leg that runs between the second 3-way switch and the light will be connected to the common terminal of the second 3-way switch. The other end of the black wire will be connected to the other side of the light (Figure 9-11).

4. *Connect the travelers.* The red and black wires of the three-conductor cable will be used to connect the two 3-way switches together. The only difference is that the 4-way switch is connected between the travelers (Figure 9-12).

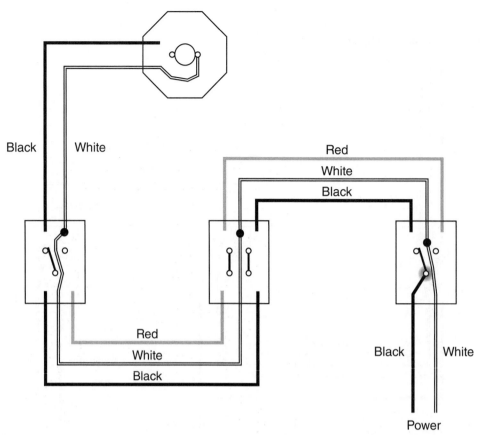

Figure 9-10 The hot conductor connects to the common terminal of one 3-way switch.

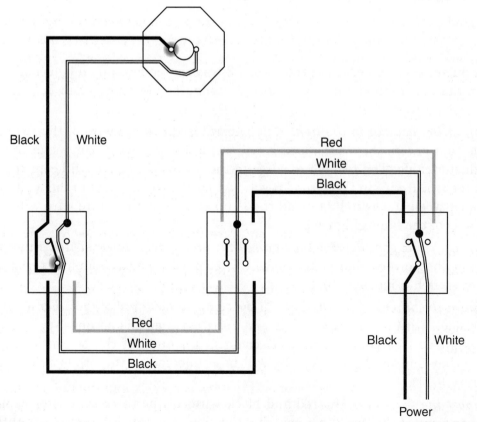

Figure 9-11 The other side of the light is connected to the common terminal of the second 3-way switch.

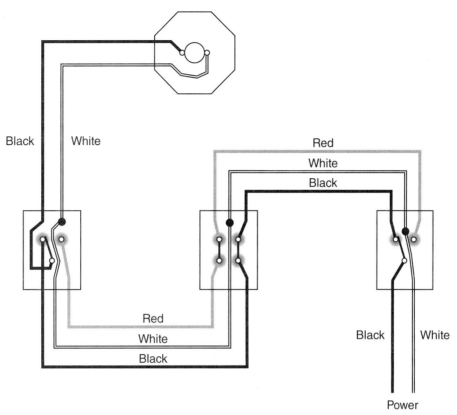

Figure 9-12 The 4-way switch is connected in the travelers.

Example Connection 2

In the second example, a two-conductor cable is connected between the light box and one 3-way switch box. A three-conductor cable runs from the 3-way switch box to the 4-way switch box and on to the second 3-way switch box. The power cable enters the 4-way switch box (Figure 9-13). Although it is seldom that a power cable will be brought to the 4-way switch box, this example is intended to illustrate how the four rules for connecting 3-way switches can be followed to make any 3- or 4-way switch connection.

1. *Connect the neutral to the light.* The white wire of the power cable is connected to the white wire of the three-conductor cable that runs to the 3-way switch box that contains the switch leg to the light (Figure 9-14). The neutral continues through the 3-way switch box and connects to the white wire of the switch leg. It is then connected to one side of the light.

2. *Connect the hot conductor to the common terminal of one 3-way switch.* The black wire of the power cable connects to the black wire of the three-conductor cable that runs to the 3-way switch box that does not contain the switch leg to the light (Figure 9-15). Notice that the common terminal of the switch will be connected to a black wire.

3. *Connect the other side of the light to the common terminal of the other 3-way switch.* The black wire of the switch leg connects to the common terminal of the second 3-way switch and to the other side of the light (Figure 9-16).

4. *Connect the travelers.* In this example, the red and white wires of one three-conductor cable connect to one set of terminals of the 4-way switch, and the red and black wires of the other three-conductor cable connect to the other set of 4-way switch terminals (Figure 9-17). The white wire that is used as a switch leg should be reidentified by marking it with colored tape or paint in each box.

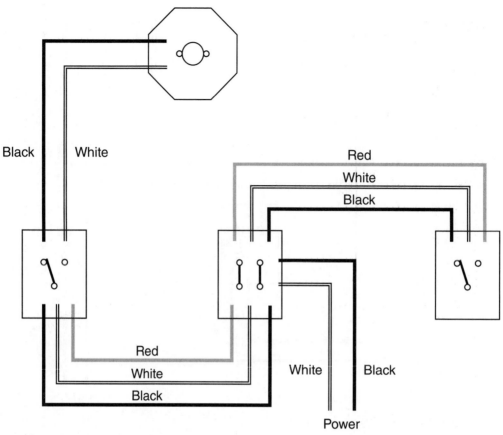

Figure 9-13 Example 2 of 4-way switch.

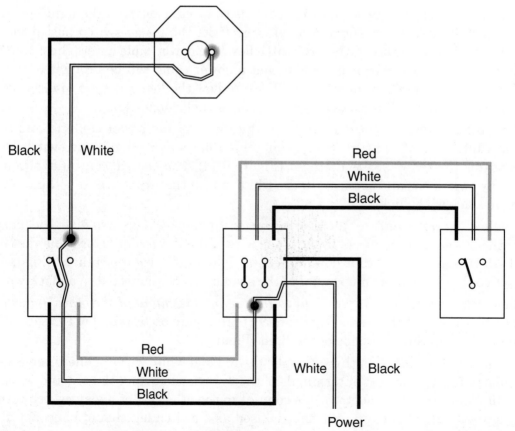

Figure 9-14 The neutral connects to the light.

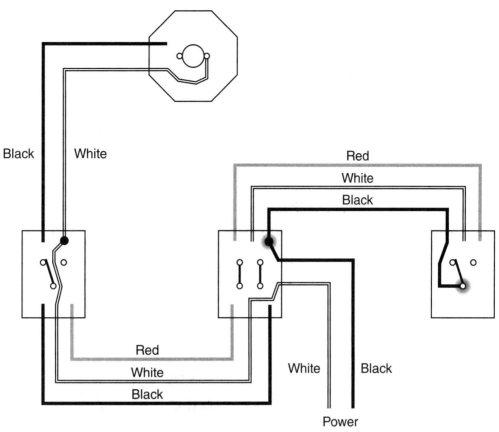

Figure 9-15 The hot conductor connects to the common terminal of one 3-way switch.

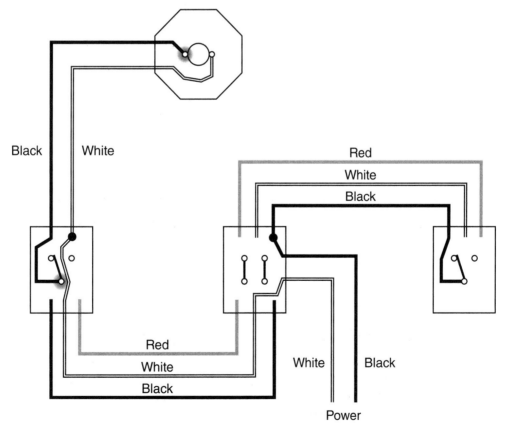

Figure 9-16 The other side of the light connects to the common terminal of the second 3-way switch.

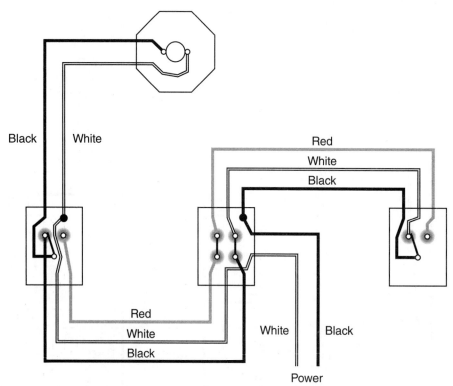

Figure 9-17 The 4-way switch is connected in the travelers.

LABORATORY EXERCISE

Materials Required

120-volt AC power supply

1 120-volt lamp (any wattage)

2 3-way switches

1 4-way switch

3 switch boxes mounted to wall studs or on a board

1 standard octagon box mounted on a rafter or on a board

1 lamp socket that will mount to the octagon box

Two-conductor cable (length will be decided by the individual laboratory)

Three-conductor cable (length will be decided by the individual laboratory)

1. **Test and verify that the power is turned off.**

2. Using the materials list, mount three switch boxes and one octagon box on a board or wall studs according to the provisions of the laboratory.

3. Connect two- and three-conductor cables between the power supply and the boxes as shown in Figure 9-18. (Note: Blueprints generally indicate electrical cables as lines with hash marks. The same notation will be used in this laboratory exercise. A line with two hash marks indicates a two-conductor cable. A line with three hash marks indicates a three-conductor cable.)

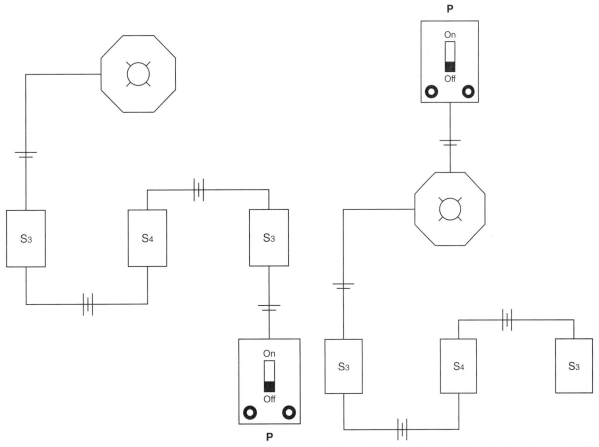

Figure 9-18 First circuit for connection.

Figure 9-19 Second circuit for connection.

4. Use the four rules for connecting 3-way switches to connect the circuit.

5. Turn on the power and test the circuit by alternately changing the position of each 3-way switch and the 4-way switch.

6. **Turn off the power** and disconnect the circuit.

7. Reposition the two- and three-conductor cables as shown in Figure 9-19.

8. Use the four rules for connecting 3-way switches to connect the circuit.

9. Turn on the power and test the circuit by alternately changing the position of each 3-way switch and the 4-way switch.

10. **Turn off the power** and disconnect the circuit.

Review Questions

1. How many terminal screws are on a 4-way switch?

_____

2. A light is to be controlled from eight different locations. How many of each type of switch are required to make this connection?

Single-pole (S_1) _____ 3-Way (S_3) _____ 4-Way (S_4) _____

_____

3. List four rules for connecting 3-way switches.

_____

4. What type of switch other than the 4-way contains four terminal screws?

_____

5. How can the two switches that contain four terminal screws be distinguished from each other?

_____

6. Refer to Figure 9-20. Connect the wires for proper switch operation. Connect the circuit so that a black conductor will supply the common terminal on each 3-way switch.

_____

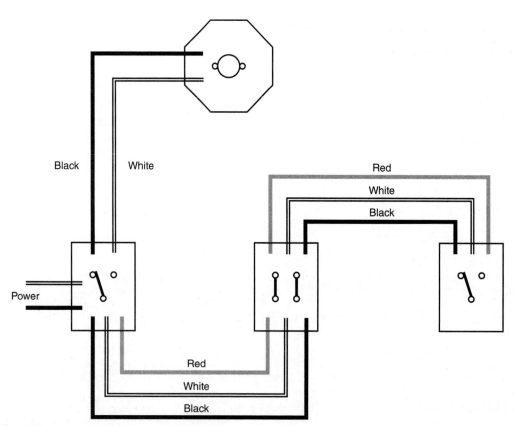

Figure 9-20 Connect the circuit.

SECTION 3

Alternating Current Loads

Unit 10 Inductance

Objectives

After studying this unit, you should be able to

- Determine the impedance of an inductor.
- Determine the inductive reactance of an inductor.
- Determine the Q of an inductor.
- Determine the inductance of an inductor.
- Determine the inductance of coils connected in parallel.
- Determine the inductive reactance of coils connected in parallel.
- Determine the inductance of coils connected in series.
- Determine the inductive reactance of coils connected in series.

Inductance is one of the major types of load in an alternating current circuit. Inductive loads exhibit different characteristics than resistive loads. Some of the characteristics of an inductive circuit are the following:

1. The current is mainly limited by inductive reactance (X_L) instead of resistance.
2. The current lags the applied voltage by 90°.
3. The pure inductive part of the circuit consumes no power.
4. Power in an inductive circuit is measured in VARs (Volt Amps Reactive) instead of watts. VARs is sometimes called wattless power. Watts is a measure of the amount of electrical energy converted into some other form, such as heat or kinetic energy. In a pure inductive circuit, power is stored in a magnetic field during part of the cycle and then returned to the circuit during another part. The electrical energy is not converted to some other form; it is stored and returned.

Impedance

Impedance (Z) is a measure of the total current-limiting effect of the circuit. It can be a combination of resistance, inductive reactance, and capacitive reactance. All inductors have some amount of resistance in the wire used to wind the inductor, but as a general rule, inductors limit current with inductive reactance instead of resistance. To determine the total current-limiting effect of an inductor, the impedance, it is necessary to add the resistance and inductive reactance together. In an AC circuit, however, the inductive part and resistive part are 90° out of phase with each other. To add these two quantities, vector addition must be used. Assume that an inductor has a resistance of 5 Ω and an inductive reactance of 8 Ω. The total current-limiting effect for this inductor can be determined using the formula:

$$Z = \sqrt{R^2 + X_L^2}$$

$$Z = \sqrt{5^2 + 8^2}$$

$$Z = \sqrt{25 + 64}$$

$$Z = \sqrt{89}$$

$$Z = 9.43 \ \Omega$$

This inductor will offer a total current-limiting effect of 9.43 Ω to the circuit.

Q of an Inductor

Q stands for quality. The Q of a coil or inductor can be determined by comparing the resistance and inductive reactance. To determine the Q of an inductor, divide the inductive reactance by the resistance. In the previous example, the coil has a wire resistance of 5 Ω and an inductive reactance of 8 Ω. To determine the Q of this coil, use this formula:

$$Q = \frac{X_L}{R}$$

$$Q = \frac{8}{5}$$

$$Q = 1.6$$

Inductors with a Q of 10 or higher are generally considered to be pure inductors and their resistance is not considered in circuit calculations. Assume that an inductor has a wire resistance of 10 Ω and an inductive reactance of 100 Ω. Now, determine the impedance of this coil:

$$Z = \sqrt{R^2 + X_L^2}$$

$$Z = \sqrt{10^2 + 100^2}$$

$$Z = \sqrt{100 + 10,000}$$

$$Z = \sqrt{10,100}$$

$$Z = 100.5 \ \Omega$$

As you can see, almost all the current-limiting effect of the coil is caused by inductive reactance. The amount of current limit due to resistance is negligible.

Inductors and Transformers

In the following laboratory experiment, a transformer winding will be used as an inductor. Inductors are also known as "chokes," "reactors," and "coils." The electrical properties of inductors and transformers are not identical, but they are similar. The greatest difference between a true reactor and a transformer is that reactors, or chokes, have the ability to limit inrush current when power is first applied to them. Transformers can exhibit extremely high inrush currents. This different characteristic between the two devices is caused by the type of magnetic core material used to make an inductor or transformer. For the purpose of this experiment, the transformer will be used as an inductor or reactor. It should be understood that although transformers and inductors are both inductive devices, the inductance of a transformer winding may vary with a change of current. True inductors or chokes will retain their inductive value as current changes.

LABORATORY EXERCISE

Name _____ Date _____

Materials Required

2 0.5-kVA control transformers with two windings rated at 240 volts and one winding rated at 120 volts

AC ammeter, in-line or clamp-on. (If a clamp-on type is employed, the use of a 10:1 scale divider is recommended.)

Connecting wires

AC voltmeter

1 120-volt AC power supply

1 208-volt AC power supply

1. With an ohmmeter, measure the resistance of the low-voltage winding marked X_1 and X_2 on the transformer.

    _____ Ω

2. Connect the circuit shown in Figure 10-1.

CAUTION

Only one winding of the transformer will be connected to power at one time, but all terminals will have voltage across them when the power is turned on. Use extreme caution NOT to touch any terminal on the transformer when the power is on, even if there is no connection made to the terminal.

---

3. Turn on the power and measure the voltage across terminals X_1 and X_2 with an AC voltmeter.

    _____ volts

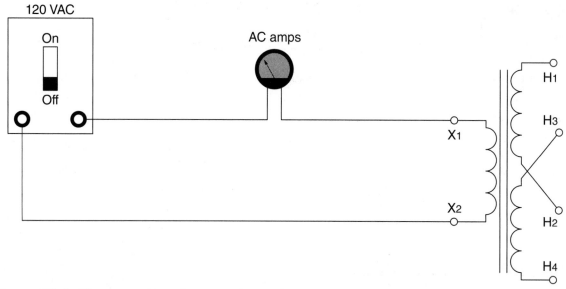

Figure 10-1 The secondary of a control transformer is used as an inductive load.

4. Measure the amount of current flow in the winding.

_____ amp(s)

5. **Turn off the power supply.**

6. Determine the total impedance of the inductor using the formula

$$Z = \frac{E}{I}$$

Z = _____ Ω

7. Now that the impedance and resistance are known, the inductive reactance of the inductor can be computed using the formula

$$X_L = \sqrt{Z^2 - R^2}$$

X_L = _____ Ω

8. Now that the inductive reactance and resistance are known, the Q of the coil can be computed using the formula

$$Q = \frac{X_L}{R}$$

Q = _____

9. The inductance (L) of the coil can be computed using the formula

$$L = \frac{X_L}{2\pi f}$$

where π = 3.1416

  f = Frequency (60)

L = _____ henrys

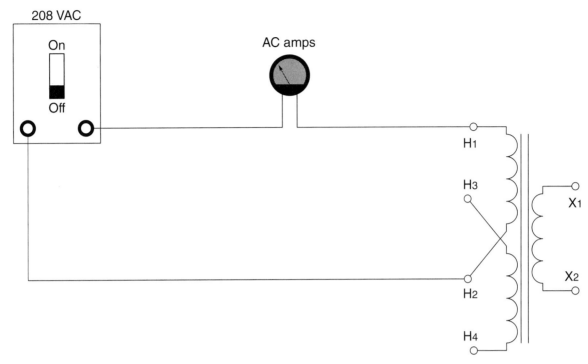

Figure 10-2 The high-voltage winding is connected to 208 volts.

10. Using an ohmmeter, measure the resistance between terminals H_1 and H_2 of the control transformer.

_____ Ω

11. Connect the circuit shown in Figure 10-2. Note that H_1 and H_2 are connected to a voltage of 208 VAC.

12. Turn on the power supply and measure the voltage applied to terminals H_1 and H_2 with an AC voltmeter.

_____ volts

13. Measure the current flow in the circuit with an AC ammeter.

_____ amp(s)

14. **Turn off the power supply.**

15. Using Ohm's law, determine the impedance of the winding.

$$Z = \frac{E}{I}$$

$Z =$ _____ Ω

16. Calculate the inductive reactance of the winding.

$$X_L = \sqrt{Z^2 - R^2}$$

$X_L =$ _____ Ω

17. Determine the Q of this inductor.

$$Q = \frac{X_L}{R}$$

$Q =$ _____

18. Calculate the inductance of the coil using the formula

$$L = \frac{X_L}{2\pi f}$$

L = _____ henry

Inductors Connected in Parallel

When inductors are connected in parallel, the reciprocals of their inductance values add in a similar manner as parallel resistors.

$$L_T = \frac{1}{\dfrac{1}{L_1} + \dfrac{1}{L_2} + \dfrac{1}{L_3} + \dfrac{1}{L_N}}$$

Since inductive reactance is directly proportional to the amount of inductance in an AC circuit, the reciprocal of the total inductive reactance will be the sum of all the reciprocals of the individual inductive reactances in the circuit.

$$X_{L_T} = \frac{1}{\dfrac{1}{X_{L_1}} + \dfrac{1}{X_{L_2}} + \dfrac{1}{X_{L_3}} + \dfrac{1}{X_{L_N}}}$$

The next step in completing this experiment is to determine the characteristics of a second 0.5 kVA control transformer. The second transformer will be referred to as transformer #2 and the first transformer will be referred to as transformer #1 for the remainder of this experiment.

19. Using control transformer #2, connect the circuit shown in Figure 10-3.

20. Turn on the power supply and measure the voltage applied to the transformer.

_____ volts

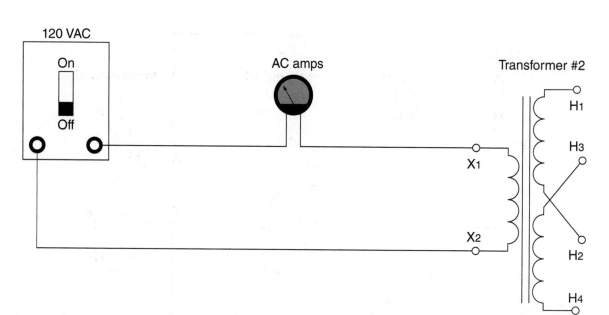

Figure 10-3 Determining the characteristics of the low-voltage winding for the second control transformer.

21. Measure the amount of current flow in the circuit.

_____ amp(s)

22. Determine the inductive reactance of this winding. (Since it has been shown earlier in this experiment that the impedance and inductive reactance are practically the same, inductive reactance will be determined by using the formula $X_L = \dfrac{E}{I}$.)

$X_L = $ _____ Ω

23. Determine the inductance of this winding.

$$L = \dfrac{X_L}{2\pi f}$$

$L = $ _____ henry

24. **Turn off the power supply.**

25. Using both transformers, connect the circuit shown in Figure 10-4.

26. Determine the total inductance of the circuit by using the inductance value of winding X_1 and X_2 of transformer #1 as determined in step 9, and the inductance value of winding X_1 and X_2 of transformer #2.

$L_T = $ _____ henry

27. Calculate the value of X_L for the circuit shown in Figure 10-4. Use the value of L_T as calculated in step 26 for the value of L in the formula.

$$X_L = 2\pi f L$$

$X_{L_T} = $ _____ Ω

28. Assuming an applied voltage of 120 volts, calculate the amount of current that should flow in this circuit.

$$I_T = \dfrac{E}{X_{L_T}}$$

$I_T = $ _____ amp(s)

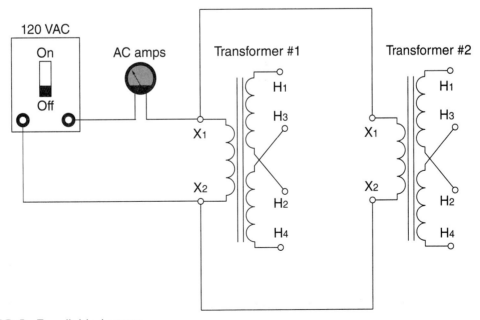

Figure 10-4 Parallel inductors.

29. Turn on the power supply and measure the current flow through the circuit.

$I_{T(measured)}$ = _____ amp(s)

30. **Turn off the power supply.**

Inductors Connected in Series

When inductors are connected in series, the total inductance is equal to the sum of the individual inductors.

$$L_T = L_1 + L_2 + L_3 + L_N$$

As with series-connected resistors, when inductors are connected in series, the inductive reactance of each inductor will be added together. Therefore, the total inductive reactance will be equal to the sum of all the inductive reactances.

$$X_{L_T} = X_{L_1} + X_{L_2} + X_{L_3} + X_{L_N}$$

31. Connect the circuit shown in Figure 10-5. Note that the power supply has changed to a 208-volt supply.

32. After the connection is completed, turn on the power supply and measure the circuit current.

I = _____ amp(s)

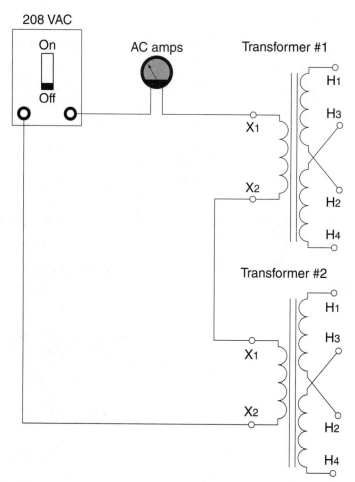

Figure 10-5   Series inductors.

33. Calculate the total inductive reactance of this connection using the current measured in step 32 and the circuit voltage.

$$X_L = \frac{E}{I}$$

$X_{L_T} =$ _____ Ω

34. Calculate the total inductance of the circuit using the calculated value of total inductive reactance.

$$L = \frac{X_L}{2\pi f}$$

$L =$ _____ henry

35. With an AC voltmeter, measure the amount of voltage across the X_1 and X_2 winding of transformer #1 (Figure 10-6).

$E_{(Transformer\ \#1)} =$ _____ volts

36. Use the measured voltage drop in step 35 and the measured current flow in step 32 to calculate the inductive reactance of this winding.

$$X_L = \frac{E}{I}$$

$X_{L(Transformer\ \#1)}$ _____ Ω

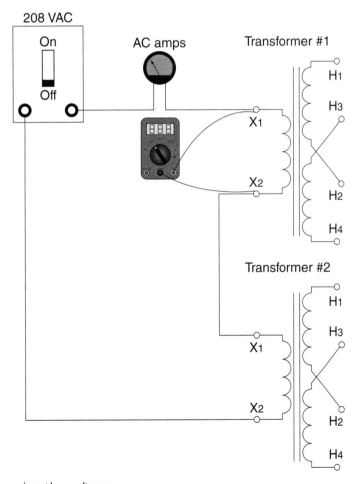

Figure 10-6 Measuring the voltage.

37. Using the value of X_L in step 36, calculate the value of inductance for this winding.

    $L_{(Transformer \#1)}$ —————————— henry

38. With the AC voltmeter, measure the voltage drop across the X_1 and X_2 winding of transformer #2.

    $E_{(Transformer \#2)}$ —————————— volts

39. **Turn off the power supply.**

40. Using the circuit current as measured in step 32, and the voltage drop measured in step 38, calculate the inductive reactance of the winding.

    $$X_L = \frac{E}{I}$$

    $X_{L(Transformer \#2)}$ —————————— Ω

41. Using the calculated value of inductive reactance for transformer #2, calculate the inductance of the winding for transformer #2.

    $$L = \frac{X_L}{2\pi f}$$

    $L_{(Transformer \#2)} =$ —————————— henry

42. Add the inductive reactance values of transformer #1 and transformer #2. Compare the sum with the calculated value of total inductive reactance in step 33. Are the two values approximately the same?

    ——————————

43. Add the values of inductance for transformer #1 and transformer #2. Compare the sum with the value of total inductance calculated in step 34. Are these two values approximately the same?

    ——————————

44. Disconnect the circuit and return the components to their proper place.

NOTE OF EXPLANATION

In some cases the values of inductance for the transformers may be different. For example, the value of inductance for transformer #1 was calculated in step 9 of this experiment. The inductance value for transformer #1 was calculated again in step 37. It is quite possible that these two values are different even if the same transformer was used to make both measurements. The reason for this difference is nonlinearity of the core material. As current flows through the windings of the transformer, the magnetic field causes "magnetic domains" or "magnetic molecules" to align. As more and more magnetic domains align themselves, the core approaches saturation. The closer the core material comes to saturation, the less magnetic effect it exhibits. The magnetic properties of the core material greatly affect the amount of inductance.

Review Questions

1. An inductor has an inductance of 0.65 henry and is connected to a 277 volt, 60 Hz power line. How much current will flow in this circuit? (Assume the wire resistance of the coil to be negligible.)

2. An ohmmeter is used to measure the wire resistance of a coil at 45 Ω. When connected to a 120 volt, 60 Hz AC circuit, there is a current flow of 1.5 amps. What is the inductive reactance of the coil?

3. A coil has an inductive reactance of 75 Ω and a wire resistance of 18 Ω. What is the Q of this coil?

4. As a general rule, for an inductor to be considered as a pure inductor, it should have a Q value of what or higher?

5. Three choke coils have inductances of 0.56 henry, 0.72 henry, and 0.43 henry. If these coils are connected in series, what would be total inductance of the circuit?

6. If the three choke coils in question 5 were to be connected in parallel, what would be the total circuit inductance?

7. An inductor is connected in a 240 volt, 60 Hz circuit. The circuit current is 2.5 amperes. If the frequency is changed to 400 Hz, how much current will flow if the voltage remains the same? (Assume the circuit to be a pure inductive circuit.)

8. An inductor has an inductive reactance of 124 Ω and a wire resistance of 44 Ω. What is the impedance of the inductor?

9. Three inductors are connected in series to a 480 volt, 60 Hz power source. The current flow in the circuit is 0.509 amp. Inductor #1 has an inductance of 1.3 henry, and inductor #2 has an inductance of 0.75 henry. What is the inductance of inductor #3? (Assume all inductors to be pure inductors.)

10. Three inductors are connected in a 208 volt, 60 Hz circuit. The circuit current is 1.2 amperes. Inductor #1 has an inductance of 1.2 henry, inductor #2 has an inductance of 1.6 henry, and inductor #3 has an inductance of 1.4 henry. Are these inductors connected in series or parallel?

Unit 11 Resistive-Inductive Series Circuits

Objectives

After studying this unit, you should be able to

- Discuss the voltage and current relationship in an RL series circuit.
- Determine the phase angle of current in an RL series circuit.
- Determine the power factor in an RL series circuit.
- Discuss the differences between apparent power, true power, and reactive power.

In an AC circuit containing pure resistance, the current and voltage are in phase with each other. This means that the voltage and current waveforms are identical as far as time is concerned (Figure 11-1). Both the current and voltage will be zero at the same time, both will reach their positive peak at the same time, and both will reach their negative peak at the same time. When the current and voltage are both positive or negative at the same time, volts times amps equals watts.

Watts

Watts is often referred to as true power. To understand true power, you must realize that electricity is a form of pure energy. Energy can be neither created nor destroyed, but its form can be changed. Watts is a measure of the amount of electrical energy changed into some other form. In the case of resistance, electrical energy is converted into thermal energy in the form of heat. In the case of a motor, electrical energy is converted into kinetic energy. There must be some form of energy conversion to have true power or watts.

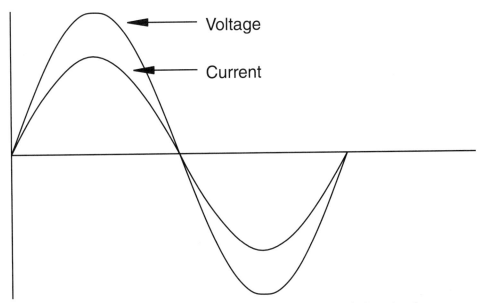

Figure 11-1 The current and voltage are in phase in a pure resistive circuit.

Inductance

In an AC circuit containing a pure inductive load, the current will lag the voltage by 90°, as shown in Figure 11-2. In a circuit of this type, there is no true power or watts because electrical energy is not converted into another form. During periods when the current and voltage are both positive or both negative, energy is stored in the form of an electromagnetic field. As current rises in the inductor, a magnetic field forms around the inductor. During periods when the voltage and current are opposite in polarity, the stored energy is given back to the circuit as the magnetic field collapses and induces a voltage back into itself. The only true power, or watts, in the circuit is caused by losses in the inductor, such as the resistance of the wire, eddy current losses caused by currents being induced into the core material, and hysteresis losses.

VARs

Another electrical term, VARs, is used to describe the power associated with a reactive load. VARs stands for "Volt Amperes Reactive." VARs is to a reactive circuit what watts is to a resistive circuit. Watts can be determined in a resistive circuit by multiplying the voltage drop across the resistor by the amount of current flow through the resistor ($E_R \times I_R$), or by the amount of current flow through the resistor by the resistance ($I_R^2 \times R$), or by dividing the voltage drop and the resistance $\left(\dfrac{E_R^2}{R} \right)$. VARs can be determined in a like manner except the inductive values are used instead of resistive values. VARs can be computed by multiplying the voltage drop across the inductor by the current flowing through the inductor ($E_L \times I_L$), or by the current and the inductive reactance of the inductor ($I_L^2 \times X_L$), or by the voltage and inductive reactance $\left(\dfrac{E_L^2}{X_L} \right)$.

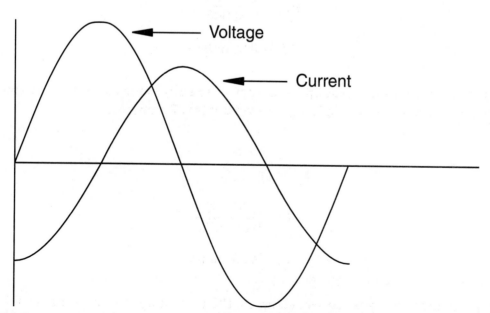

Figure 11-2   In a pure inductive circuit, the current lags the voltage by 90°.

Apparent Power (Volt Amps)

Another electrical quantity that is very similar to watts and VARs is volt amps (VA). Volt amps is generally referred to as apparent power because it is computed in a similar manner as watts and VARs, except that the applied or total circuit values are used instead of resistive or inductive value. Volt amps can be determined by multiplying the total or applied voltage by the total circuit current ($E_T \times I_T$), or by using the total current and circuit impedance ($I_T^2 \times Z$), or by using the applied voltage and impedance $\left(\dfrac{E_T^{\,2}}{Z} \right)$.

Angle Theta (∅)

When an AC circuit contains elements of both resistance and inductance, the voltage and current will be out of phase by some amount between 0° and 90°. The amount of out-of-phase condition is determined by the values of resistance and inductance and is expressed as angle theta (∅). The power factor is the cosine of angle theta. Angle theta can be computed using the formula cos ∠∅ = PF.

Power Factor

Power factor is a ratio of the amount of true power or watts as compared to apparent power or volt amps. Power factor is expressed as a percent. Power companies become upset when the power factor drops to a low percent because they must supply more current than is actually being consumed. Assume that an industrial plant operates on 480 volts three-phase and that the apparent power is 250 kVA. Also assume that a watt meter indicates a true power of 200 kW. The power factor can be determined by dividing the true power by the apparent power.

$$PF = \frac{P}{VA}$$

$$PF = \frac{200{,}000}{250{,}000}$$

$$PF = 0.80 \text{ or } 80\%$$

This indicates that 80% of the supplied energy is actually being consumed. At the present time, the power company is supplying a current of 300.7 amperes.

$$I = \frac{VA}{E \times \sqrt{3}}$$

$$I = \frac{250{,}000}{480 \times 1.732}$$

$$I = 300.7 \text{ amps}$$

If the power factor were to be corrected to 100% or unity, the current would drop to 240.6 amperes.

$$I = \frac{P}{E \times \sqrt{3}}$$

$$I = \frac{200,000}{480 \times 1.732}$$

$$I = 240.6 \text{ amps}$$

Power factor correction will be discussed in a later unit.

Example Problem

A series circuit containing a resistor and inductor is shown in Figure 11-3. It is assumed the circuit is connected to 120 VAC with a frequency of 60 Hz. The resistor has a resistance of 24 Ω and the inductor has an inductive reactance of 32 Ω. The following values will be computed:

Z - Total impedance of the circuit

I_T - Total circuit current

E_R - Voltage drop across the resistor

P - True power or watts

E_L - Voltage drop across the inductor

L - Inductance of the inductor

$VARs_L$ - Reactive power

VA - Volt amps or apparent power

PF - Power factor

$\angle\varnothing$ - Angle theta (the angle or degree amount that the current and voltage are out of phase with each other)

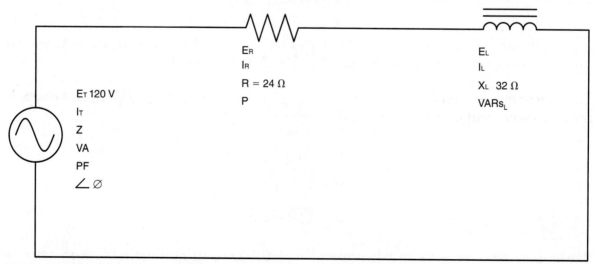

Figure 11-3   RL series circuit.

The first step in determining the missing values is to compute the total circuit impedance. One of the rules for series circuits states that the total resistance is equal to the sum of the individual resistances. This basic rule is still true, but it must be bent a bit to fit this circuit. It is true that the ohmic values still add, but the inductive and resistive parts of the circuit are out of phase with each other by 90°. Vector addition must be used to determine the total impedance. The formula for determining impedance in a series circuit containing resistance and inductive reactance is $Z = \sqrt{R^2 + X_L^2}$.

$$Z = \sqrt{24^2 + 32^2}$$

$$Z = \sqrt{276 + 1{,}024}$$

$$Z = \sqrt{1{,}600}$$

$$Z = 40 \ \Omega$$

The total circuit current can now be computed using Ohm's law.

$$I_T = \frac{E_T}{Z}$$

$$I_T = \frac{120}{40}$$

$$I_T = 3 \text{ amps}$$

The current is the same at any point in a series circuit. The values of I_R and I_L are, therefore, the same as I_T.

Now that the amount of current flowing through the resistor is known, the voltage drop across the resistor can be determined using Ohm's law.

$$E_R = I_R \times R$$

$$E_R = 3 \times 24$$

$$E_R = 72 \text{ volts}$$

The true power, or watts, can be computed using any of the power formulas. In this example the true power will be computed using E × I.

$$P = E_R \times I_R$$

$$P = 72 \times 3$$

$$P = 216 \text{ watts}$$

The voltage drop across the inductor can be determined using Ohm's law and reactive values.

$$E_L = I_L \times X_L$$

$$E_L = 3 \times 32$$

$$E_L = 96 \text{ volts}$$

Note that the resistor has a voltage drop of 72 volts and the inductor has a voltage drop of 96 volts. One of the rules for series circuit states that the total or applied voltage is equal to the sum of the voltage drop in the circuit. The circuit has 120 volts applied. Therefore, the sum of 72 and 96 should equal 120. In order for these two values to equal the applied voltage, vector addition must be employed. In a series circuit, the current is the same through all parts of the circuit. Since the voltage drop across the resistor is in phase with the current and the voltage drop across the inductor leads the current by 90°, the voltage drops across the resistor and inductor are 90° out of phase with each other. Total voltage can be determined using the formula

$$E_T = \sqrt{E_R^2 + E_L^2}$$

$$E_T = \sqrt{72^2 + 96^2}$$

$$E_T = \sqrt{14,400}$$

$$E_T = 120 \text{ volts}$$

The reactive VARs can be computed in a manner similar to determining the value for watts or volt amps, except that reactive values are used.

$$VARs_L = E_L \times I_L$$

$$VARs_L = 96 \times 3$$

$$VARs_L = 288$$

The value of inductance of the inductor can be computed using the formula

$$L = \frac{X_L}{2\pi f}$$

$$L = \frac{32}{377}$$

$$L = 0.0849 \text{ henry}$$

The apparent power or volt amps can be computed using formulas similar to those for determining watts or VARs, except that total circuit values are used in the formula

$$VA = E_T + I_T$$

$$VA = 120 \times 3$$

$$VA = 360$$

The circuit power factor can be computed using the formula

$$PF = \frac{P}{VA}$$

$$PF = \frac{216}{360}$$

$$PF = 0.6 \text{ or } 60\%$$

The power factor is the cosine of angle theta. In this circuit, the decimal power factor is 0.6. The cosine of angle theta is 0.6. To determine angle theta, find the angle that corresponds to a cosine of 0.6. Most scientific calculators contain trigonometric functions. To find what angle corresponds to one of the sin, cos, or tan functions, it is generally necessary to use the invert key, the arc key, or one of the keys marked $\sin^{-1}$, $\cos^{-1}$, or $\tan^{-1}$.

$$\cos \varnothing = 0.6$$

$$\varnothing = 53.13°$$

The current and voltage are 53.13° out of phase with each other in this circuit.

The circuit, with all completed values, is shown in Figure 11-4.

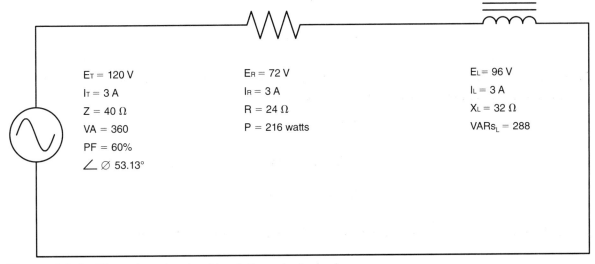

Figure 11-4 RL series circuit with all the missing values.

LABORATORY EXERCISE

Name _____ Date _____

Materials Required

1 0.5 kVA control transformer with two windings rated at 240 volts and one winding rated at 120 volts

1 100 ohm resistor

1 250 ohm resistor

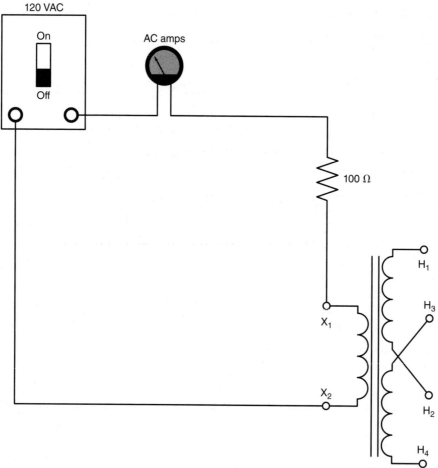

Figure 11-5 Connecting an RL series circuit.

AC ammeter (In-line or Clamp-on may be used. If a clamp-on type is employed, the use of a 10:1 scale divider is recommended)

Connecting wires

AC voltmeter

1 120 Volt AC power supply

Formulas for resistive-inductive series circuits shown in Figure 11-6

1. Connect the circuit shown in Figure 11-5. NOTE: During this experiment the H terminals of the transformer will provide a step-up in voltage. Be careful not to come in contact with these terminals.

2. Turn on the power and measure the current flow in the circuit with an AC ammeter.
   I = _____ amps

3. Measure the voltage drop across the 100 ohm resistor an AC voltmeter.
   E_R = _____ volts

4. Compute the true power in the circuit using the resistive values of voltage and current.
   $P = E_R \times I_R$
   P = _____ watts

5. Measure the voltage drop across winding X_1 and X_2 of the transformer with an AC voltmeter. **Turn off the power.**

   E_L = _____ volts

6. Compute the inductive reactance of the inductor (X_1 to X_2) using Ohm's law.

   $$X_L = \frac{E_L}{I}$$

   X_L = _____ Ω

7. Compute the inductive VARs in the circuit using the following formula: $VARS_L = E_L \times I_L$.

   $VARS_L$ = _____

8. Compute the apparent power (VA) using the total circuit values. $VA = E_T \times I_T$

   VA = _____

9. Compute the circuit power factor using the following formula: $PF = \dfrac{P}{VA}$

   PF = _____ %

10. Using the decimal value of the power factor, determine the phase angle difference between the voltage and current in this circuit. $\text{Cos } \angle\varnothing = PF$

    $\angle\varnothing$ = _____ °

11. Replace the 100 ohm resistor with a 250 ohm resistor.

12. Turn on the power and measure the current in the circuit using an AC ammeter.

    I = _____ A

13. Measure the voltage drop across the 250 ohm resistor with an AC voltmeter.

    E_R = _____ volts

14. Compute the true power in the circuit using Ohm's law.

    P = _____ watts

15. Measure the voltage drop across the inductor (winding X_1 and X_2 of the transformer) with an AC voltmeter. **Turn off the power.**

    E_L = _____ volts

16. Compute the inductive reactance of the inductor using Ohm's law.

    X_L = _____ Ω

17. Compute the reactive VARs for the inductor using Ohm's law.

    $VARS_L$ = _____

18. Compute the inductance of the inductor.

    L = _____ henry

19. Compute the apparent power of the circuit using Ohm's law.

    VA = _____

20. Compute the power factor of the circuit.

    PF = _____ %

21. Determine the value of angle theta.

    $\angle\varnothing$ = _____ °

22. In an RL series circuit the voltage drop across the resistor and inductor are out of phase with each other. Add the voltage drop across the resistor in step #3 and the voltage drop across the inductor in step #5.

E = _____ volts

Is the sum of the two voltages greater than the voltage source?

_____

23. Because the voltage across the resistor and the voltage across the indictor are out of phase with each other, vector addition must be used. If the components were a true 90° out of phase with each other, the vector sum should be equal to the voltage applied to the circuit. However, a transformer winding is not a true indictor or choke, and the winding does contain some amount of resistance which will prevent the voltage across the transformer winding from being 90° out of phase with the voltage across the resistor. Calculate the vector sum of the two voltage values using the following formula:

$$E_T = \sqrt{E_R^2 + E_L^2}$$

$E_T =$ _____ volts

24. Disconnect the circuit and return the components to their proper place.

Review Questions

Refer to the formulas shown in Figure 11-6 to answer some of the following questions.

1. What is the phase angle difference between current and voltage in a pure resistive circuit?

_____

2. What is the phase angle difference between current and voltage in a pure inductive circuit?

_____

3. An inductor and resistor are connected in series. The resistor has a resistance of 26 Ω and the inductor has an inductive reactance of 16 Ω. What is the impedance of the circuit?

_____

4. An RL series circuit is connected to a 120 volt, 60 Hz line. The inductor has a voltage drop of 54 volts. What is the voltage drop across the resistor?

_____

5. An inductor is using 1.6 kVARs and has an inductive reactance of 14 Ω. How much current is flowing through the inductor?

_____

6. An RL series circuit is connected to a 208 volt, 60 Hz line. The resistor has a power dissipation of 46 watts and the inductor is operating at 38 VARs. How much current is flowing in the circuit?

_____

7. An RL series circuit is connected to a 480 volt, 60 Hz line. The apparent power of the circuit is 82 kVA. What is the impedance of the circuit?

_____

8. An RL series circuit has a power factor of 82%. Determine angle theta.

_____

9. An RL series circuit is connected to a 240 volt, 60 Hz line. The inductor has a current of 8 amperes flowing through it. The X_L of the inductor is 12 Ω. The resistor has a resistance of 9 Ω. How much current is flowing through the resistor?

10. An RL series circuit has an apparent power of 650 VA and a true power of 375 watts. What is the reactive power in the circuit?

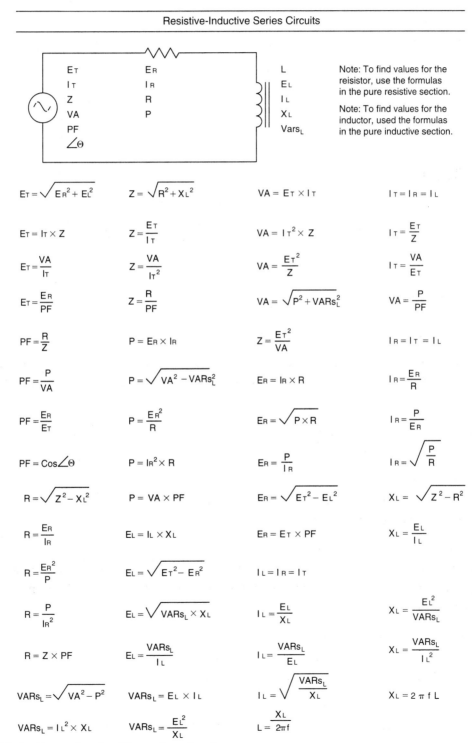

Figure 11-6 Formulas for RL series circuits.

Unit 12 RL Parallel Circuits

Objectives

After studying this unit, you should be able to

- Discuss the voltage and current relationship in an RL parallel circuit.
- Determine the phase angle of current in an RL parallel circuit.
- Determine the power factor in an RL parallel circuit.
- Discuss the differences between apparent power, true power, and reactive power.
- Find the impedance in an RL parallel circuit.

In any parallel circuit, the voltage must be the same across all branches. Since the current is in phase with the voltage in a pure resistive circuit, and the current lags the voltage by 90° in a pure inductive circuit, the current flow through the inductive branch will be out of phase with the current through the resistive branch. The total current will be out of phase with the applied voltage by some amount between 0° and 90° depending on the relative values of resistance and inductance.

Determining electrical values in an RL parallel circuit is very similar to determining values in a series circuit with a few exceptions. Probably the greatest difference is calculating the value of impedance when the values of R and X_L are known. Recall that vector addition can be used with the ohmic values of an RL series circuit to determine the impedance.

$$Z = \sqrt{R^2 + X_L{}^2}$$

The same basic concept is true for an RL parallel circuit, except that the reciprocal value of R and X_L must be used.

$$Z = \frac{1}{\sqrt{\left(\dfrac{1}{R}\right)^2 + \left(\dfrac{1}{X_L}\right)^2}}$$

Example: A resistor and inductor are connected in parallel. The resistor has a resistance of 50 Ω and the inductor has an inductive reactance of 60 Ω. Find the impedance of the circuit.

Solution:

$$Z = \frac{1}{\sqrt{\left(\dfrac{1}{50}\right)^2 + \left(\dfrac{1}{60}\right)^2}}$$

$$Z = \frac{1}{\sqrt{(0.02)^2 + (0.01667)^2}}$$

$$Z = \frac{1}{\sqrt{0.0004 + 0.0002779}}$$

$$Z = \frac{1}{\sqrt{0.0006779}}$$

$$Z = \frac{1}{0.02604}$$

$$Z = 38.4 \ \Omega$$

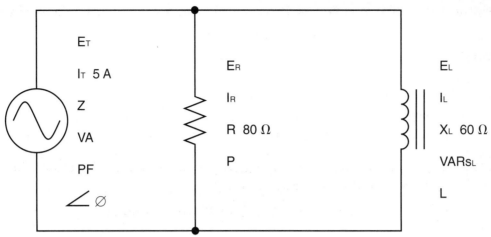

Figure 12-1 RL parallel circuit.

Example Circuit

An RL parallel circuit is connected to a 60 Hz line (see Figure 12-1). The resistor has a resistance of 80 Ω and the inductor has an inductive reactance of 60 Ω. The circuit has total current flow of 5 amps. The following values will be computed:

E_T - Total circuit voltage

Z - Circuit impedance

VA - Apparent power

I_R - Current flow through the resistor

P - True power

I_L - Current flow through the inductor

$VARs_L$ - Reactive power

L - Inductance of the inductor

PF - Power factor

$\angle\emptyset$ - Angle theta

The first step in determining the missing values for this problem is to determine the circuit impedance using the values of R and X_L.

$$Z = \frac{1}{\sqrt{\left(\frac{1}{R}\right)^2 + \left(\frac{1}{X_L}\right)^2}}$$

$$Z = \frac{1}{\sqrt{\left(\frac{1}{80}\right)^2 + \left(\frac{1}{60}\right)^2}}$$

$$Z = \frac{1}{\sqrt{0.00015625 + 0.00027778}}$$

$$Z = \frac{1}{0.0208333}$$

$$Z = 48 \ \Omega$$

Now that the value of impedance is known, the total voltage can be determined using Ohm's law.

$$E_T = I_T \times Z$$

$$E_T = 5 \times 48$$

$$E_T = 240 \text{ volts}$$

In a parallel circuit, the voltage is the same across all branches. Therefore, the voltage drops across the resistor and inductor are also 240 volts.

The apparent power can be computed using the following formula:

$$VA = E_T \times I_T$$

$$VA = 240 \times 5$$

$$VA = 1,200$$

The current flowing through the resistor can be computed using Ohm's law.

$$I_R = \frac{E_R}{R}$$

$$I_R = \frac{240}{80}$$

$$I_R = 3 \text{ amps}$$

The true power in the circuit can be computed using resistive values.

$$P = E_R \times I_R$$

$$P = 240 \times 3$$

$$P = 720 \text{ watts}$$

The current flow through the inductor can be computed using Ohm's law.

$$I_L = \frac{E_L}{X_L}$$

$$I_L = \frac{240}{60}$$

$$I_L = 4 \text{ amps}$$

The inductive VARs can be computed using the inductive values and Ohm's law.

$$VARs_L = E_L \times I_L$$

$$VARs_L = 240 \times 4$$

$$VARs_L = 960$$

The inductance of the inductor can be computed using the following formula:

$$L = \frac{X_L}{2\pi f}$$

$$L = \frac{60}{377}$$

$$L = 0.159 \text{ henry}$$

The power factor can be determined by comparing the true power and apparent power.

$$PF = \frac{P}{VA}$$

$$PF = \frac{720}{1,200}$$

$$PF = 0.6 \text{ or } 60\%$$

The cosine of angle theta is equal to the decimal power factor value.

$$\cos \angle\varnothing = PF$$

$$\cos \angle\varnothing = 0.6$$

$$\angle\varnothing = 53.13°$$

The circuit with all the missing values is shown in Figure 12-2.

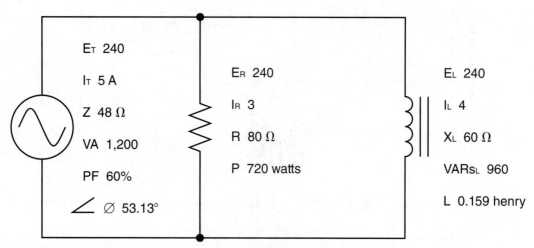

Figure 12-2 RL parallel circuit with all the missing values.

LABORATORY EXERCISE

Name _____ Date _____

Materials Required

Formulas for resistive–inductive parallel circuits shown in Figure 12-6

1 0.5 kVA control transformer with two windings rated at 240 volts and one winding rated at 120 volts

1 100 ohm resistor

1 150 ohm resistor

AC ammeter (In-line or clamp-on types may be used. If a clamp-on type is employed, the use of a 10:1 scale divider is recommended.)

Connecting wires

AC voltmeter

1 120 volt AC power supply

1. Connect the circuit shown in Figure 12-3.
2. Turn on the power supply and measure the total circuit current with an AC ammeter.

   I_T _____ amps
3. **Turn off the power supply.**
4. Connect an AC ammeter in series with the 100 ohm resistor as shown in Figure 12-4.
5. Turn on the power supply and measure the current flowing through the lamp.

   I_R _____ amps
6. **Turn off the power supply.**
7. Connect the AC ammeter in series with the inductor as shown in Figure 12-5.
8. Turn on the power supply and measure the current flow through the transformer winding.

   I_L _____ amps
9. **Turn off the power supply**
10. Calculate the true power in the circuit using resistive value of voltage and current.
    $P = E_R \times I_R$
    $P =$ _____ watts
11. Calculate the value of inductive reactance using inductive values of voltage and current.

    $X_L = \dfrac{E_L}{I_L}$

    X_L _____ Ω

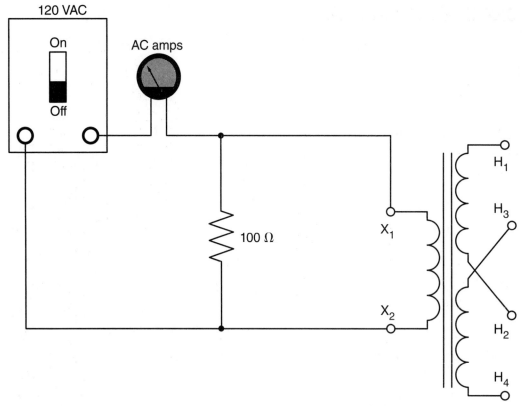

Figure 12-3 The ammeter is connected to measure the total circuit current.

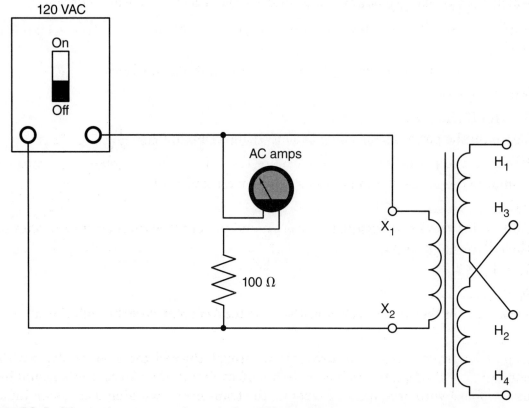

Figure 12-4 Measuring current through the resistive load.

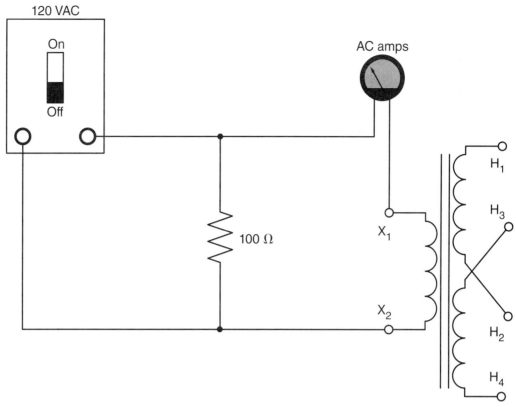

Figure 12-5 Measuring current through the inductive load.

12. Calculate the reactive power using inductive values of voltage and current.

    $VARS_L = E_L \times I_L$

    $VARS_L$ _____

13. Compute the apparent power in the circuit using the total values.

    $VA = E_T \times I_T$.

    VA _____

14. Compute the power factor using the following formula: $PF = \dfrac{P}{VA}$

    PF _____ %

15. Compute the angle theta using this formula: $\cos \angle \varnothing = PF$

    $\angle \varnothing$ _____ °

16. Add the current flow through the resistor and current flow through the inductor using the following formula:

    $I_T = I_R + I_L$

    $I_T =$ _____ A

    Is the sum of the currents greater than the total current flow through the circuit?

    _____

17. In an RL parallel circuit the current flow through the resistor is out of phase with the current flow through the inductor. In a perfect circuit, the two currents would be 90° out of phase with each other. However, the transformer winding used as an inductor is not a perfect inductor and the current flow will not be at a 90° angle. Vector addition

must be employed to determine the total current value. Calculate the total circuit current using the following formula:

$$I_T = \sqrt{I_R^2 + I_L^2}$$

$I_T =$ _____ A

18. Replace the 100 ohm resistor with a 150 ohm resistor in the RL parallel circuit.

19. Reconnect the AC ammeter to measure the total circuit current. Turn on the power supply and measure the total current flow in the circuit.

    I_T _____ amps

20. **Turn off the power supply** and reconnect the ammeter to measure the current flow through the 150 ohm resistor.

21. Turn on the power supply and measure the resistive current.

    I_R _____ amps

22. **Turn off the power supply** and reconnect the ammeter to measure the current flow through the transformer winding.

23. Turn on the power supply and measure the inductive current.

    I_L _____ amps

24. **Turn off the power supply.**

25. Compute the total impedance of the circuit using Ohm's law.

    Z _____ Ω

26. Compute the true power in the circuit using Ohm's law.

    P _____ watts

27. Compute the inductive reactance of the inductor.

    X_L _____ Ω

28. Compute the reactive power using Ohm's law.

    $VARS_L$ _____

29. Compute the apparent power in the circuit using total values of voltage and current.

    VA _____

30. Compute the circuit power factor using the values of true power and apparent power.

    PF _____ %

31. Compute angle theta.

    $\angle\emptyset$ _____ °

32. Disconnect the circuit and return the components to their proper place.

Review Questions

To answer the following questions, it may be necessary to refer to the formulas shown in Figure 12-6.

1. A 5 mh inductor is connector to a 400 Hz line. What is the inductive reactance of the inductor?

    _____

2. A resistor and inductor are connected in parallel to a 120 volt, 60 Hz line. The circuit has a current flow of 3 amperes. The resistor has a resistance of 72 Ω. What is the inductance of the inductor?

    _____

3. A resistor with a resistance of 50 Ω is connected in parallel with an inductor with an inductance of 0.175 henry. The power source is 60 Hz. What is the impedance of the circuit?

    _____

4. A resistor and inductor are connected in parallel to a 277 volt, 60 Hz line. The resistor has a current of 12 amperes flowing through it, and the inductor has a current flow of 8 amperes flowing through it. What is the total current flow in the circuit?

    _____

5. A resistor and inductor are connected in parallel to a 400 Hz line. The inductor has a voltage drop of 136 volts across it. What is the voltage drop across the resistor?

    _____

6. An inductor has a current flow of 2 amperes when connected to a 240 volt, 50 Hz line. How much current will flow through the inductor if it is connected to a 240 volt, 60 Hz line?

    _____

7. An RL parallel circuit has an apparent power of 2,400 VA. The true power is 1,860 watts. What is the circuit power factor?

    _____

8. How many degrees out of phase are the voltage and current in question 7?

    _____

9. An RL parallel circuit has a power factor of 64%. The circuit voltage is 480 volts and the total current is 25.6 amperes. What is the true power in the circuit?

    _____

10. The voltage and current are 44° out of phase with each other in an RL parallel circuit. What is the circuit power factor?

    _____

Resistive-Inductive Parallel Circuits

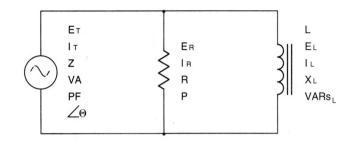

$$Z = \cfrac{1}{\sqrt{\left(\cfrac{1}{R}\right)^2 + \left(\cfrac{1}{X_L}\right)^2}} \qquad Z = \frac{VA}{I_T^2} \qquad I_T = \sqrt{I_R^2 + I_L^2} \qquad I_T = \frac{VA}{E_T} \qquad I_T = \sqrt{\frac{VA}{Z}}$$

$$Z = \frac{E_T}{I_T} \qquad Z = \frac{E_T^2}{VA} \qquad Z = R \times PF \qquad I_T = \frac{E_T}{Z} \qquad I_T = \frac{I_R}{PF} \qquad E_T = E_R = E_L$$

$$VA = E_T \times I_T \qquad PF = \frac{Z}{R} \qquad E_L = I_L \times X_L \qquad I_L = \sqrt{I_T^2 - I_R^2} \qquad E_T = I_T \times Z$$

$$VA = I_T^2 \times Z \qquad PF = \frac{P}{VA} \qquad E_L = E_T = E_R \qquad I_L = \frac{E_L}{X_L} \qquad E_T = \frac{VA}{I_T}$$

$$VA = \frac{E_T^2}{Z} \qquad PF = \frac{I_R}{I_T} \qquad E_L = \sqrt{VARs_L \times X_L} \qquad I_L = \frac{VARs_L}{E_L} \qquad E_T = \sqrt{VA \times Z}$$

$$VA = \sqrt{P^2 + VARs_L^2} \qquad PF = Cos\angle\Theta \qquad E_L = \frac{VARs_L}{I_L} \qquad I_L = \sqrt{\frac{VARs_L}{X_L}} \qquad L = \frac{X_L}{2\pi f}$$

$$VA = \frac{P}{PF} \qquad VARs_L = \sqrt{VA^2 - P^2} \qquad VARs_L = E_L \times I_L \qquad VARs_L = \frac{E_L^2}{X_L}$$

$$VARs_L = I_L^2 \times X_L \qquad E_R = I_R \times R \qquad I_R = \sqrt{I_T^2 - I_L^2}$$

$$X_L = \frac{E_L}{I_L} \qquad X_L = \cfrac{1}{\sqrt{\left(\cfrac{1}{R}\right)^2 - \left(\cfrac{1}{Z}\right)^2}} \qquad E_R = \sqrt{P \times R} \qquad I_R = \frac{E_R}{R}$$

$$X_L = \frac{E_L^2}{VARs_L} \qquad X_L = 2\pi f L \qquad E_R = \frac{P}{I_R} \qquad I_R = \frac{P}{E_R}$$

$$X_L = \frac{VARs_L}{I_L^2} \qquad R = \frac{E_R}{I_R} \qquad E_R = E_T = E_L \qquad I_R = \sqrt{\frac{P}{R}}$$

$$R = \frac{E_R^2}{P} \qquad R = \frac{P}{I_R^2} \qquad I_R = I_T \times PF$$

$$R = \cfrac{1}{\sqrt{\left(\cfrac{1}{Z}\right)^2 - \left(\cfrac{1}{X_L}\right)^2}} \qquad P = \frac{E_R^2}{R} \qquad R = \frac{Z}{PF} \qquad P = \sqrt{VA^2 - VARs_L^2}$$

$$P = E_R \times I_R \qquad P = I_R^2 \times R \qquad P = VA \times PF$$

Figure 12-6 RL parallel circuit formulas.

Unit 13 Capacitance

Objectives

After studying this unit, you should be able to

- Describe polarized and nonpolarized capacitors.
- Perform an ohmmeter test on a capacitor.
- Calculate values of capacitance and capacitive reactance.
- Determine the value of a capacitor by making electrical measurements.

Capacitors are the third major type of electrical load to be discussed. A basic capacitor is constructed by separating two metal plates with an insulating material called a dielectric (Figure 13-1). There are three factors that determine the amount of capacitance a capacitor will have:

1. *The surface area of the plates.* The greater the surface area, the more capacitance the capacitor will exhibit.
2. *The distance between the plates.* The closer the two plates are together, the greater the amount of capacitance.
3. *The type of dielectric.* Different types of dielectric can produce more or less capacitance.

Dielectric materials have a rating called the dielectric constant. Some common dielectric materials and their dielectric constants are shown in Figure 13-2. The dielectric constant of a material is determined by measuring the amount of increased capacitance when a particular material is employed. Air has a dielectric constant of 1 and is used as the base reference. Assume that a dielectric material is inserted between the plates of the capacitor without changing the distance between the plates. Now assume that the capacitor exhibits 10 times more capacitance with the dielectric material inserted between the plates instead of air. The material will have a dielectric constant of 10.

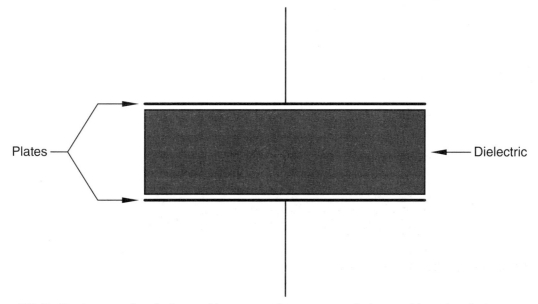

Figure 13-1 Basic capacitor is formed by separating two metal plates with an insulating material.

Material	Dielectric Constant
Air	1
Bakelite	4.0 to 10.0
Castor Oil	4.3 to 4.7
Cellulose Acetate	7.0
Ceramic	1,200
Dry Paper	3.5
Hard Rubber	2.8
Insulating Oils	2.2 to 4.6
Lucite	2.4 to 3.0
Mica	6.4 to 7.0
Mycalex	8.0
Paraffin	1.9 to 2.2
Porcelain	5.5
Pure Water	81
Pyrex Glass	4.1 to 4.9
Rubber Compounds	3.0 to 7.0
Teflon	2
Titanium Dioxide Compounds	90 to 170

Figure 13-2 Dielectric constant of different materials.

The dielectric or insulating material has the ability to change the amount of capacitance because of the manner in which a capacitor stores an electric charge. When a capacitor is charged, electrons are deposited on one plate and removed from the other, as seen in Figure 13-3. As electrons flow away from the capacitor plate connected to the positive battery terminal and to the capacitor plate connected to the negative battery terminal, a voltage is developed across the two capacitor plates. This electron flow will continue until the voltage across the plates is equal to the battery voltage. When the two voltages become equal, the current flow stops. If the battery is disconnected from the circuit, the capacitor is left in a charged state. The voltage across the capacitor plates produces stress on the dielectric material. Since one plate is now more positive and the other plate is more negative, the atoms of the dielectric become stressed

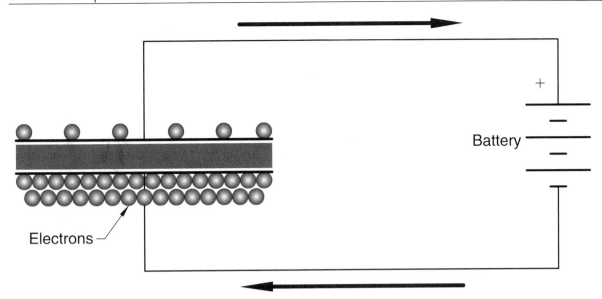

Figure 13-3 Charging a capacitor.

due to the attraction and repulsion of charged particles. The negative electrons in the atoms of the dielectric are attracted to the positive plate and repelled from the negative plate, as seen in Figure 13-4. This molecular stretching is called dielectric stress. The greater the potential difference between the two plates, the greater the dielectric stress. This is the reason that the voltage rating of a capacitor should never be exceeded. This molecular stretching is similar to stretching a rubber band. If the stress becomes too great (i.e., too much voltage), the dielectric will break down and short the capacitor. The dielectric stress produces an electrostatic field. Most of the capacitor's energy is stored in the electrostatic field.

The dielectric stress gives the capacitor the ability to produce almost infinite current. This is the reason that capacitors are one of the most dangerous components in the electrical field.

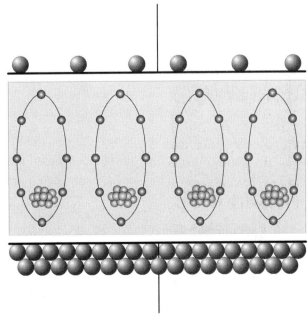

Figure 13-4 Atoms of the dielectric in a charged capacitor.

Never charge a capacitor and hand it to another person! Many people believe this to be a comical act, but in reality it is extremely dangerous. One of the basic characteristics of a capacitor is that it opposes a change of voltage. When a charged capacitor is connected across a load, it will produce any current necessary to prevent a voltage change. Assume that a capacitor has been charged to 500 volts. Now assume that the capacitor is connected across a 100 Ω load resistor. The capacitor will produce an initial current of 5 amps in an effort to prevent a voltage change (I = 500/100). If a person should contact the leads of a capacitor charged to 500 volts, the effect is the same as making contact with a 500 volt power line.

> **Caution**
>
> Never charge a capacitor and hand it to another person!

Capacitor Charge and Discharge

Capacitors charge and discharge at an exponential rate. This is the same rate of increase and decrease as the current flow through an inductor. Recall that each time constant of the exponential curve has a value of 63.2% of the whole. It is assumed that there are five time constants in the curve. An exponential charging curve is shown in Figure 13-5. The curve assumes the capacitor to be connected to a 100 volt source. The capacitor charge and

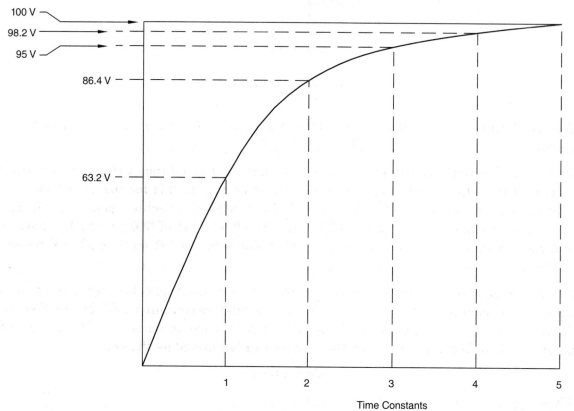

Figure 13-5 Capacitors charge at an exponential rate.

discharge time is determined by the amount of capacitance and resistance in the circuit. The formula (T = RC) can be used to determine the charge or discharge time of a capacitor where:

T = Time in seconds

R = Resistance in ohms

C = Capacitance in farads

The value of T is for one time constant.

Example: Determine the amount of time needed to charge a 100 μF capacitor connected in series with a 50 kΩ resistor.

$$T = 50,000 \times 0.000100 \ (100 \ \mu F)$$

$$T = 5 \text{ sec per time constant}$$

$$\text{Total time} = 5 \text{ sec} \times 5 \text{ time constants}$$

$$\text{Total time} = 25 \text{ sec}$$

Capacitance Values

Capacitance is measured in units called farads. A farad is the amount of capacitance necessary to produce a charge of 1 coulomb with a voltage change of 1 volt across its plates. The farad is an extremely large amount of capacitance. A 1 farad capacitor with air as a dielectric and a spacing of 1 inch between the plates would have a plate area of approximately 1,100 square miles. Since the farad is such a large amount of capacitance, capacitance values are generally expressed as microfarads (μF), nanofarads (nF), and picofarads (pF).

$$\mu F = \frac{1}{1,000,000} \text{ or } \times 10^{-6}$$

$$nF = \frac{1}{1,000,000,000} \text{ or } \times 10^{-9}$$

$$pF = \frac{1}{1,000,000,000,000} \text{ or } \times 10^{-12}$$

The picofarad is sometimes called micro-microfarad because it is one-millionth of a microfarad.

Most formulas that use capacitance assume that the capacitance value is in farads. As a general rule, the value given must be converted into farads for use in a formula. A value of 25 μF can be converted to farads by moving the decimal point six places to the left so that 25 μF becomes 0.000025 farads. Or a value of 300 pF can be converted to farads by moving the decimal point twelve places to the left so that 300 pF becomes 0.000000000300 farads.

Another method of entering capacitance values with a scientific calculator is to enter the value using scientific notation. Most scientific calculators contain an EE key or EXP key, depending on the manufacturer. These keys can be used to enter a value in scientific notation. Micro is one-millionth or 10^{-6}. 25 μF can be entered as shown:

$$25 \text{ EE } 6 \ +/-$$

300 pF can be entered as shown:

$$300 \text{ EE } 12 \ +/-$$

Polarized and Nonpolarized Capacitors

Capacitors can be divided into two major categories: polarized and nonpolarized. Nonpolarized capacitors are often called AC capacitors because they can be connected to alternating current circuits. These capacitors are not polarity sensitive. Either of the capacitor terminals can be connected to a positive or negative voltage.

Polarized capacitors can be connected to direct current circuits only. One of their terminals will be marked with a + or − sign to indicate the proper polarity. If the polarity connection is reversed, the capacitor will be damaged and possibly explode. Polarized capacitors are often called electrolytic capacitors. The greatest advantage of an electrolytic capacitor is that it can exhibit a large amount of capacitance in a small case size. Electrolytic capacitors are generally used in electronic equipment.

Capacitor Voltage Rating

As stated previously, the voltage rating of a capacitor should never be exceeded. The voltage ratings are marked in different ways, often making it difficult for a student to determine exactly what the voltage rating is. Some common capacitor voltage ratings are:

VDC (Volts DC)

WVDC (Working Volts DC)

VAC (Volts AC)

The VDC and WVDC ratings are essentially the same. Polarized capacitors will be rated in DC volts because they must be connected to direct current. Nonpolarized capacitors are generally marked VAC or WVAC. If the voltage rating is given as VAC, it is an RMS value. If a nonpolarized capacitor is marked WVDC, it is a peak value, not an RMS value.

Example: Assume that an AC capacitor is marked 600 WVDC. What is the maximum RMS value of voltage that can be connected to the capacitor? To find the answer, assume 600 volts to be a peak value and multiply by 0.707 to find the RMS value.

$$E_{RMS} = 600 \times 0.707$$

$$E_{RMS} = 424.2 \text{ volts}$$

Capacitance in AC Circuits

When a capacitor is connected in a DC circuit, current will flow during the time the capacitor is charging or discharging. Once the capacitor has become charged, it becomes an open circuit. When a capacitor is connected into an AC circuit, the capacitor charges and discharges each time the current changes direction. Since the current is continually changing direction, current will appear to flow through the capacitor.

As the capacitor charges, the impressed voltage across the capacitor opposes the applied voltage, acting like a resistance to the flow of current. This opposition to current flow is called capacitive reactance and is symbolized X_C. The formula for determining capacitive reactance is:

$$X_C = \frac{1}{2\pi fC}$$

where

X_C = Capacitive reactance in ohms

π = 3.1416

f = Frequency in Hz

C = Capacitance in farads

Example: A 10 μF capacitor is connected to a 120 volt, 60 Hz line. How much current will flow in this circuit? The first step is to determine the capacitive reactance.

$$X_C = \frac{1}{2 \times \pi \times 60 \times 0.000010 \text{ farad}}$$

$$X_C = 265.26 \ \Omega$$

The current flow can now be determined by replacing the value of resistance (R) with the value of capacitive reactance (X_C) in an Ohm's law formula:

$$I = \frac{E}{X_C}$$

$$I = \frac{120}{265.26}$$

$$I = 0.452 \text{ amp}$$

If the value of capacitive reactance is known, the capacitance can be determined using the formula:

$$C = \frac{1}{2\pi f X_C}$$

Example: A capacitor is connected in a 480 volt, 60 Hz circuit. The circuit has a current of 14 amperes. What is the capacitance of the capacitor? The first step is to determine the capacitive reactance using Ohm's law.

$$X_C = \frac{E}{I}$$

$$X_C = \frac{480}{14}$$

$$X_C = 34.29 \ \Omega$$

Now that the value of X_C is known, the capacitance can be determined.

$$C = \frac{1}{2 \times \pi \times 60 \times 34.29}$$

$$C = 0.00007736 \text{ farad or } 77.36 \ \mu\text{F}$$

Capacitors Connected in Series

When capacitors are connected in series, the total capacitance is reduced because it has the effect of increasing the distance between the plates. The formula for determining the total

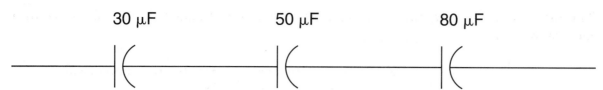

Figure 13-6 Capacitors connected in series.

capacitance of capacitors in series is similar to determining total resistance in a parallel circuit.

$$C_T = \frac{1}{\dfrac{1}{C_1} + \dfrac{1}{C_2} + \dfrac{1}{C_3} + \dfrac{1}{C_N}} \quad \text{(reciprocal method)}$$

or

$$C_T = \frac{C_1 \times C_2}{C_1 + C_2} \quad \text{(product over sum method)}$$

or

$$C_T = \frac{C}{N}$$

This formula can be employed when all capacitor values are the same. The product over sum method can be employed to determine the value of two capacitors at a time.

Assume that a circuit contains three capacitors with values of 30 μF, 50 μF, and 80 μF connected in series, as shown in Figure 13-6. What is the total capacitance of this circuit?

$$C_T = \frac{1}{\dfrac{1}{30} + \dfrac{1}{50} + \dfrac{1}{80}}$$

$$C_T = 15.19 \ \mu F$$

The total capacitive reactance is equal to the sum of the individual capacitive reactances in the circuit.

$$X_{C_T} = X_{C_1} + X_{C_2} + X_{C_3} + X_{C_N}$$

The advantage of connecting capacitors in series is that it increases the voltage rating of the connection. Assume that three capacitors rated at 100 μF and 100 volts each are connected in series. The total capacitance of the connection would be 33.33 μF ($C_T = \frac{100}{3}$), but the voltage rating would now be 300 volts instead of 100 volts.

Capacitors Connected in Parallel

When capacitors are connected in parallel, the total capacitance value will be the sum of the individual capacitors. Connecting capacitors in parallel has the effect of increasing the plate area of one capacitor. The formula for finding the total capacitance of parallel-connected capacitors is:

$$C_T = C_1 + C_2 + C_3 + C_N$$

The total capacitive reactance can be determined in a manner similar to determining parallel resistance.

$$X_{C_T} = \cfrac{1}{\cfrac{1}{X_{C_1}} + \cfrac{1}{X_{C_2}} + \cfrac{1}{X_{C_3}} + \cfrac{1}{X_{C_N}}} \quad \text{(reciprocal method)}$$

or

$$X_{C_T} = \frac{X_{C_1} \times X_{C_2}}{X_{C_1} + X_{C_2}} \quad \text{(product over sum method)}$$

or

$$X_{C_T} = \frac{X_C}{N}$$

This formula can be employed when all values of capacitive reactance are the same. The product over sum method can be employed when two values of capacitive reactance are known.

Voltage and Current Relationships

In a pure capacitive circuit, the voltage and current are 90° out of phase with each other, as shown in Figure 13-7. In a pure capacitive circuit, the current *leads* the applied voltage instead of lagging the applied voltage. Since the current and voltage are 90° out of phase, there is no true power or watts produced in a pure capacitive circuit. The electrical energy is stored in the form of an electrostatic field during part of the cycle and returned to the circuit during another part. Like inductive circuits, the power in a capacitive circuit is in VARs and not watts. To distinguish between inductive and capacitive VARs, capacitive VARs will be noted as VARs$_C$. The formulas for determining capacitive VARs are the same basic formulas used for watts and inductive VARs.

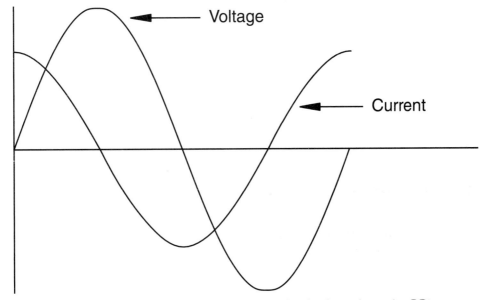

Figure 13-7   In a pure capacitive circuit, the current leads the voltage by 90°.

$$VARs_C = E_C \times I_C$$

$$VARs_C = I_C^2 \times X_C$$

$$VARs_C = \frac{E_C^2}{X_C}$$

Capacitor Testing

Testing capacitors is difficult at best. To really test a capacitor, two conditions must be assessed. One is the condition of the dielectric, and the other is the condition of the plate area. The first test can be made with an ohmmeter. The ohmmeter is used to check for a short circuit between the plates. If an analog ohmmeter is used, when the meter leads are connected to the capacitor the needle should swing upscale and then return to infinity. If the ohmmeter leads are reversed, the needle should move approximately twice as far upscale and then return to infinity. If a digital ohmmeter is employed, it should indicate either infinity or a very high resistance in the meg-ohm range.

The ohmmeter test can indicate a short circuit, but it cannot test the condition of the dielectric. To test the dielectric, rated voltage must be applied to the capacitor. As a general rule, a test instrument called a "HiPot" is used to test the dielectric. HiPot is an abbreviation for high potential. The HiPot permits rated voltage to be applied to the capacitor while measuring the leakage current through the dielectric. In theory, there should be no current flow through the dielectric. Leakage current is an indication that the dielectric is breaking down under load.

The second test involves measuring the amount of capacitance of the capacitor. If the plates are in good condition, the capacitance should be within a few percent of the marked capacitor rating. The capacitance value can be determined in several ways. One is to use a capacitance tester designed to measure the amount of capacitance. Another is to connect the capacitor in an AC circuit and measure the current flow. The capacitive reactance can then be computed using Ohm's law, and the capacitance value can be computed once the capacitive reactance is known.

LABORATORY EXERCISE

Name _____ Date _____

Materials Required

1 120-VAC power supply

1 AC voltmeter

1 AC ammeter, in-line or clamp-on. (If a clamp-on meter is employed, the use of a 10:1 scale divider is recommended.)

1 ohmmeter

1 7.5-μF capacitor 240 VAC or more

1 10-μF capacitor 240 VAC or more

1 25-μF capacitor 240 VAC or more

1. Use an ohmmeter to test the resistance of the following capacitors. Be sure to reverse the ohmmeter leads when checking each capacitor. List the amount of resistance for each capacitor.

   7.5 μF _____ Ω   10 μF _____ Ω   25 μF _____ Ω

Caution

If any of the capacitors tested indicate a low value of resistance (less than 100 kΩ), inform your instructor before continuing with this experiment.

---

2. Calculate the amount of capacitive reactance for the 7.5 μF capacitor. Assume a frequency of 60 Hz.

$$X_C = \frac{1}{2\pi fC}$$

   (Note: A shortcut is to use a value of 377 for $(2\pi f)$ if the frequency is 60 Hz.)

   $X_C =$ _____ Ω

3. Using the calculated value of capacitive reactance, compute the amount of current that should flow if the capacitor is connected to 120 VAC.

   I = _____ amp(s)

4. Connect the circuit shown in Figure 13-8. Turn on the power and measure the amount of current flow. Compare the measured value with the computed value. Are the two answers approximately the same?

   I = _____ amp(s)   Yes/no _____

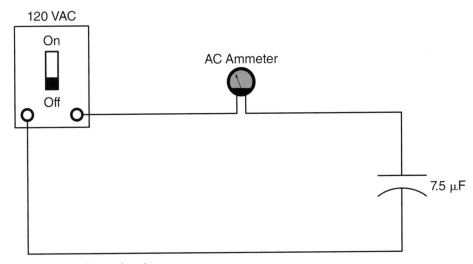

Figure 13-8 A capacitive circuit.

5. **Turn off the power supply.**

6. Replace the 7.5 µF capacitor with a 10 µF capacitor.

7. Turn on the power supply and measure the amount of current flow in the circuit.

I = _____ amp(s)

8. Using the measure value of current, compute the capacitive reactance for this circuit.

$$X_C = \frac{E}{I}$$

$X_C =$ _____ Ω

9. Using the computed value of capacitive reactance, compute the capacitance of the capacitor. (Note: The computed value will be in farads. Change your answer to microfarads.) Is the computed value approximately the same as the marked value on the capacitor?

$$C = \frac{1}{2\pi f X_C}$$

C = _____ µF   Yes/no _____

10. **Turn off the power supply.**

11. Replace the 10 µF capacitor with the 25 µF capacitor.

12. Compute the amount of current that should flow if a voltage of 120 volts is applied to the circuit.

I = _____ amp(s)

13. Turn on the power supply and measure the current in the circuit. Is the measured current approximately the same as the computed value?

I = _____ amp(s)   Yes/no _____

14. **Turn off the power supply.**

15. Connect the three capacitors to form a series circuit as shown in Figure 13-9.

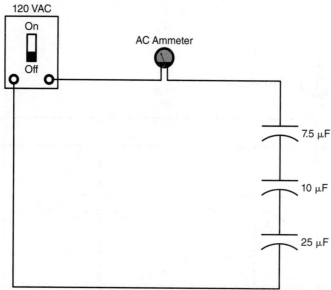

Figure 13-9 Capacitors connected in series.

16. Turn on the power supply and measure the current in the circuit.

    I = _____ amp(s)

17. Use an AC voltmeter to measure the voltage drop across each capacitor.

    7.5 μF _____ volts   10 μF _____ volts   25 μF _____ volts

18. Using the measured values of voltage and current, compute the capacitive reactance of each capacitor. Recall that in a series circuit the current is the same in all parts of the circuit.

    7.5 μF _____ Ω   10 μF _____ Ω   25 μF _____ Ω

19. Using the formula ($X_{C_T} = X_{C_1} + X_{C_2} + X_{C_3} + X_{C_N}$), compute the capacitive reactance for the circuit.

    X_{C_T} = _____ Ω

20. Use Ohm's law and the values of applied voltage and circuit current to compute the total capacitive reactance for this circuit. Is this computed value approximately the same as the capacitive reactance value in step 19?

    X_{C_T} = _____ Ω   Yes/no _____

21. **Turn off the power supply.**

22. Reconnect the three capacitors to form a parallel connection as shown in Figure 13-10.

23. Turn on the power supply and measure the current flow in the circuit.

    I = _____ amp(s)

24. **Turn off the power supply.**

25. Calculate the capacitive reactance of the circuit using Ohm's law.

    X_{C_T} = _____ Ω

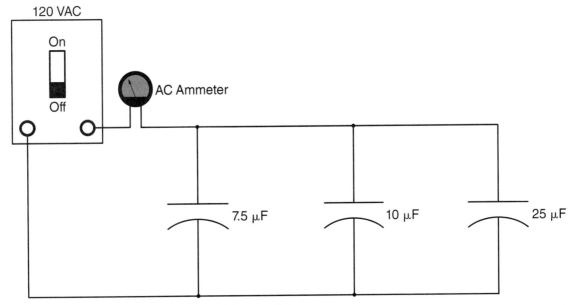

Figure 13-10 Capacitors connected in parallel.

26. Use the values of capacitive reactance computed in step 18 for each of the capacitors to compute the total amount of capacitive reactance for the circuit. Is the computed value approximately the same as the value computed in step 25?

$$X_{C_T} = \cfrac{1}{\cfrac{1}{X_{C_1}} + \cfrac{1}{X_{C_2}} + \cfrac{1}{X_{C_3}} + \cfrac{1}{X_{C_N}}}$$

$X_{C_T} = $ _____ Ω   Yes/no _____

27. Using the computed value of capacitive reactance in step 26, compute the total amount of capacitance in the circuit.

$C_T = $ _____ μF

28. Add the values of capacitance listed on the three capacitors. Is the sum of the listed values approximately the same as the computed value in step 27?

$C_T = $ _____ μF   Yes/no _____

29. Disconnect the circuit and return the components to their proper place.

Review Questions

1. Name three factors that determine the capacitance of a capacitor.

_____

_____

_____

2. A nonpolarized capacitor is connected in a DC circuit. Is there a danger of damaging the capacitor when the power is turned on?

_____

3. A polarized capacitor is connected to an AC circuit. Is there a danger of damaging the capacitor when the power is turned on?

_____

4. Four capacitors having values of 100 μF, 175 μF, 75 μF, and 50 μF are connected in series. What is the total capacitance of the circuit?

_____

5. Assume that the circuit in question 4 is connected to a 240 volt AC, 60 Hz power supply. How much voltage would be dropped across the 100 μF capacitor?

_____

6. Assume that the 175 μF capacitor in question 4 has a voltage rating of 35 WVDC. If the circuit is connected to 240 VAC, will the voltage rating of the capacitor be exceeded?

_____

7. A 500 pF capacitor is connected to a 60 Hz circuit. What is the capacitive reactance of this capacitor?

_____

8. Assume that a 500 pF capacitor is connected in a circuit with a frequency of 100 kHz. What is the capacitive reactance of the capacitor?

_____

9. Assume that three capacitors with values of 5 μF, 12 μF, and 22 μF are connected in parallel. What is the total capacitance of the circuit?

_____

10. Assume that the circuit in question 9 is connected to a 208-volt, 60 Hz power line. How much current will flow in the circuit?

_____

Unit 14 Resistive-Capacitive Series Circuits

Objectives

After studying this unit, you should be able to

- Discuss the voltage and current relationship in an RC series circuit.
- Determine the phase angle of current in an RC series circuit.
- Determine the power factor in an RC series circuit.

In an AC circuit containing pure resistance, the current and voltage are in phase with each other. In an AC circuit containing a pure capacitive load, the current will lead the voltage by 90°. Since the current is the same at any point in a series circuit, the voltage drop across the resistor and the voltage drop across the capacitor will be 90° out of phase with each other, as shown in Figure 14-1.

Example Problem

A series circuit containing a resistor and capacitor is shown in Figure 14-2. It is assumed the circuit is connected to 120 VAC with a frequency of 60 Hz. The resistor has a resistance of 30 Ω and the capacitor has a capacitive reactance of 46 Ω. The following values will be computed:

Z - Total impedance of the circuit

I_T - Total circuit current

E_R - Voltage drop across the resistor

P - True power or watts

E_C - Voltage drop across the capacitor

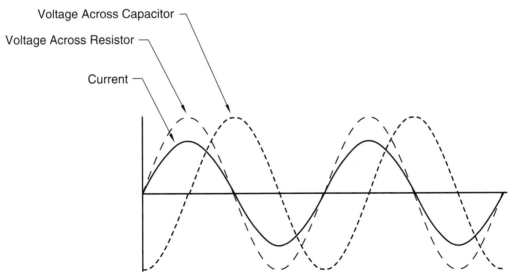

Figure 14-1 The voltage across the resistor and the voltage across the capacitor are 90° out of phase with each other.

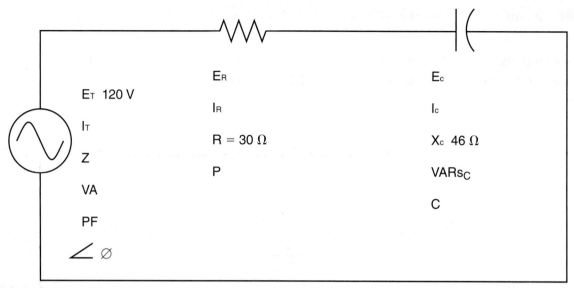

Figure 14-2 RC series circuit.

C - Capacitance of the capacitor

VARs$_C$ - Reactive power

VA - Volt amps or apparent power

PF - Power factor

$\angle\varnothing$ - Angle theta (the angle's degree amount that the current and voltage are out of phase with each other)

The first step in determining the missing values is to compute the total circuit impedance. Since the capacitive and resistive parts of the circuit are out of phase with each other by 90°, vector addition must be used to determine the total impedance. The formula for impedance determining the impedance in a series circuit containing resistance and capacitive reactance is $Z = \sqrt{R^2 + X_c^{\,2}}$.

$$Z = \sqrt{30^2 + 46^2}$$

$$Z = \sqrt{900 + 2{,}116}$$

$$Z = \sqrt{3{,}016}$$

$$Z = 54.92\ \Omega$$

The total circuit current can now be computed using Ohm's law.

$$I_T = \frac{E_T}{Z}$$

$$I_T = \frac{120}{54.92}$$

$$I_T = 2.18\ \text{amps}$$

The current is the same at any point in a series circuit. The values of I_C and I_L are, therefore, the same as I_T.

Now that the amount of current flowing through the resistor is known, the voltage drop across the resistor can be determined using Ohm's law.

$$E_R = I_R \times R$$

$$E_R = 2.18 \times 30$$

$$E_R = 65.4 \text{ volts}$$

The true power or watts can be computed using any of the power formulas. In this example the true power will be computed using $E \times I$.

$$P = E_R \times I_R$$

$$P = 65.4 \times 2.18$$

$$P = 142.57 \text{ watts}$$

The voltage drop across the capacitor can be determined using Ohm's law and reactive values.

$$E_c = I_c \times X_c$$

$$E_c = 2.18 \times 46$$

$$E_c = 100.28 \text{ volts}$$

The reactive VARs can be computed in a manner similar to determining the value for watts or volt amps, except that reactive values are used.

$$VARs_C = E_c \times I_c$$

$$VARs_C = 100.28 \times 2.18$$

$$VARs_C = 218.61$$

The capacitance value of the capacitor can be computed using the formula:

$$C = \frac{1}{2\pi f X_c}$$

$$C = \frac{1}{2 \times 3.1416 \times 60 \times 46}$$

$$C = \frac{1}{377 \times 46}$$

$$C = 0.00005766 \text{ farad or } 57.66 \text{ }\mu\text{F}$$

The apparent power or volt amps can be computed using formulas similar to those for determining watts or VARs, except that total circuit values are used in the formula:

$$VA = E_T \times I_T$$

$$VA = 120 \times 2.18$$

$$VA = 261.6$$

The circuit power factor can be computed using the formula:

$$PF = \frac{P}{VA}$$

$$PF = \frac{142.57}{261.6}$$

$$PF = 0.545 \text{ or } 54.5\%$$

The power factor is the cosine of angle theta. In this circuit, the decimal power factor is 0.545. The cosine of angle theta is 0.545. To determine angle theta, find the angle that corresponds to a cosine of 0.545. Most scientific calculators contain trigonometric functions. To find what angle corresponds to one of the sin, cos, or tan functions, it is generally necessary to use the invert key, the arc key, or one of the keys marked $\sin^{-1}$, $\cos^{-1}$, or $\tan^{-1}$.

$$\cos \angle \varnothing = 0.545$$

$$\angle \varnothing = 56.97°$$

The current and voltage are 56.97° out of phase with each other in this circuit.

The circuit with all completed values is shown in Figure 14-3.

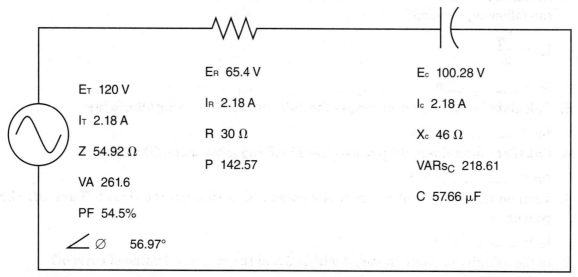

Figure 14-3 RC series circuit with all missing values.

LABORATORY EXERCISE

Name _____ Date _____

Materials Required

1 25 μF AC capacitor with a minimum voltage rating of 240 VAC

1 10 μF AC capacitor with a minimum voltage rating of 240 VAC

1 100 ohm resistor

1 150 ohm resistor

AC ammeter (In-line or clamp-on type may be used. If a clamp-on type is employed, the use of a 10:1 scale divider is recommended.)

Connecting wires

AC voltmeter

1 120 volt AC power supply

Formulas for resistive–capacitive series circuits shown in Figure 14-6

1. Connect the circuit shown in Figure 14-4.

2. Calculate the capacitive reactance of the 25 μF capacitor assuming a frequency of 60 Hz.

    $X_C = $ _____ Ω

3. Calculate the impedance of the circuit using the following formula:

    $$Z = \sqrt{R^2 + X_C^2}$$

    $Z = $ _____ Ω

4. Assuming a voltage of 120 volts, calculate the total current flow in the circuit using the following formula:

    $$I_T = \frac{E_T}{Z}$$

    $I_T = $ _____ A

5. Calculate the voltage drop across the 100 ohm resistor using Ohm's law.

    $E_R = $ _____ volts

6. Calculate the voltage drop across the 25 μF capacitor using Ohm's law.

    $E_C = $ _____ volts

7. Turn on the power and measure the current flow through the circuit. **Turn off the power**.

    $I_T = $ _____ A

8. Is the calculated value in step 4 within 5% of the measured value of current?

    _____

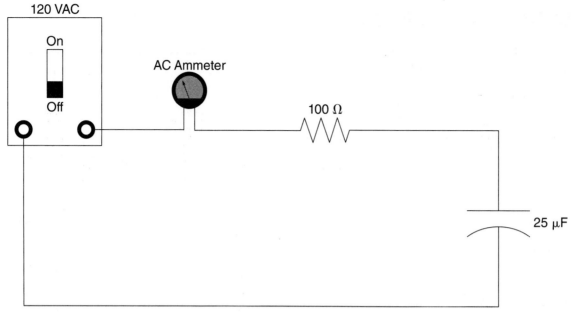

Figure 14-4 A 100 ohm resistor is connected in series with a 25 µF capacitor.

9. Turn on the power and measure the voltage drop across the resistor and capacitor. **Turn off the power**.

$E_R =$ _____ volts

$E_C =$ _____ volts

10. Compare the measured values of voltage drop across the resistor and capacitor with the calculated values in steps 5 and 6. Are the values within 5% of each other?

_____

11. In a series circuit the current in all parts of the circuit must be the same. Therefore, the voltage drop across the resistor is 90° out of phase with the voltage drop across the capacitor. When vector addition is used, the sum of the two voltage drops should equal the applied voltage. Calculate the total or applied voltage using the following formula:

$E_T = \sqrt{E_R^2 + E_C^2}$

$E_T =$ _____ volts

12. Is the computed value of voltage within 5% of the voltage that was applied to the circuit?

_____

13. Determine the true power of the circuit using the circuit current and the voltage drop across the resistor.

$P = E_R \times I$

$P =$ _____ watts

14. Determine the capacitive VARs using the circuit current and the voltage drop across the capacitor.

$VARs_C = E_C \times I$

$VARs_C =$ _____

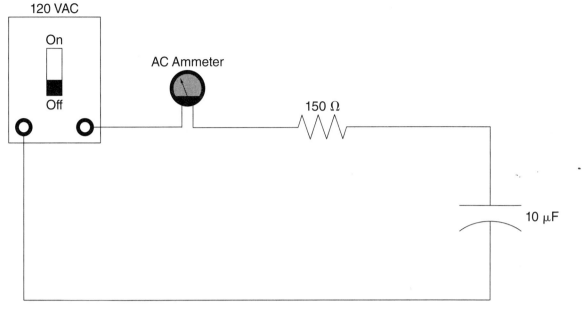

Figure 14-5 Replacing the circuit components.

15. Determine the apparent power in the circuit using the applied voltage and the circuit current.

$$VA = E_T \times I$$

$$VA = \underline{\hspace{2cm}}$$

16. Calculate the circuit power factor using the true power and apparent power.

$$PF = \frac{P}{VA}$$

$$PF = \underline{\hspace{2cm}} \%$$

17. Calculate angle theta.

$$\angle\varnothing = Cos\ PF$$

$$\angle\varnothing = \underline{\hspace{2cm}}°$$

18. Connect the circuit shown in Figure 14-5.

19. Calculate the capacitive reactance of the 10 µF capacitor assuming a frequency of 60 Hz.

$$X_C = \underline{\hspace{2cm}} \Omega$$

20. Calculate the impedance of the circuit using the following formula:

$$Z = \sqrt{R^2 + X_C^2}$$

$$Z = \underline{\hspace{2cm}} \Omega$$

21. Assuming a voltage of 120 volts, calculate the total current flow in the circuit using the following formula:

$$I_T = \frac{E_T}{Z}$$

$$I_T = \underline{\hspace{2cm}} A$$

22. Calculate the voltage drop across the 150 ohm resistor using Ohm's law.

    $E_R =$ _____ volts

23. Calculate the voltage drop across the 25 μF capacitor using Ohm's law.

    $E_C =$ _____ volts

24. Turn on the power and measure the current flow through the circuit. **Turn off the power**.

    $I_T =$ _____ A

25. Is the calculated value in step #21 within 5% of the measured value of current?

    _____

26. Turn on the power and measure the voltage drop across the resistor and capacitor. **Turn off the power**.

    $E_R =$ _____ volts

    $E_C =$ _____ volts

27. Compare the measured values of voltage drop across the resistor and capacitor with the calculated values in steps 22 and 23. Are the values within 5% of each other?

    _____

28. In a series circuit the current is the same in any part of the circuit must be the same. Therefore, the voltage drop across the resistor is 90° out of phase with the voltage drop across the capacitor. When vector addition is used, the sum of the two voltage drops should equal the applied voltage. Calculate the total or applied voltage using the following formula:

    $$E_T = \sqrt{E_R^2 + E_C^2}$$

    $E_T =$ _____ volts

29. Is the computed value of voltage within 5% of the voltage that was applied to the circuit?

    _____

30. Determine the true power of the circuit using the circuit current and the voltage drop across the resistor.

    $P = E_R \times I$

    $P =$ _____ watts

31. Determine the capacitive VARs using the circuit current and the voltage drop across the capacitor.

    $VARs_C = E_C \times I$

    $VARs_C =$ _____

32. Determine the apparent power in the circuit using the applied voltage and the circuit current.

    $VA = E_T \times I$

    $VA =$ _____

33. Calculate the circuit power factor using the true power and apparent power.

    $$PF = \frac{P}{VA}$$

    $PF =$ _____ %

34. Calculate angle theta.

$\angle \varnothing$ = Cos PF

$\angle \varnothing$ = _____ °

35. Disconnect the circuit and return the components to their proper place.

Review Questions

Refer to the formulas shown in Figure 14-6 to answer some of the following questions:

1. The current and voltage are 64° out of phase in an RC series circuit. The apparent power is 260 VA. What is the true power in the circuit?

_____

2. The circuit in question 1 has a current flow of 1.25 amps. What is the applied voltage?

_____

3. A capacitor and resistor are connected in series. The resistor has a resistance of 26 Ω and the capacitor has a capacitive reactance of 16 Ω. What is the impedance of the circuit?

_____

4. An RC series circuit is connected to a 120 volt, 60 Hz line. The capacitor has a voltage drop of 84 volts. What is the voltage drop across the resistor?

_____

5. A capacitor is using 1.6 kVARs and has a capacitive reactance of 24 Ω. How much current is flowing through the inductor?

_____

6. An RC series circuit is connected to a 277 volt, 60 Hz line. The resistor has a power dissipation of 146 watts and the capacitor is using 138 VARs. How much current is flowing in the circuit?

_____

7. An RC series circuit is connected to a 480 volt, 60 Hz line. The apparent power of the circuit is 8.2 kVA. What is the impedance of the circuit?

_____

8. An RC series circuit has a power factor of 72%. How many degrees are the voltage and current out of phase with each other?

_____

9. An RL series circuit is connected to a 240 volt, 60 Hz line. The capacitor has a current of 1.25 amps flowing though it. The X_C of the inductor is 22 Ω. The resistor has a resistance of 16 Ω. How much voltage is dropped across the resistor?

_____

10. An RC series circuit has an apparent power of 650 VA and a true power of 475 watts. What is the reactive power in the circuit?

_____

Resistive-Capacitive Series Circuits

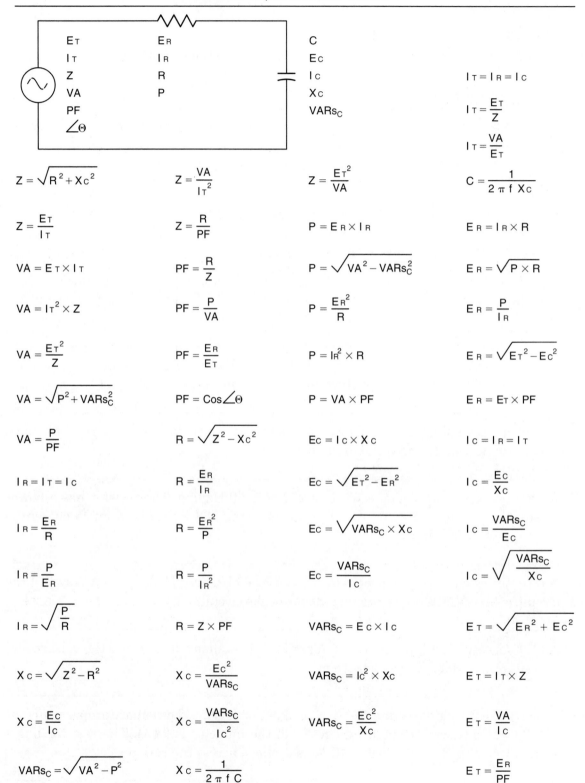

$$Z = \sqrt{R^2 + X_C^2}$$

$$Z = \frac{VA}{I_T^2}$$

$$Z = \frac{E_T^2}{VA}$$

$$C = \frac{1}{2 \pi f X_C}$$

$$Z = \frac{E_T}{I_T}$$

$$Z = \frac{R}{PF}$$

$$P = E_R \times I_R$$

$$E_R = I_R \times R$$

$$VA = E_T \times I_T$$

$$PF = \frac{R}{Z}$$

$$P = \sqrt{VA^2 - VARs_C^2}$$

$$E_R = \sqrt{P \times R}$$

$$VA = I_T^2 \times Z$$

$$PF = \frac{P}{VA}$$

$$P = \frac{E_R^2}{R}$$

$$E_R = \frac{P}{I_R}$$

$$VA = \frac{E_T^2}{Z}$$

$$PF = \frac{E_R}{E_T}$$

$$P = I_R^2 \times R$$

$$E_R = \sqrt{E_T^2 - E_C^2}$$

$$VA = \sqrt{P^2 + VARs_C^2}$$

$$PF = \cos \angle \Theta$$

$$P = VA \times PF$$

$$E_R = E_T \times PF$$

$$VA = \frac{P}{PF}$$

$$R = \sqrt{Z^2 - X_C^2}$$

$$E_C = I_C \times X_C$$

$$I_C = I_R = I_T$$

$$I_R = I_T = I_C$$

$$R = \frac{E_R}{I_R}$$

$$E_C = \sqrt{E_T^2 - E_R^2}$$

$$I_C = \frac{E_C}{X_C}$$

$$I_R = \frac{E_R}{R}$$

$$R = \frac{E_R^2}{P}$$

$$E_C = \sqrt{VARs_C \times X_C}$$

$$I_C = \frac{VARs_C}{E_C}$$

$$I_R = \frac{P}{E_R}$$

$$R = \frac{P}{I_R^2}$$

$$E_C = \frac{VARs_C}{I_C}$$

$$I_C = \sqrt{\frac{VARs_C}{X_C}}$$

$$I_R = \sqrt{\frac{P}{R}}$$

$$R = Z \times PF$$

$$VARs_C = E_C \times I_C$$

$$E_T = \sqrt{E_R^2 + E_C^2}$$

$$X_C = \sqrt{Z^2 - R^2}$$

$$X_C = \frac{E_C^2}{VARs_C}$$

$$VARs_C = I_C^2 \times X_C$$

$$E_T = I_T \times Z$$

$$X_C = \frac{E_C}{I_C}$$

$$X_C = \frac{VARs_C}{I_C^2}$$

$$VARs_C = \frac{E_C^2}{X_C}$$

$$E_T = \frac{VA}{I_C}$$

$$VARs_C = \sqrt{VA^2 - P^2}$$

$$X_C = \frac{1}{2 \pi f C}$$

$$E_T = \frac{E_R}{PF}$$

Figure 14-6 Formulas for RC series circuits.

Unit 15 RC Parallel Circuits

Objectives

After studying this unit, you should be able to

- Discuss the voltage and current relationship in an RC parallel circuit.
- Determine the phase angle of current in an RC parallel circuit.
- Determine the power factor in an RC parallel circuit.
- Discuss the differences between apparent power, true power, and reactive power.
- Find the impedance in an RC parallel circuit.

In any parallel circuit, the voltage must be the same across all branches. Since the current is in phase with the voltage in a pure resistive circuit, and the current leads the voltage by 90° in a pure capacitive circuit, the current flow through the capacitive branch will be out of phase with the current through the resistive branch. The total current will be out of phase with the applied voltage by some amount between 0° and 90° depending on the relative values of resistance and capacitive reactance.

Determining electrical values in an RC parallel circuit is very similar to determining values in a series circuit with a few exceptions. Probably the greatest difference is calculating the value of impedance when the values of R and X_C are known. Recall that vector addition can be used with the ohmic values of an RC series circuit to determine the impedance.

$$Z = \sqrt{R^2 + X_c^2}$$

The same basic concept is true for an RL parallel circuit, except that the reciprocal value of R and X_C must be used.

$$Z = \cfrac{1}{\sqrt{\left(\dfrac{1}{R}\right)^2 + \left(\dfrac{1}{X_c}\right)^2}}$$

Example: A resistor and capacitor are connected in parallel. The resistor has a resistance of 50 Ω and the capacitor has a capacitive reactance of 60 Ω. Find the impedance of the circuit, seen in Figure 15-1.

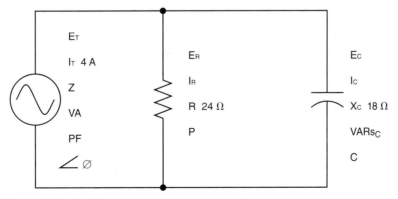

Figure 15-1 RC parallel circuit.

Solution:

$$Z = \frac{1}{\sqrt{\left(\frac{1}{50}\right)^2 + \left(\frac{1}{60}\right)^2}}$$

$$Z = \frac{1}{\sqrt{(0.02)^2 + (0.01667)^2}}$$

$$Z = \frac{1}{\sqrt{0.0004 + 0.0002779}}$$

$$Z = \frac{1}{\sqrt{0.0006779}}$$

$$Z = \frac{1}{0.02604}$$

$$Z = 38.4 \ \Omega$$

Example Circuit

An RC parallel circuit is connected to a 60 Hz line. The resistor has a resistance of 24 Ω and the capacitor has a capacitive reactance of 18 Ω. The circuit has total current flow of 4 amps. The following values will be computed:

E_T - Total circuit voltage

Z - Circuit impedance

VA - Apparent power

I_R - Current flow through the resistor

P - True power

I_C - Current flow through the inductor

$VARs_C$ - Reactive power

C - Capacitance of the capacitor

PF - Power factor

$\angle\varnothing$ - Angle theta

The first step in determining the missing values for this problem is to determine the circuit impedance using the values of R and X_C.

$$Z = \frac{1}{\sqrt{\left(\frac{1}{R}\right)^2 + \left(\frac{1}{X_c}\right)^2}}$$

$$Z = \frac{1}{\sqrt{\left(\frac{1}{24}\right)^2 + \left(\frac{1}{18}\right)^2}}$$

$$Z = \frac{1}{\sqrt{0.001736 + 0.003086}}$$

$$Z = \frac{1}{0.06944}$$

$$Z = 14.4 \ \Omega$$

Now that the value of impedance is known, the total voltage can be determined using Ohm's law.

$$E_T = I_T \times Z$$

$$E_T = 4 \times 14.4$$

$$E_T = 57.6 \ \text{volts}$$

In a parallel circuit, the voltage is the same across all branches. Therefore, the voltage drops across the resistor and capacitor are 57.6 volts.

The apparent power can be computed using the formula:

$$VA = E_T \times I_T$$

$$VA = 57.6 \times 4$$

$$VA = 230.4$$

The current flowing through the resistor can be computed using Ohm's law.

$$I_R = \frac{E_R}{R}$$

$$I_R = \frac{57.6}{24}$$

$$I_R = 2.4 \ \text{amps}$$

The true power in the circuit can be computed using resistive values.

$$P = E_R \times I_R$$

$$P = 57.6 \times 2.4$$

$$P = 138.24 \ \text{watts}$$

The current flow through the capacitor can be computed using Ohm's law.

$$I_c = \frac{E_c}{X_c}$$

$$I_c = \frac{57.6}{18}$$

$$I_c = 3.2 \, \text{amps}$$

The capacitive VARs can be computed using the inductive values and Ohm's law.

$$\text{VARs}_C = E_c \times I_c$$

$$\text{VARs}_C = 57.6 \times 3.2$$

$$\text{VARs}_C = 184.32$$

The capacitance of the capacitor can be computed using the formula:

$$C = \frac{1}{2\pi f X_c}$$

$$C = \frac{1}{377 \times 18}$$

$$C = 0.00014736 \, \text{farad or} \, 147.36 \, \mu\text{F}$$

The power factor can be determined by comparing the true power and apparent power.

$$PF = \frac{P}{VA}$$

$$PF = \frac{138.24}{230.4}$$

$$PF = 0.6 \, \text{or} \, 60\%$$

The cosine of angle theta is equal to the decimal power factor value.

$$\cos \angle\varnothing = PF$$

$$\cos \angle\varnothing = 0.6$$

$$\angle\varnothing = 53.13°$$

The circuit with all the missing values is shown in Figure 15-2.

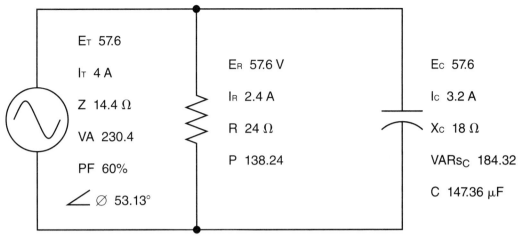

Figure 15-2 RC parallel circuit with all the missing values.

LABORATORY EXERCISE

Name _____ Date _____

Materials Required

1 25 µF AC capacitor rated at 240 VAC or greater

1 7.5 µF AC capacitor rated at 240 VAC or greater

1 100 ohm resistor

1 150 ohm resistor

AC ammeter (In-line or clamp-on may be used. If a clamp-on type is employed, the use of a 10:1 scale divider is recommended.)

Connecting wires

AC voltmeter

1 120 volt AC power supply

Formulas for resistive–capacitive parallel circuits shown in Figure 15-6

1. Connect the circuit shown in Figure 15-3.
2. Calculate the capacitive reactance of the capacitor assuming a frequency of 60 Hz.

   $X_C =$ _____

3. Calculate the impedance of the circuit using the following formula:

   $$Z = \frac{1}{\sqrt{\left(\dfrac{1}{R}\right)^2 + \left(\dfrac{1}{X_c}\right)^2}}$$

   $Z =$ _____ Ω

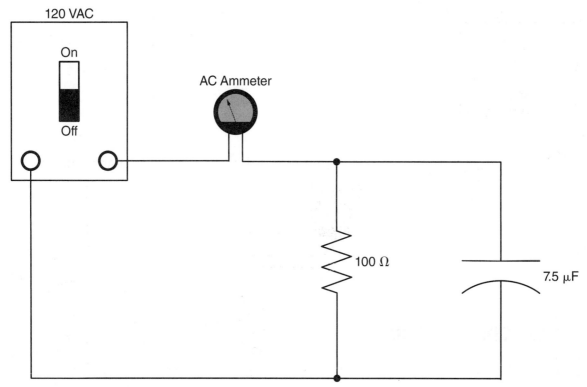

Figure 15-3 The AC ammeter measures the total circuit current.

4. Assume an applied voltage of 120 volts. Calculate the total circuit current using the following formula:

$$I_T = \frac{E}{Z}$$

$I_T = $ _____ A

5. Assume an applied voltage of 120 volts. Calculate the current flow through the 100 ohm resistor using the following formula:

$$I_R = \frac{E}{R}$$

$I_R = $ _____ A

6. Assume an applied voltage of 120 volts. Calculate the current that appears to flow through the capacitor using the following formula:

$$I_C = \frac{E}{X_C}$$

$I_C = $ _____ A

7. Turn on the power supply and measure the total current in the circuit. **Turn off the power.**

$I_C = $ _____ A

8. Compare the measured value with the calculated value in step 4. Are the two values within 5% of each other?

9. Reconnect the circuit as shown in Figure 15-4.

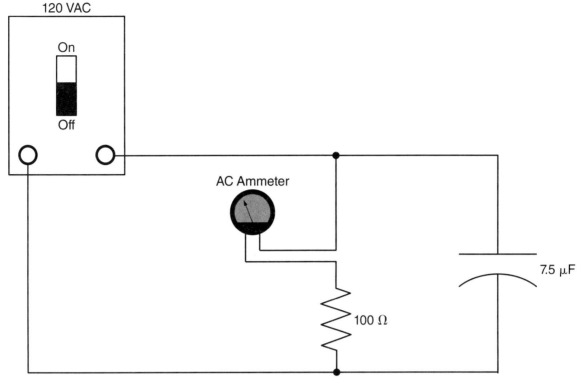

Figure 15-4 Measuring current through the resistive load.

10. Turn on the power and measure the current flow through the resistor. **Turn off the power.**

    I_R = _____ A

11. Compare the measured value with the calculated value in step 5. Are the two values within 5% of each other?

    _____

12. Reconnect the circuit as shown in Figure 15-5.

13. Turn on the power and measure the current that appears to flow through the capacitor. **Turn off the power.**

    I_C = _____ A

14. Compare the measured value with the calculated value in step 6. Are the two values within 5% of each other?

    _____

15. Calculate the true power in the circuit using the values of voltage and current that apply to the resistor.

    P = _____ watts

16. Calculate the reactive power using the values of voltage and current that apply to the capacitor.

    $VARs_C$ = _____

17. Calculate the apparent power in the circuit using the values of voltage and current that apply to the entire circuit.

    VA = _____

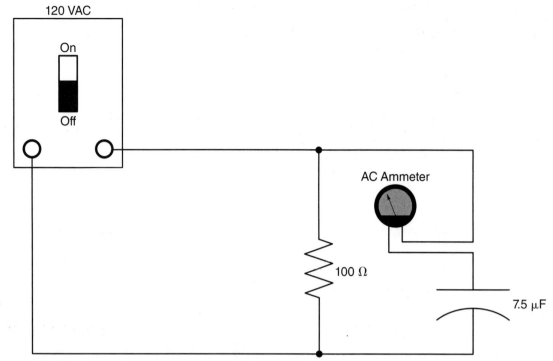

Figure 15-5 Measuring current through the capacitive load.

18. Calculate the circuit power factor using the following formula. Be sure to express the answer as a percentage.

$$PF = \frac{P}{VA}$$

PF = _____ %

19. Determine the degree of out-of-phase condition between the voltage and total current in the circuit by calculating the value of angle theta.

Cos $\angle\varnothing$ = PF

$\angle\varnothing$ = _____ °

20. Replace the 100 ohm resistor in the circuit with a 150 ohm resistor, and replace the 7.5 µF capacitor with a 25 µF capacitor. Reconnect the ammeter to measure the total circuit current as shown in Figure 15-4.

21. Calculate the capacitive reactance of the capacitor assuming a frequency of 60 Hz.

X_C = _____ Ω

22. Calculate the impedance of the circuit using the following formula:

$$Z = \frac{1}{\sqrt{\left(\frac{1}{R}\right)^2 + \left(\frac{1}{X_c}\right)^2}}$$

Z = _____ Ω

23. Assume an applied voltage of 120 volts. Calculate the total circuit current using the following formula:

$$I_T = \frac{E}{Z}$$

I_T = _____ A

24. Assume an applied voltage of 120 volts. Calculate the current flow through the 150 ohm resistor using the following formula:

$$I_R = \frac{E}{R}$$

$I_R =$ _____ A

25. Assume an applied voltage of 120 volts. Calculate the current that appears to flow through the capacitor using the following formula:

$$I_C = \frac{E}{X_C}$$

$I_C =$ _____ A

26. Turn on the power supply and measure the total current in the circuit. **Turn off the power.**

$I_T =$ _____ A

27. Compare the measured value with the calculated value in step 4. Are the two values within 5% of each other?

28. Reconnect the circuit to measure the current through the resistor as shown in Figure 15-4.

29. Turn on the power and measure the current flow through the resistor. **Turn off the power.** $I_R =$ _____ A

30. Compare the measured value with the calculated value in step #5. Are the two values within 5% of each other?

_____

31. Reconnect the circuit to measure the current that appears to flow through the capacitor as shown in Figure 15-5.

32. Turn on the power and measure the current that appears to flow through the capacitor. **Turn off the power.**

$I_C =$ _____ A

33. Compare the measured value with the calculated value in step 6. Are the two values within 5% of each other?

_____

34. Calculate the true power in the circuit using the values of voltage and current that apply to the resistor.

P = _____ watts

35. Calculate the reactive power using the values of voltage and current that apply to the capacitor.

$VARs_C =$ _____

36. Calculate the apparent power in the circuit using the values of voltage and current that apply to the entire circuit.

VA = _____

37. Calculate the circuit power factor using this formula: Be sure to express the answer as a percent.

$$PF = \frac{P}{VA}$$

PF = _____ %

38. Determine the degree of out of phase condition between the voltage and total current in the circuit by calculating the value of angle theta.

Cos ∠∅ = PF

∠∅ = _____ °

39. Disconnect the circuit and return the components to their proper place.

Review Questions

To answer the following questions, it may be necessary to refer to the formulas shown in Figure 15-6.

1. A 50 µF capacitor is connected to a 400 Hz line. What is the capacitive reactance of the capacitor?

_____

2. A resistor and capacitor are connected in parallel to a 120 volt, 60 Hz line. The circuit has a current flow of 3 amperes. The resistor has a resistance of 68 Ω. What is the capacitance of the capacitor?

_____

3. A resistor with a resistance of 50 Ω is connected in parallel with a capacitor with a capacitance of 35 µF. The power source is 60 Hz. What is the impedance of the circuit?

_____

4. A resistor and capacitor are connected in parallel to a 277 volt, 60 Hz line. The resistor has a current of 16 amperes flowing through it, and the capacitor has a current flow of 28 amperes flowing through it. What is the total current flow in the circuit?

_____

5. A resistor and capacitor are connected in parallel to a 1,000 Hz line. The capacitor has a voltage drop of 200 volts across it. What is the voltage drop across the resistor?

_____

6. A capacitor has a current flow of 1.6 amperes when connected to a 240 volt, 50 Hz line. How much current will flow through the capacitor if it is connected to a 240 volt, 60 Hz line?

_____

7. An RC parallel circuit has an apparent power of 21 kVA. The true power is 18.6 kW. What is the circuit power factor?

_____

8. How many degrees out of phase are the voltage and current in question 7?

_____

9. An RC parallel circuit has a power factor of 84%. The circuit voltage is 480 volts and the total current is 28 amperes. What is the true power in the circuit?

_____

10. The voltage and current are 24° out of phase with each other in an RC parallel circuit. What is the circuit power factor?

_____

Resistive-Capacitive Parallel Circuits

$$E_T = E_R = E_C$$

$$E_T = I_T \times Z$$

$$E_T = \frac{VA}{I_T}$$

$$E_T = \sqrt{VA \times Z}$$

$$I_T = \sqrt{I_R^2 + I_C^2} \qquad VA = E_T \times I_T \qquad PF = \frac{Z}{R} \qquad P = E_R \times I_R$$

$$I_T = \frac{E_T}{Z} \qquad VA = I_T^2 \times Z \qquad PF = \frac{P}{VA} \qquad P = \sqrt{VA^2 - VARs_C^2}$$

$$I_T = \frac{VA}{E_T} \qquad VA = \frac{E_T^2}{Z} \qquad PF = \frac{E_R}{E_T} \qquad P = \frac{E_R^2}{R}$$

$$I_T = \frac{I_R}{PF} \qquad VA = \sqrt{P^2 + VARs_C^2} \qquad PF = Cos \angle \Theta \qquad P = I_R^2 \times R$$

$$I_T = \sqrt{\frac{VA}{Z}} \qquad VA = \frac{P}{PF} \qquad R = \frac{E_R}{I_R} \qquad P = VA \times PF$$

$$E_R = I_R \times R \qquad I_R = \sqrt{I_T^2 - I_C^2} \qquad R = \sqrt{\frac{1}{\left(\frac{1}{Z}\right)^2 - \left(\frac{1}{X_C}\right)^2}} \qquad E_C = I_C \times X_C$$

$$E_R = \sqrt{P \times R} \qquad I_R = \frac{E_R}{R} \qquad R = \frac{E_R^2}{P} \qquad E_C = E_T = E_R$$

$$E_R = \frac{P}{I_R} \qquad I_R = \frac{P}{E_R} \qquad R = \frac{P}{I_R^2} \qquad E_C = \sqrt{VARs_C \times X_C}$$

$$E_R = E_T = E_C \qquad I_R = \sqrt{\frac{P}{R}} \qquad R = \frac{Z}{PF} \qquad E_C = \frac{VARs_C}{I_C}$$

$$E_R = E_T \times PF \qquad X_C = \frac{E_C}{I_C} \qquad X_C = \frac{E_C^2}{VARs_C} \qquad VARs_C = I_C^2 \times X_C$$

$$I_C = \sqrt{I_T^2 - I_R^2} \qquad X_C = \sqrt{\frac{1}{\left(\frac{1}{Z}\right)^2 - \left(\frac{1}{R}\right)^2}} \qquad X_C = \frac{VARs_C}{I_C^2} \qquad VARs_C = \frac{E_C^2}{X_C}$$

$$I_C = \frac{E_C}{X_C} \qquad X_C = \frac{1}{2 \pi f C} \qquad C = \frac{1}{2 \pi f X_C} \qquad VARs_C = E_C \times I_C$$

$$I_C = \frac{VARs_C}{E_C} \qquad I_C = \sqrt{\frac{VARs_C}{X_C}} \qquad Z = \frac{E_T^2}{VA} \qquad VARs_C = \sqrt{VA^2 - P^2}$$

$$Z = \sqrt{\frac{1}{\left(\frac{1}{R}\right)^2 + \left(\frac{1}{X_C}\right)^2}} \qquad Z = \frac{VA}{I_T^2} \qquad Z = \frac{E_T}{I_T} \qquad Z = R \times PF$$

Figure 15-6 RC parallel circuit formulas.

Unit 16 Resistive-Inductive-Capacitive Series Circuits

Objectives

After studying this unit, you should be able to

- Determine values of watts and VARs for circuit components.
- Determine circuit power factor.
- Measure values of current and voltage in an RLC series circuit.
- Compute the phase angle difference between current and voltage in an RLC series circuit.
- Connect an RLC series circuit.

In a pure resistive circuit, the current and voltage are in phase with each other. In a pure inductive circuit, the voltage leads the current by 90°, and in a pure capacitive circuit, the voltage lags the current by 90°. Since the current is the same at any point in a series circuit, the voltage drops across the different components are out of phase with each other, as shown in Figure 16-1. Since the voltage across the capacitor lags the current by 90° and the voltage across the inductor leads the current by 90°, they are 180° out of phase with each other. Since these two voltages are 180° out of phase with each other, one tends to cancel the effects of the other. The result is that the larger voltage is reduced and the smaller is eliminated as far as the circuit is concerned. Assume that an RLC series circuit contains a resistor with a voltage drop of 100 volts, an inductor with a voltage drop of 150 volts, and a capacitor with a voltage drop of 120 volts. The capacitive voltage reduces the inductive voltage to the point that the circuit is essentially an RL series circuit that has a resistor with a voltage drop of 100 volts and an inductor with a voltage drop of 30 volts (Figure 16-2).

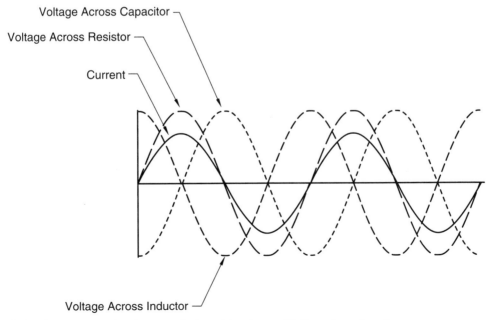

Voltage Across Capacitor

Voltage Across Resistor

Current

Voltage Across Inductor

Figure 16-1 Current and voltage relationships in an RLC series circuit.

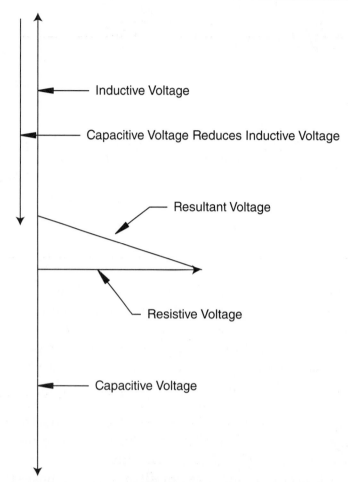

Figure 16-2 Inductive and capacitive voltage drops oppose each other.

Since the inductive and capacitive voltages oppose each other, it is possible for these components to have a greater voltage drop than the voltage applied to the circuit. The circuit shown in Figure 16-3 contains a resistor, inductor, and capacitor connected in series. The resistor has a resistance of 24 Ω, the inductor has an inductive reactance of 40 Ω, and the capacitor has a capacitive reactance of 35 Ω. The circuit has a total voltage of 208 volts. The following values will be computed:

Z - Total circuit impedance

I - Circuit current

E_R - Voltage drop across the resistor

P - True power or watts

E_L - Voltage drop across the inductor

$VARs_L$ - Reactive power component of the inductor

E_C - Voltage drop across the capacitor

$VARs_C$ - Reactive power component of the capacitor

VA - Apparent power

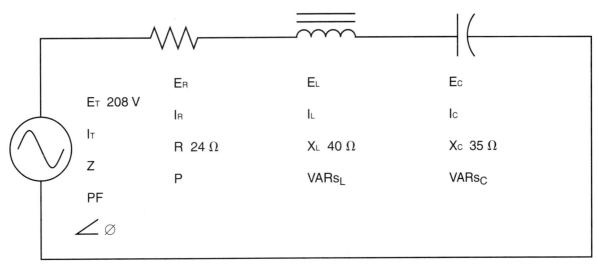

Figure 16-3 RLC series circuit.

PF - Power factor

$\angle\emptyset$ - Angle theta

The first step is to determine the total circuit impedance using the formula:

$$Z = \sqrt{R^2 + (X_L - X_c)^2}$$

$$Z = \sqrt{24^2 + (40 - 35)^2}$$

$$Z = \sqrt{601}$$

$$Z = 24.51 \ \Omega$$

Now that the impedance is known, the total circuit current can be computed using Ohm's law.

$$I_T = \frac{E_T}{Z}$$

$$I_T = \frac{208}{24.51}$$

$$I_T = 8.49 \ \text{amps}$$

In a series circuit, current is the same through all parts. Therefore, I_R, I_L, and I_C have a value of 8.49 amps.

The voltage drop across the resistor can be computed using Ohm's law.

$$E_R = I_R \times R$$

$$E_R = 8.49 \times 24$$

$$E_R = 203.76 \ \text{volts}$$

True power can be determined using the formula:

$$P = E_R \times I_R$$

$$P = 203.76 \times 8.49$$

$$P = 1{,}729.92 \text{ watts}$$

The voltage drop across the inductor can be computed using Ohm's law and inductive values.

$$E_L = I_L \times X_L$$

$$E_L = 8.49 \times 40$$

$$E_L = 339.6 \text{ volts}$$

Note that the voltage across the inductor is greater than the applied voltage to the circuit. The reactive power for the inductor can be computed using the formula:

$$VARs_L = E_L \times I_L$$

$$VARs_L = 339.6 \times 8.49$$

$$VARs_L = 2{,}883.2$$

The voltage drop across the capacitor can be determined using Ohm's law and capacitive values.

$$E_c = I_c \times X_c$$

$$E_c = 8.49 \times 35$$

$$E_c = 297.15 \text{ volts}$$

The reactive power for the capacitor can be computed using the formula:

$$VARs_C = E_c \times I_c$$

$$VARs_C = 297.15 \times 8.49$$

$$VARs_C = 2{,}522.8$$

The apparent power can be computed using the total values of voltage and current.

$$VA = E_T \times I_T$$

$$VA = 208 \times 8.49$$

$$VA = 4{,}288.72$$

The power factor can be determined using the formula:

$$PF = \frac{P}{VA}$$

$$PF = \frac{1,729.92}{4,288.72}$$

$$PF = 0.4034 \text{ or } 40.34\%$$

The power factor is the cosine of angle theta.

$$\angle\varnothing = \cos 0.0434$$

$$\angle\varnothing = 66.21°$$

The circuit with all the missing values is shown in Figure 16-4. Formulas for RLC series circuits are shown in Figure 16-6.

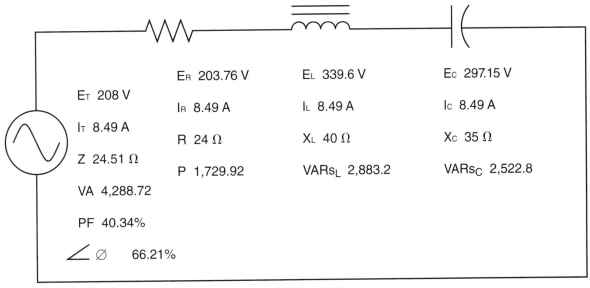

Figure 16-4 RLC series circuit with all the missing values.

LABORATORY EXERCISE

Name _____ Date _____

Materials Required

Formulas for RLC series circuits as shown in Figure 16-6

1 120 volt power supply

1 100 ohm resistor

1 150 ohm resistor

1 0.5 kVA control transformer with two windings rated at 240 volts each and one winding rated at 120 volts

1 25 μF AC capacitor with a voltage rating not less than 240 volts

1 10 μF AC capacitor with a voltage rating not less than 240 volts

1 AC ammeter (An in-line or clamp-on type meter may be used. If a clamp-on type is employed the use of a 10:1 scale divider is recommended.)

1 AC voltmeter

1. Connect the circuit shown in Figure 16-5.

2. Turn on the power supply and measure the current flow in the circuit.

   I_T _____ amps

3. Measure the voltage drop across the 100 ohm resistor, terminals X_1 and X_2 of the transformer, and the 25 μF capacitor with an AC voltmeter. **Turn off the power.**

   $E_{R \ (Resistor)}$ ————— $E_{L \ (Transformer)}$ ————— $E_{C \ (Capacitor)}$ —————

4. Does any component exhibit a greater voltage drop than the amount of voltage being applied to the circuit? If yes, which component?

   Yes/no _____   Component _____

5. Using the measured value of voltage and current, compute the apparent power, true power, $VARs_L$, and $VARs_C$.

   VA _____   P _____

   $VARs_L$ _____ $VARs_C$ _____

6. Compute the circuit power factor.

   PF = _____ %

7. How many degrees are the voltage and current out of phase with each other?

   $\angle\varnothing$ = _____ °

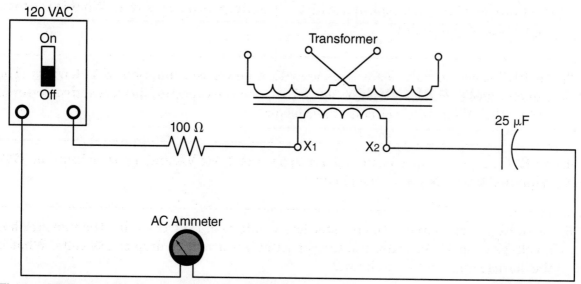

Figure 16-5 Connecting an RLC series circuit.

8. Replace the 25 μF AC capacitor with a 10 μF AC capacitor, and replace the 100 ohm resistor with a 150 ohm resistor.

9. Turn on the power supply and measure the current flow in the circuit.

    I_T _____ amps

10. Measure the voltage drop across the 150 ohm resistor, terminals X_1 and X_2 of the transformer, and the 10 μF capacitor with an AC voltmeter. **Turn off the power**.

    $E_{R\ (Lamp)}$ _____ $E_{L\ (Transformer)}$ _____ $E_{C\ (Capacitor)}$ _____

11. Does any component$_{(s)}$ exhibit a greater voltage drop than the amount of voltage being applied to the circuit? If yes which component?

    Yes/no _____   Component$_{(s)}$ _____

12. Using the measured value of voltage and current, compute the apparent power, true power, VARs$_L$, and VARs$_C$.

    VA _____   P _____

    VARs$_L$ _____   VARs$_C$ _____

13. Compute the circuit power factor.

    PF = _____ %

14. How many degrees are the voltage and current out of phase with each other?

    $\angle\varnothing$ = _____ °

15. Disconnect the circuit and return the components to their proper place.

Review Questions

Refer to the formula sheet in Figure 16-6 to answer the following questions:

1. A 120 volt, 60 Hz circuit contains a resistor, inductor, and capacitor connected in series. The total impedance of the circuit is 16.5 Ω. How much current is flowing through the inductor?

    _____

2. A circuit has a 48 Ω resistor connected in series with an inductor with an inductive reactance of 96 Ω and a capacitor with a capacitive reactance of 76 Ω. What is the total impedance of the circuit?

    _____

3. An RLC series circuit has a resistor with a power consumption of 124 watts. The inductor has a reactive power of 366 VARs$_L$ and the capacitor has a reactive power of 288 VARs$_C$. What is the circuit power factor?

    _____

4. An RLC series circuit has an apparent power of 1,564 VA and a power factor of 77%. What is the true power in the circuit?

    _____

5. In an RLC series circuit, the resistor has a voltage drop of 121 volts, the inductor has a voltage drop of 216 volts, and the capacitor has a voltage drop of 189 volts. What is the voltage applied to the circuit?

    _____

6. What is the power factor of the circuit in question 5?

   _____

7. An RLC series circuit has an applied voltage of 277 volts. The inductor has a voltage drop of 355 volts and the capacitor has a voltage drop of 124 volts. What is the voltage drop across the resistor?

   _____

8. The resistor of an RLC series circuit has a resistance of 16 Ω and a voltage drop of 56 volts. The capacitor has a capacitive reactance of 24 Ω. What is the reactive power of the capacitor?

   _____

9. An RLC series circuit has a power factor of 84%. The resistor has a voltage drop of 96 volts. What is the total voltage applied to the circuit?

   _____

10. The resistor of an RLC series circuit has a resistance of 48 Ω. The inductor has an inductive reactance of 88 Ω and the capacitor has a capacitive reactance of 62 Ω. The apparent power of the circuit is 780 VA. What is the total current in the circuit?

   _____

Resistive-Inductive-Capacitive Series Circuits

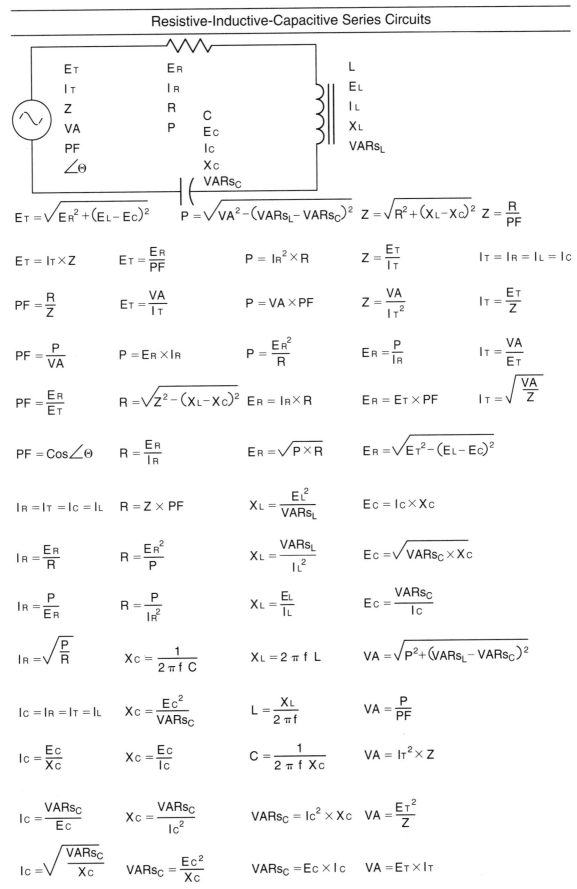

$$E_T = \sqrt{E_R^2 + (E_L - E_C)^2} \qquad P = \sqrt{VA^2 - (VARs_L - VARs_C)^2} \quad Z = \sqrt{R^2 + (X_L - X_C)^2} \quad Z = \frac{R}{PF}$$

$$E_T = I_T \times Z \qquad E_T = \frac{E_R}{PF} \qquad P = I_R^2 \times R \qquad Z = \frac{E_T}{I_T} \qquad I_T = I_R = I_L = I_C$$

$$PF = \frac{R}{Z} \qquad E_T = \frac{VA}{I_T} \qquad P = VA \times PF \qquad Z = \frac{VA}{I_T^2} \qquad I_T = \frac{E_T}{Z}$$

$$PF = \frac{P}{VA} \qquad P = E_R \times I_R \qquad P = \frac{E_R^2}{R} \qquad E_R = \frac{P}{I_R} \qquad I_T = \frac{VA}{E_T}$$

$$PF = \frac{E_R}{E_T} \qquad R = \sqrt{Z^2 - (X_L - X_C)^2} \quad E_R = I_R \times R \qquad E_R = E_T \times PF \qquad I_T = \sqrt{\frac{VA}{Z}}$$

$$PF = Cos \angle\Theta \qquad R = \frac{E_R}{I_R} \qquad E_R = \sqrt{P \times R} \qquad E_R = \sqrt{E_T^2 - (E_L - E_C)^2}$$

$$I_R = I_T = I_C = I_L \quad R = Z \times PF \qquad X_L = \frac{E_L^2}{VARs_L} \qquad E_C = I_C \times X_C$$

$$I_R = \frac{E_R}{R} \qquad R = \frac{E_R^2}{P} \qquad X_L = \frac{VARs_L}{I_L^2} \qquad E_C = \sqrt{VARs_C \times X_C}$$

$$I_R = \frac{P}{E_R} \qquad R = \frac{P}{I_R^2} \qquad X_L = \frac{E_L}{I_L} \qquad E_C = \frac{VARs_C}{I_C}$$

$$I_R = \sqrt{\frac{P}{R}} \qquad X_C = \frac{1}{2 \pi f \, C} \qquad X_L = 2 \pi f \, L \qquad VA = \sqrt{P^2 + (VARs_L - VARs_C)^2}$$

$$I_C = I_R = I_T = I_L \quad X_C = \frac{E_C^2}{VARs_C} \qquad L = \frac{X_L}{2 \pi f} \qquad VA = \frac{P}{PF}$$

$$I_C = \frac{E_C}{X_C} \qquad X_C = \frac{E_C}{I_C} \qquad C = \frac{1}{2 \pi f \, X_C} \qquad VA = I_T^2 \times Z$$

$$I_C = \frac{VARs_C}{E_C} \qquad X_C = \frac{VARs_C}{I_C^2} \qquad VARs_C = I_C^2 \times X_C \quad VA = \frac{E_T^2}{Z}$$

$$I_C = \sqrt{\frac{VARs_C}{X_C}} \quad VARs_C = \frac{E_C^2}{X_C} \qquad VARs_C = E_C \times I_C \quad VA = E_T \times I_T$$

Figure 16-6 Formulas for RLC series circuits.

Unit 17 Resistive-Inductive-Capacitive Parallel Circuits

Objectives

After studying this unit, you should be able to

- Calculate the impedance in an RLC parallel circuit.
- Determine the phase angle difference between voltage and current.
- Determine the circuit power factor.
- Connect an RLC parallel circuit.

In a parallel circuit, the voltage is the same across all branches. In a resistive circuit, the voltage and current are in phase with each other. In an inductive circuit, the current lags the voltage, and in a capacitive circuit, the current leads the voltage. Since the voltage is the same across all components, the currents through the resistive, inductive, and capacitive branches are out of phase with each other.

Determining the Impedance of an RLC Parallel Circuit

The impedance of a parallel RLC circuit can be determined in a manner similar to determining the impedance of an RLC series circuit. The difference lies in the fact that the reciprocal values of resistance, inductive reactance, and capacitive reactance must be used.

Example: An RLC parallel circuit has a resistor valued at 40 Ω, an inductor with an inductive reactance of 60 Ω, and a capacitor with a capacitive reactance of 30 Ω. What is the total circuit impedance?

The impedance can be determined using the formula:

$$Z = \frac{1}{\sqrt{\left(\dfrac{1}{R}\right)^2 + \left(\dfrac{1}{X_L} - \dfrac{1}{X_c}\right)^2}}$$

$$Z = \frac{1}{\sqrt{\left(\dfrac{1}{40}\right)^2 + \left(\dfrac{1}{60} - \dfrac{1}{30}\right)^2}}$$

$$Z = \frac{1}{\sqrt{0.000625 + 0.0002778}}$$

$$Z = \frac{1}{\sqrt{0.0009028}}$$

$$Z = 33.282 \ \Omega$$

An RLC parallel circuit contains a 75 Ω resistor, a 0.32 henry inductor, and a 15 μF capacitor (Figure 17-1). The circuit has an applied voltage of 240 volts and the frequency is 60 Hz.

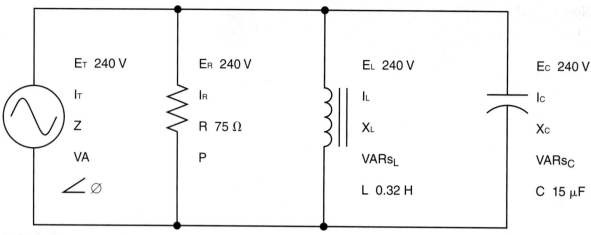

Figure 17-1 IRLC parallel circuit.

The following values will be determined:

X_L - Inductive reactance

X_C - Capacitive reactance

Z - Circuit impedance

I_T - Total circuit current

VA - Apparent power

I_R - Current flow through the resistor

P - True power

I_L - Current flow through the inductor

$VARs_L$ - Reactive power of the inductor

I_C - Current flow through the capacitive branch

$VARs_C$ - Reactive power of the capacitor

PF - Circuit power factor

∠∅ - Angle theta

The first step is to determine the values of inductive reactance and capacitive reactance. Recall that in a circuit with a frequency of 60 Hz, $2\pi f$ has a value of 377.

$$X_L = 2\pi fL$$

$$X_L = 377 \times 0.32$$

$$X_L = 120.64 \ \Omega$$

$$X_c = \frac{1}{2\pi fC}$$

$$X_c = \frac{1}{377 \times 15EE-6}$$

$$X_c = 176.83 \ \Omega$$

Now that the values of inductive reactance and capacitive reactance are known, the total circuit impedance can be determined.

$$Z = \frac{1}{\sqrt{\left(\dfrac{1}{R}\right)^2 + \left(\dfrac{1}{X_L} - \dfrac{1}{X_c}\right)^2}}$$

$$Z = \frac{1}{\sqrt{\left(\dfrac{1}{75}\right)^2 + \left(\dfrac{1}{120.64} - \dfrac{1}{176.83}\right)^2}}$$

$$Z = \frac{1}{\sqrt{0.000453 + 0.00000694}}$$

$$Z = \frac{1}{\sqrt{0.00046}}$$

$$Z = 46.625 \ \Omega$$

The total circuit current can now be computed using Ohm's law.

$$I_T = \frac{E_T}{Z}$$

$$I_T = \frac{240}{46.625}$$

$$I_T = 5.147 \ \text{amps}$$

The apparent power can be determined using the total values for voltage and current.

$$VA = E_T \times I_T$$

$$VA = 240 \times 5.147$$

$$VA = 1{,}235.28$$

Since all the branches of a parallel circuit have the same voltage applied across them, the current flow through the resistor can be determined using Ohm's law.

$$I_R = \frac{E_R}{R}$$

$$I_R = \frac{240}{75}$$

$$I_R = 3.2 \ \text{amps}$$

True power can be computed using the formula:

$$P = E_R \times I_R$$

$$I = \frac{E}{R}$$

$$P = 240 \times 3.2$$

$$P = 768 \text{ watts}$$

The current through the inductive branch of the circuit can be determined using the following formula:

$$I_L = \frac{E_L}{X_L}$$

$$I_L = \frac{240}{120.64}$$

$$I_L = 1.989 \text{ amps}$$

Reactive power for the inductive branch can be computed using the inductive values of voltage and current.

$$\text{VARs}_L = 240 \times 1.989$$

$$\text{VARs}_L = 477.36$$

The current through the capacitive branch can be calculated using the following formula:

$$I_c = \frac{E_c}{X_c}$$

$$I_c = \frac{240}{176.83}$$

$$I_c = 1.357 \text{ amps}$$

The reactive power for the capacitive branch can be determined using capacitive values of voltage and current.

$$\text{VARs}_C = 240 \times 1.357$$

$$\text{VARs}_C = 325.68$$

The power factor can be computed using the values for apparent power and true power.

$$PF = \frac{P}{VA}$$

$$PF = \frac{768}{1,235.28}$$

$$PF = 0.6217 \text{ or } 62.17\%$$

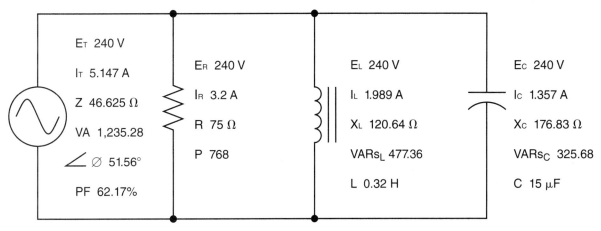

Figure 17-2 RLC parallel circuit with all the missing values.

The power factor is the cosine of angle theta.

$$\angle\varnothing = \text{inv}\cos 0.6217$$

$$\angle\varnothing = 51.56°$$

The circuit with all missing values is shown in Figure 17-2. Formulas for RLC parallel circuits are shown in Figure 17-7.

LABORATORY EXERCISE

Name _____ Date _____

Materials Required

Formulas for RLC parallel circuits as shown in Figure 17-7

1 120-volt power supply

1 0.5-kVA control transformer with 2 windings rated at 240 volts each and one winding rated at 120 volts.

1 100-ohm resistor

1 150-ohm resistor

1 25-μF AC capacitor rated not less than 240 volts

1 10-μF AC capacitor rated not less than 240 volts

1 AC voltmeter

1 AC ammeter (in-line or clamp-on type may be used. If a clamp-on type meter is used, a 10:1 scale divider is recommended.)

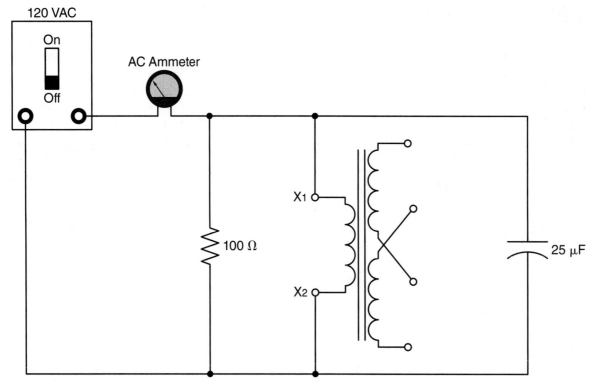

Figure 17-3   RLC parallel circuit.

1. Connect the circuit shown in Figure 17-3.
2. Turn on the power supply and measure the voltage with an AC voltmeter.

   E_T _____ volts
3. Measure the total current flow in the circuit.

   I_T _____ amps
4. Compute the apparent power in the circuit. ($VA = E_T \times I_T$)

   VA _____
5. **Turn off the power supply** and reconnect the circuit to measure the current through the resistive branch as shown in Figure 17-4.
6. Turn on the power and measure the current flow through the resistive branch. **Turn off the power.**

   I_R _____ amps
7. Compute the true power in the circuit using Ohm's law. ($P = E_R \times I_R$)

   P _____ watts
8. Reconnect the circuit to measure the current flow through the inductive part of the circuit (Figure 17-5).
9. Turn on the power supply and measure the current through the inductive branch of the circuit. **Turn off the power.**

   I_L _____ amps
10. Compute the reactive power in the inductive branch. ($VARs_L = E_L \times I_L$)

    $VARs_L$ _____

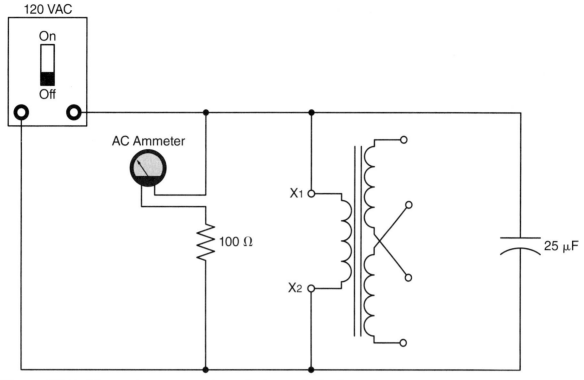

Figure 17-4 Measuring current through the resistive branch.

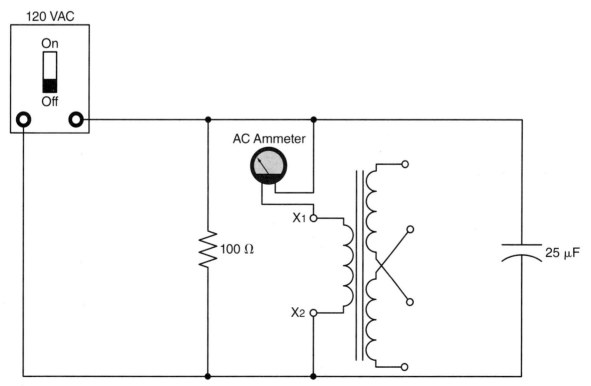

Figure 17-5 Measuring current through the inductive branch.

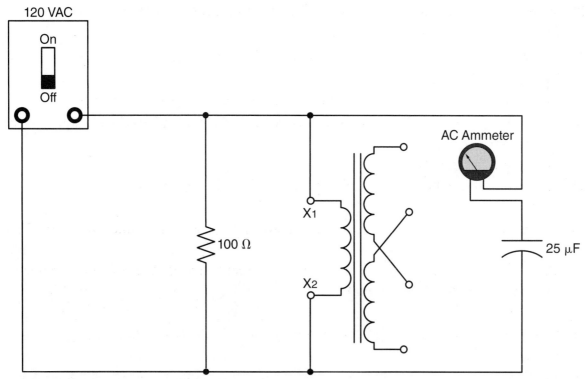

Figure 17-6 Measuring current through the capacitive branch.

11. Reconnect the ammeter to measure the current flow through the capacitive branch (Figure 17-6).

12. Turn on the power supply and measure the current flow through the capacitive branch. **Turn off the power.**

    I_C _____ amps

13. Compute the reactive power through the capacitive branch. (VARs$_C$ = E_C × I_C)

    VARs$_C$ _____

14. Using the values determined for apparent power and true power compute the circuit power factor. (PF = $\frac{P}{VA}$)

    PF _____ %

15. Does this circuit have a leading or lagging power factor? Explain you answer.

    Leading/lagging _____

    _____

    _____

    _____

16. How many degrees are the voltage and current out of phase with each other in this circuit? (Cos ∠∅ = PF)

    ∠∅ _____ °

17. Replace the 100-ohm resistor in the circuit with a 150-ohm resistor.

18. Replace the 25-μF AC capacitor with a 10-μF capacitor.

19. Turn on the power supply and measure the voltage with an AC voltmeter.

    E_T _____ volts

20. Measure the total current flow in the circuit. **Turn off the power**.

    I_T _____ amps

21. Compute the apparent power in the circuit. (VA = $E_T \times I_T$)

    VA _____

22. Reconnect the circuit to measure the current through the resistive branch.

23. Turn on the power and measure the current flow through the resistive branch. **Turn off the power**.

    I_R _____ amps

24. Compute the true power in the circuit using Ohm's law. (P = $E_R \times I_R$)

    P _____ watts

25. Reconnect the circuit to measure the current flow through the inductive part of the circuit.

26. Turn on the power supply and measure the current through the inductive branch of the circuit. **Turn off the power**.

    I_L _____ amps

27. Compute the reactive power in the inductive branch. (VARs$_L$ = $E_L \times I_L$)

    VARs$_L$ _____

28. Reconnect the ammeter to measure the current flow through the capacitive branch.

29. Turn on the power supply and measure the current flow through the capacitive branch. **Turn off the power**.

    I_C _____ amps

30. Compute the reactive power through the capacitive branch. (VARs$_C$ = $E_C \times I_C$)

    VARs$_C$ _____

31. Using the values determined for apparent power and true power compute the circuit power factor. (PF = $\frac{P}{VA}$)

    PF _____ %

32. Does this circuit have a leading or lagging power factor? Explain you answer.

    Leading/lagging _____

    _____

    _____

    _____

33. How many degrees are the voltage and current out of phase with each other in this circuit? (Cos $\angle\varnothing$ = PF)

    $\angle\varnothing$ _____ °

34. Disconnect the circuit and return the components to their proper place.

Review Questions

Use the formulas shown in Figure 17-7 to answer some of the following questions:

1. An RLC parallel circuit has a resistive current of 4.8 amps, an inductive current of 7.6 amps, and a capacitive current of 6.6 amps. What is the total current flow in the circuit?

    _____

2. Assume that the circuit in question 1 is connected to a 480 volt, 60 Hz line. What is the inductance of the inductor?

_____

3. What is the impedance of an RLC parallel circuit if the resistive branch has a resistance of 112 Ω, the inductive branch has an inductive reactance of 86 Ω, and the capacitive branch has a capacitive reactance of 94 Ω?

_____

4. An RLC parallel circuit has a total current of 15.8 amps. The inductive branch has a current flow of 9.6 amps, and the capacitive branch has a current flow of 12.4 amps. How much current is flowing in the resistive branch?

_____

5. An RLC parallel circuit has a true power of 138 watts. The inductive VARs are 218 and the capacitive VARs are 86. What is the circuit power factor?

_____

6. An RLC parallel circuit has a total current of 12 amps and an apparent power of 6,280 VA. How much voltage is across the capacitive branch of the circuit?

_____

7. The capacitive branch of an RLC parallel circuit has a voltage drop of 208 volts. The reactive power is 2,268 VARs. What is the capacitive reactance of this branch?

_____

8. An RLC parallel circuit has a total current flow of 23.8 amperes. The circuit power factor is 56.5%. How much current is flowing through the resistive branch of the circuit?

_____

9. An RLC parallel circuit has an apparent power of 22.5 kVA. The total current is 46.6 amperes. What is the circuit impedance?

_____

10. An RLC parallel circuit is connected to a 240 volt, 60 Hz line. The apparent power of the circuit is 3,865 VA. The inductive VARs are 1,892 and the capacitive VARs are 1,563. What is the resistance of the resistive branch?

_____

Resistive-Inductive-Capacitive Parallel Circuits

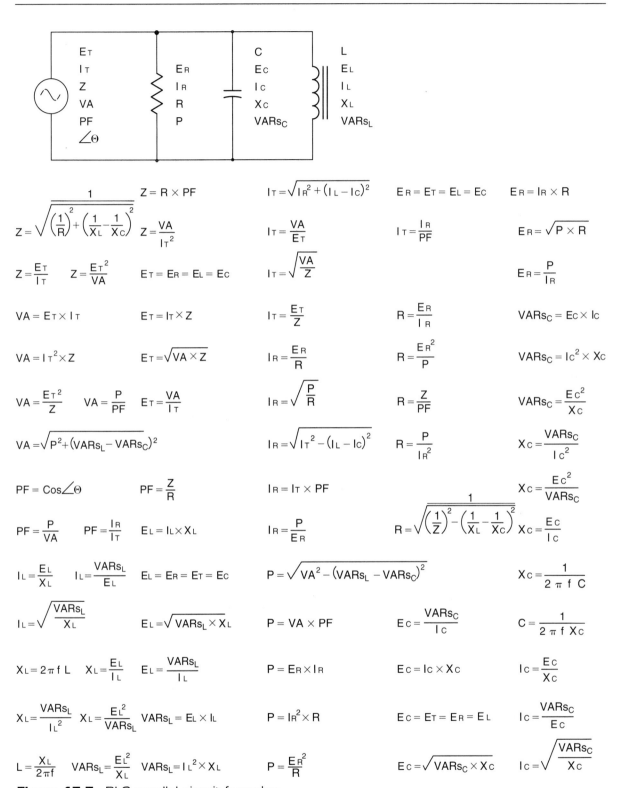

Figure 17-7 RLC parallel circuit formulas.

Unit 18 Power Factor Correction

Objectives

After studying this unit, you should be able to:

- Discuss the relationship between apparent power and true power.
- Determine the power factor of a circuit.
- Discuss the importance of power factor correction.
- Determine the amount of capacitance needed to correct a lagging power factor.

Power factor is a ratio of the apparent power and the true power or watts. In an alternating current circuit, the voltage and current can become out of phase with each other. Inductive loads cause the current to lag the applied voltage, and capacitive loads cause the current to lead the applied voltage. Power factor is always expressed as a percent. Basically, it indicates the portion of the power supplied by the power utility that is actually being utilized by the connected load. The utility company must supply the apparent power or volt-amperes. Wattmeters measure the true power or watts. Electric power is purchased in units of kilowatt hours. When the power factor falls below a certain percent, the electric utility generally charges a surcharge to make up the difference.

Example: Assume that a company has a power factor of 56%. Now assume that the wattmeter is indicating that power is being consumed at a rate of 50 kilowatts per hour. With a power factor of 56%, the power utility must actually supply 89.3 kVA (50 / 0.56). Assuming a three-phase voltage of 480 volts, the utility company is supplying 140.94 amperes, computed using following formula:

$$I = \frac{VA}{E \times \sqrt{3}}$$

If the power factor were to be correct to 100%, the apparent power and true power would become the same and the power company would have to furnish only 78.9 amperes to operate the load.

Although a power factor of 100% or unity is the ideal situation, as a general rule industrial customers generally try to correct to about 95%. The amount of capacitance needed to correct the power factor to 100% is much greater than that need for 95% correction.

Example: Assume that a single-phase motor is operating with a 60% lagging power factor. Also assume that the voltage is 240 volts single-phase at 60 Hz and that an ammeter indicates a current of 12.5 amperes. Determine the following:

- Apparent power (VA)
- True power (P)
- Reactive power (VARs$_L$). The VARs are inductive because the power factor is lagging.
- The amount of capacitance needed to correct the power factor to 100%
- The amount of capacitance needed to correct the power factor to 95%

Determining Capacitance for 100% PF

The apparent power can be computed using the formula:

$$VA = E_{Applied} \times I$$

$$VA = 240 \times 12.5$$

$$VA = 3,000$$

The true power can be determined using the formula:

$$P = E \times I \times PF$$

$$P = 240 \times 12.5 \times 0.60$$

$$P = 1,800 \text{ watts}$$

The inductive VARs can be found using the formula:

$$VARs_L = \sqrt{VA^2 - P^2}$$

$$VARs_L = \sqrt{3,000^2 - 1,800^2}$$

$$VARs_L = 2,400$$

To correct the power factor to unity or 100%, an equal amount of capacitive VARs would have to be added to the circuit. To determine the amount of capacitance needed, first determine the amount of capacitive current necessary to produce 2,400 capacitive VARs.

$$I_c = \frac{VARs_C}{E}$$

$$I_c = \frac{2,400}{240}$$

$$I_c = 10 \text{ amperes}$$

Now that the amount of current needed is known, the amount of capacitive reactance needed to produce that much current can be calculated.

$$X_c = \frac{E}{I_c}$$

$$X_c = \frac{240}{10}$$

$$X_c = 24 \ \Omega$$

The amount of capacitance needed to produce a capacitive reactance of 24 Ω at 60 Hz can be computed using the formula:

$$C = \frac{1}{2\pi f X_c}$$

$$C = \frac{1}{377 \times 24}$$

$$C = 0.0001105 \text{ farad or } 110.5 \ \mu F$$

Determining Capacitance for 95% PF

To determine the amount of capacitance needed to produce a 95% power factor, it is first necessary to determine what the apparent power should be to produce a 95% power factor. This can be computed using the formula:

$$VA = \frac{P}{PF}$$

$$VA = \frac{1,800}{0.95}$$

$$VA = 1,894.75$$

The next step is to determine the amount of reactive power needed to produce an apparent power of 1,894.75 VA. When correcting power factor, it is generally desirable to leave the power factor lagging as opposed to leading.

$$VARs_L = \sqrt{VA^2 - P^2}$$

$$VARs_L = \sqrt{1,894.75^2 - 1,800^2}$$

$$VARs_L = 591.67$$

At present, the reactive power is 2,400 $VARs_L$. To reduce the circuit to 591.6 $VARs_L$, 1,808.4 capacitive VARs will have to be added to the circuit (2,400 − 591.6).

The current needed to produce 1,808.4 $VARs_C$ can be determined using the formula:

$$I_c = \frac{VARs_C}{E}$$

$$I_c = \frac{1,808.4}{240}$$

$$I_c = 7.53 \text{ amperes}$$

Now that the amount of current needed is known, the amount of capacitive reactance needed to produce that much current can be calculated.

$$X_c = \frac{E}{I_c}$$

$$X_c = \frac{240}{7.53}$$

$$X_c = 31.87 \ \Omega$$

The amount of capacitance needed to produce a capacitive reactance of 31.87 Ω at 60 Hz can be computed using the formula:

$$C = \frac{1}{2\pi f X_c}$$

$$C = \frac{1}{377 \times 31.87}$$

$$C = 0.00008328 \text{ farad or } 83.28 \text{ } \mu F$$

Note that substantially less capacitance is needed to correct the power factor to 95%. The current will now drop from 12.5 amperes to 7.89 amperes (1,894.75 VA / 240 V).

LABORATORY EXERCISE

Name _____ Date _____

Materials Required

1 120-volt AC power supply

1 0.5-kVA control transformer with 2 windings rated at 240 volts each and one winding rated at 120 volts

1 100-ohm resistor

1 150-ohm resistor

1 25-μf AC capacitor rated at not less than 240 volts

1 10-μf AC capacitor rated at not less than 240 volts

1 7.5-μf AC capacitor rated at not less than 240 volts

1 AC voltmeter

1 AC ammeter (an in-line or clamp-on type meter may be used. If a clamp-on type is employed, the use of a 10:1 scale divider is recommended.)

In this experiment, the low-voltage winding (X_1 and X_2) of a transformer will be connected in series with 60 ohms of resistance to produce a load with a lagging power factor. A 100 ohm resistor is connected in parallel with a 150 ohm resistor to produce a total of 60 ohms. The apparent power, true power, and reactive power of the circuit will then be calculated. After these values have been determined, the amount of capacitance needed to correct the power factor will be computed.

1. Connect the circuit shown in Figure 18-1.

2. Turn on the power and measure the total current.

   I_T _____ amps.

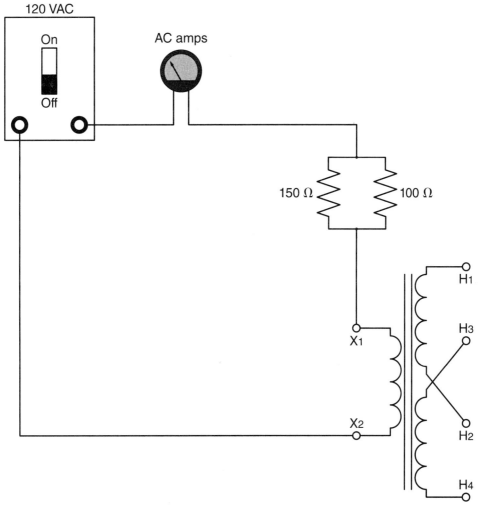

Figure 18-1 The 100 ohm and 150 ohm resistors are connected in parallel to produce a resistance of 60 ohms.

3. Measure the voltage applied to the circuit.

   E_T = _____ volts

4. Measure the voltage drop across the 60 ohm resistance.

   E_R _____ volts

5. Measure the voltage across the X_1 and X_2 winding of the transformer. **Turn off the power.**

   E_L _____ volts

6. Compute the apparent power in the circuit using the total voltage and the total circuit current. ($VA = E_T \times I$)

   VA _____

7. Compute the true power in the circuit using the voltage drop across the 60 watt lamp and the circuit current. ($P = E_R \times I$)

   P _____ watts

8. Compute the inductive VARs in the circuit using the voltage drop across the transformer winding and the circuit current. ($VARs_L = E_L \times I$)

   $VARs_L$ _____

9. Compute the circuit power factor. ($PF = \frac{P}{VA}$)

   PF _____ %

10. Determine the phase angel difference between the voltage and current. ($\cos PF = \angle\varnothing$)

    $\angle\varnothing$ _____ °

11. To determine the amount of capacitance needed to correct the power factor to 95%, first calculate the apparent power needed to produce a 95% power factor. (*Note:* A value of 0.95 should be used for power factor in the formula, because the power factor desired is 95%.)

    ($VA = \frac{P}{PF}$)

    VA _____

12. Determine the inductive VARs necessary to produce the apparent power determined in step 12. ($VARs = \sqrt{VA^2 - P^2}$)

    $VARs_L$ _____

13. To determine the capacitive VARs needed, subtract the needed VARs in step 13 from the calculated VARs in step 8.

    $VARs_C$ _____

14. Compute the amount of capacitive necessary to produce the capacitive VARs determined in step 13. ($X_C = \frac{E^2}{VARs_C}$)

    X_C _____ Ω

15. Compute the amount of capacitance needed to produce the capacitive reactance determined in step 14. ($C = \frac{1}{2\pi fX}$)

    C _____ μF

16. From the capacitors provided in this experiment, select the capacitor that is the nearest value to the capacitor value determined in step 16. Do not go over the computed value. Connect the capacitor across the circuit as shown in Figure 18-2.

17. Turn on the power supply and measure the total circuit current. **Turn off the power.**

    I_T _____ amps

18. Compute the apparent power of the circuit using the current value measured in step 17.

    VA _____

19. Assuming that the true power has remained constant, compute the circuit power factor using the apparent power value determined in step 18.

    PF _____ %

20. Was there an improvement in the circuit power factor?

    Yes/no _____

21. Did the total current decrease after the capacitor was added to the circuit?

    Yes/no _____

22. Disconnect the circuit and return the components to their proper place.

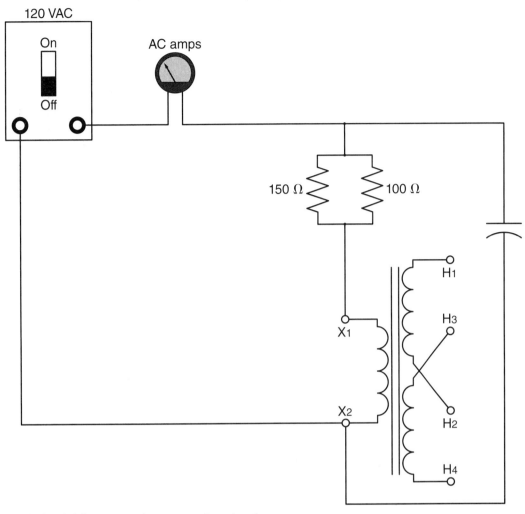

Figure 18-2 Adding capacitance to the circuit.

Review Questions

1. A single-phase motor is connected to a 120 volt, 60 Hz line. The motor has a current draw of 15.7 amperes. A watt meter indicated the motor is actually using 1,340 watts. Find the following:

   Apparent power:

   _____

   Power factor of the motor:

   _____

   Reactive power of the motor:

   _____

   Capacitance needed to correct the power factor to 100%:

   _____

   Capacitance needed to correct the power factor to 95%:

   _____

2. An inductive load is connected to a 208 volt, 60 Hz line. The circuit has a current draw of 6.25 amperes. The circuit power factor is 45%. If a 10 μF capacitor is connected in parallel with the inductive load, what will the new power factor be? Determine the following values:

Apparent power:

True power:

Reactive power before the capacitor is connected:

New power factor:

SECTION 4

Transformers and Motors

Objectives

After studying this unit, you should be able to:

- Discuss the construction of an isolation transformer.
- Determine the winding configuration with an ohmmeter.
- Connect a transformer and make voltage measurements.
- Compute the turns-ratio of the windings.

A transformer is a magnetically operated machine that can change values of voltage, current, and impedance without a change of frequency. Transformers are the most efficient machines known. Their efficiencies commonly range from 90% to 99% at full load. Transformers can be divided into several classifications such as:

- Isolation
- Auto
- Current

A basic law concerning transformers is that all values of a transformer are proportional to its turns-ratio. This does not mean that the exact number of turns of wire on each winding must be known to determine different values of voltage and current for a transformer. What must be known is the ratio of turns. For example, assume a transformer has two windings. One winding, the primary, has 1,000 turns of wire and the other, the secondary, has 250 turns of wire, as shown in Figure 19-1. The turns-ratio of this transformer is 4 to 1 or 4:1 (1,000/250 = 4). This indicates there are four turns of wire on the primary for every one turn of wire on the secondary.

Helpful Hint

A basic law concerning transformers is that all values of a transformer are proportional to its turns-ratio.

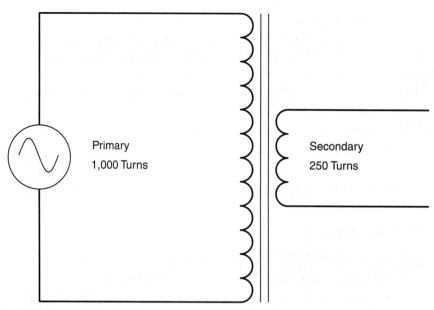

Figure 19-1 All values of a transformer are proportional to its turns-ratio.

Transformer Formulas

There are different formulas that can be used to find the values of voltage and current for a transformer. The following is a list of standard formulas:

where:

N_P = Number of turns in the primary

N_S = Number of turns in the secondary

E_P = Voltage of the primary

E_S = Voltage of the secondary

I_P = Current in the primary

I_S = Current in the secondary

$$\frac{E_P}{E_S} = \frac{N_P}{N_S}$$

$$\frac{E_P}{S_P} = \frac{I_S}{I_P}$$

$$\frac{N_P}{N_S} = \frac{I_S}{I_P}$$

or

$$E_P \times N_S = E_S \times N_P$$

$$E_P \times I_P = E_S \times I_S$$

$$N_P \times I_P = N_S \times I_S$$

The primary winding of a transformer is the power input winding. It is the winding that is connected to the incoming power supply. The secondary winding is the load winding or output winding. It is the side of the transformer that is connected to the driven load, seen in Figure 19-2. Any winding of a transformer can be used as a primary or secondary winding provided its voltage or current rating is not exceeded. Transformers can also be operated at a lower voltage than their rating indicates, but they cannot be connected to a higher voltage. Assume the transformer shown in Figure 19-2, for example, has a primary voltage rating of 480 volts and the secondary has a voltage rating of 240 volts. Now assume that the primary winding is connected to a 120 volt source. No damage would occur to the transformer, but the secondary winding would produce only 60 volts.

Isolation Transformers

The transformers shown in Figures 19-1 and 19-2 are *isolation* transformers. This means that the secondary winding is physically and electrically isolated from the primary winding. There is no electrical connection between the primary and secondary winding. *This transformer is magnetically coupled, not electrically coupled.* This "line isolation" is often a very desirable characteristic. Since there is no electrical connection between the load and power supply, the transformer becomes a filter between the two. The isolation transformer will attenuate any voltage spikes that originate on the supply side before they are transferred to the load side. Some isolation transformers are built with a turns-ratio of 1:1. A transformer of this type will have the same input and output voltage and is used for the purpose of isolation only.

The reason that the transformer can greatly reduce any voltage spikes before they reach the secondary is because of the rise time of current through an inductor. The current in an inductor rises at an exponential rate, as shown in Figure 19-3. As the current increases in value, the expanding magnetic field cuts through the conductors of the coil and induces a voltage that is opposed to the applied voltage. The amount of induced voltage is proportional to the rate of change of current. This simply means that the faster current attempts to increase, the greater the opposition to that increase will be. Spike voltages and currents are generally of very short duration, which means that they increase in value very rapidly (Figure 19-4). This rapid change of value causes the opposition to the change to increase just as rapidly. By the time the spike has been transferred to the secondary winding of the transformer, it has been eliminated or greatly reduced (Figure 19-5).

Another purpose of isolation transformers is to remove or isolate some piece of electrical equipment from circuit ground. It is sometimes desirable that a piece of electrical equipment not be connected directly to circuit ground. This is often done as a safety

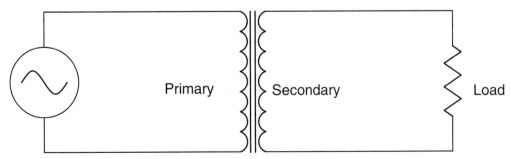

Figure 19-2 Isolation transformer.

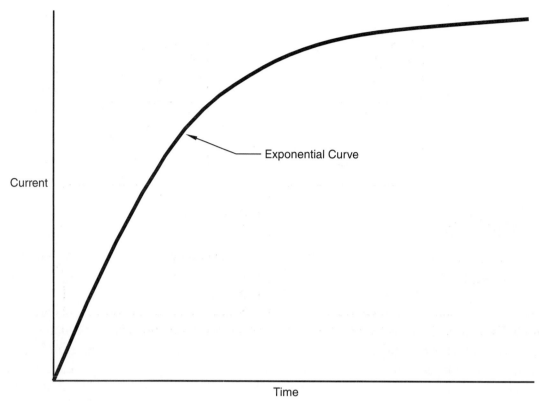

Figure 19-3 The current through an inductor rises at an exponential rate.

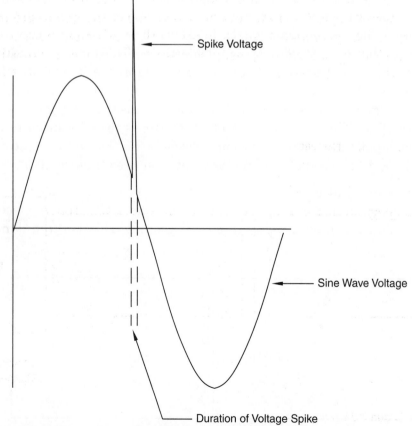

Figure 19-4 Voltage spikes are generally of very short duration.

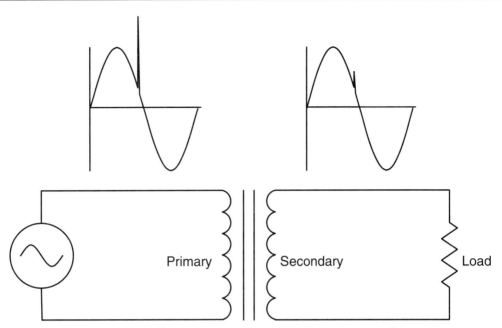

Figure 19-5   The isolation transformer greatly reduces the voltage spike.

precaution to eliminate the hazard of an accidental contact between a person at ground potential and the ungrounded conductor. If the case of the equipment should come in contact with the ungrounded conductor, the isolation transformer would prevent a circuit being completed to ground through a person touching the case of the equipment. Many alternating current circuits have one side connected to ground. A familiar example of this is the common 120 volt circuit with a grounded neutral conductor, as seen in Figure 19-6. An isolation transformer can be used to remove or isolate a piece of equipment from circuit ground.

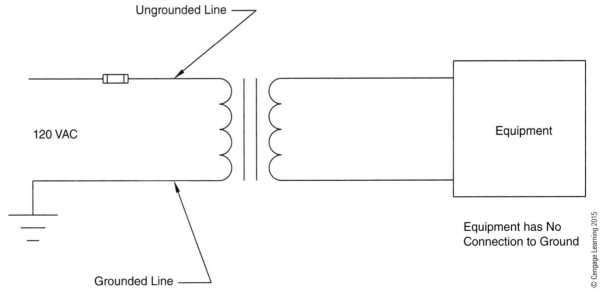

Figure 19-6   Isolation transformer used to remove a piece of electrical equipment from ground.

Excitation Current

There will always be some amount of current flow in the primary of a transformer even if there is no load connected to the secondary. This is called the *excitation current* of the transformer. The excitation current is the amount of current required to magnetize the core of the transformer. The excitation current remains constant from no load to full load. As a general rule, the excitation current is such a small part of the full load current it is often omitted when making calculations.

Transformer Calculations

In the following examples, values of voltage, current, and turns for different transformers will be computed.

Example #1: Assume the isolation transformer shown in Figure 19-2 has 240 turns of wire on the primary and 60 turns of wire on the secondary. This is a ratio of 4:1 (240/60 = 4). Now assume that 120 volts is connected to the primary winding. What is the voltage of the secondary winding?

$$\frac{E_P}{E_S} = \frac{N_P}{N_S}$$

$$\frac{120}{E_S} = \frac{240}{60}$$

$$E_S = 30 \text{ volts}$$

The transformer in this example is known as a step-down transformer because it has a lower secondary voltage than primary voltage.

Now assume that the load connected to the secondary winding has an impedance of 5 Ω. The next problem is to calculate the current flow in the secondary and primary windings. The current flow of the secondary can be computed using Ohm's law since the voltage and impedance are known.

$$I = \frac{E}{Z}$$

$$I = \frac{30}{5}$$

$$I = 6 \text{ amps}$$

Now that the amount of current flow in the secondary is known, the primary current can be computed using the following formula:

$$\frac{E_P}{E_S} = \frac{I_S}{I_P}$$

$$\frac{120}{30} = \frac{6}{I_P}$$

$$120 I_P = 180$$

$$I_P = 1.5 \text{ amps}$$

Notice that the primary voltage is higher than the secondary voltage, but the primary current is much less than the secondary current. A good rule for transformers is that power in must equal power out. If the primary voltage and current are multiplied together, the result should equal the product of the voltage and current of the secondary.

Primary	Secondary
$120 \times 1.5 = 180$ volt-amps	$30 \times 6 = 180$ volt-amps

---

Helpful Hint

A good rule for transformers is that power in must equal power out.

Example #2: In the next example, assume that the primary winding contains 240 turns of wire and the secondary contains 1,200 turns of wire. This is a turns-ratio of 1:5 (1,200/240 = 5). Now assume that 120 volts is connected to the primary winding. Compute the voltage output of the secondary winding.

$$\frac{E_P}{E_S} = \frac{N_P}{N_S}$$

$$\frac{120}{E_S} = \frac{240}{1,200}$$

$$240E_S = 144,000$$

$$E_S = 600 \text{ volts}$$

Notice that the secondary voltage of this transformer is higher than the primary voltage. This type of transformer is known as a *step-up* transformer.

Now assume that the load connected to the secondary has an impedance of 2,400 Ω. Find the amount of current flow in the primary and secondary windings. The current flow in the secondary winding can be computed using Ohm's law.

$$I = \frac{E}{Z}$$

$$I = \frac{600}{2,400}$$

$$I = 0.25 \text{ amp}$$

Now that the amount of current flow in the secondary is known, the primary current can be computed using the formula:

$$\frac{E_P}{E_S} = \frac{I_S}{I_P}$$

$$\frac{120}{600} = \frac{0.25}{I_P}$$

$$120I_P = 150$$

$$I_P = 1.25 \text{ amps}$$

Notice that the amount of power input equals the amount of power output.

Primary	Secondary
$120 \times 1.25 = 150$ volt-amps	$600 \times 0.25 = 150$ volt-amps

Calculating Transformer Values Using the Turns-Ratio

As illustrated in the previous examples, transformer values of voltage, current, and turns can be computed using formulas. It is also possible to compute these same values using the turns-ratio. There are several ways in which turns-ratios can be expressed. One method is to use a whole number value such as 13:5 or 6:21. The first ratio indicates that one winding has 13 turns of wire for every 5 turns of wire in the other winding. The second ratio indicates that there are 6 turns of wire in one winding for every 21 turns in the other.

A second method is to use the number 1 as a base. When using this method, the number 1 is always assigned to the winding with the lowest voltage rating. The ratio is found by dividing the higher voltage by the lower voltage. The number on the left side of the ratio represents the primary winding and the number on the right of the ratio represents the secondary winding. For example, assume a transformer has a primary rated at 240 volts and a secondary rated at 96 volts, as shown in Figure 19-7. The turns-ratio can be computed by dividing the higher voltage by the lower voltage.

$$\text{Ratio} = \frac{240}{96}$$

$$\text{Ratio} = 2.5:1$$

Notice in this example that the primary winding has the higher voltage rating and the secondary has the lower. Therefore, the 2.5 is placed on the left and the base unit, 1, is placed on the right. This ratio indicates that there are 2.5 turns of wire in the primary winding for every 1 turn of wire in the secondary.

Now assume that there is a resistance of 24 Ω connected to the secondary winding. The amount of secondary current can be found using Ohm's law.

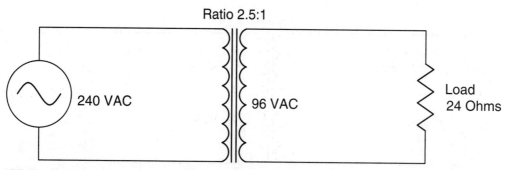

Figure 19-7 Computing transformer values using the turns-ratio.

$$I_s = \frac{96}{24}$$

$$I_s = 4 \text{ amps}$$

The primary current can be found using the turns-ratio. Recall that the volt-amps of the primary must equal the volt-amps of the secondary. Since the primary voltage is greater, the primary current will have to be less than the secondary current. Therefore, the secondary current will be divided by the turns-ratio.

$$I_P = \frac{I_S}{\text{Turns-ratio}}$$

$$I_P = \frac{4}{2.5}$$

$$I_P = 1.6 \text{ amps}$$

To check the answer, find the volt-amps of the primary and secondary.

Primary	Secondary
$240 \times 1.6 = 384$	$96 \times 4 = 384$

Now assume that the secondary winding contains 150 turns of wire. The primary turns can be found by using the turns-ratio, also. Since the primary voltage is higher than the secondary voltage, the primary must have more turns of wire. Since the primary must contain more turns of wire, the secondary turns will be multiplied by the turns-ratio.

$$N_P = N_S \times \text{Turns-ratio}$$

$$N_P = 150 \times 2.5$$

$$N_P = 375 \text{ turns}$$

In the next example, assume a transformer has a primary voltage of 120 volts and a secondary voltage of 500 volts. The secondary has a load impedance of 1,200 Ω. The secondary contains 800 turns of wire (Figure 19-8). The turns-ratio can be found by dividing the higher voltage by the lower voltage.

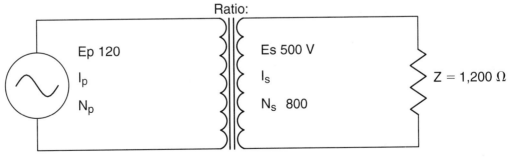

Figure 19-8 Calculating transformer values.

$$\text{Ratio} = \frac{500}{120}$$

$$\text{Ratio} = 1{:}4.17$$

The secondary current can be found using Ohm's law.

$$I_S = \frac{500}{1{,}200}$$

$$I_S = 0.417 \text{ amps}$$

In this example the primary voltage is lower than the secondary voltage. Therefore, the primary current must be higher. To find the primary current, multiply the secondary current by the turns-ratio.

$$I_P = I_S \times \text{Turns-ratio}$$

$$I_P = 0.417 \times 4.17$$

$$I_P = 1.74 \text{ amps}$$

To check this answer, compute the volt-amps of both windings.

Primary	Secondary
$120 \times 1.74 = 208.8$	$500 \times 0.417 = 208.5$

The slight difference in answers is caused by rounding off of values.

Since the primary voltage is less than the secondary voltage, the turns of wire in the primary will also be less. The primary turns will be found by dividing the turns of wire in the secondary by the turns-ratio.

$$N_P = \frac{N_S}{\text{Turns-ratio}}$$

$$N_P = \frac{800}{4.17}$$

$$N_P = 192 \text{ turns}$$

LABORATORY EXERCISE

Name _____ Date _____

Materials Required

480-240/120-volt, 0.5-kVA control transformer

Ohmmeter

AC voltmeter, in-line or clamp-on. (If a clamp-on type is used, a 10:1 scale divider is recommended.)

These experiments are intended to provide the electrician with hands-on experience dealing with transformers. The transformers used in these experiments are standard control transformers with two high-voltage windings rated at 240 volts each generally used to provide primary voltages of 480/240, and one low-voltage winding rated at 120 volts. The transformers have a rating of 0.5 kVA. The loads are standard 100 watt lamps that may be connected in parallel or series. It is assumed that the power supply is 208/120 volt three-phase four wire. It is also possible used with a 240/120 volt three-phase high leg system, provided adjustments are made in the calculations.

As in industry, these transformers will be operated with full voltage applied to the windings. The utmost caution must be exercised when dealing with these transformers. These transformers can provide enough voltage and current to seriously injure or kill anyone. The power should be disconnected before attempting to make or change any connections.

Caution

These transformers can provide enough voltage and current to seriously injure or kill anyone.

The transformer used in this experiment contains two high-voltage windings and one low-voltage winding. The high-voltage windings are labeled H_1 - H_2 and H_3 - H_4. The low-voltage winding is labeled X_1 - X_2.

1. Set the ohmmeter to the Rx1 range and measure the resistance between the following terminals:

    H_1 - H_2 _____ Ω
    H_1 - H_3 _____ Ω
    H_1 - H_4 _____ Ω
    H_1 - X_1 _____ Ω
    H_1 - X_2 _____ Ω
    H_2 - H_3 _____ Ω
    H_2 - H_4 _____ Ω
    H_2 - X_1 _____ Ω
    H_2 - X_2 _____ Ω
    H_3 - H_4 _____ Ω
    H_3 - X_1 _____ Ω
    H_3 - X_2 _____ Ω
    H_4 - X_1 _____ Ω
    H_4 - X_2 _____ Ω
    X_1 - X_2 _____ Ω

2. Using the information provided by the measurements from step 1, which sets or terminals form complete circuits within the transformer?

    _____

    _____

    _____

    These circuits represent the connections to the three separate windings within the transformer.

3. Which of the windings exhibits the lowest resistance and why?

_____

_____

_____

4. The H_1 - H_2 terminals are connected to one of the high-voltage windings and the H_3 - H_4 terminals are connected to the second high-voltage winding. Each of these windings is rated at 240 volts. When this transformer will be connected for 240 volt operation, the two high-voltage windings are connected in parallel to form one winding by connecting H_1 to H_3 and H_2 to H_4, as shown in Figure 19-9. This will provide a 2:1 turns-ratio with the low-voltage winding.

   When this transformer is operated with 480 volts connected to the primary, the high-voltage windings are connected in series by connecting H_2 to H_3 and connecting power to H_1 and H_4, as shown in Figure 19-10. This effectively doubles the primary turns, providing a 4:1 turns-ratio with the low-voltage winding.

5. Connect the two high-voltage windings for parallel operation as shown in Figure 19-9. Assume a voltage of 208 volts is applied to the high-voltage windings. Compute the voltage that should be seen on the low-voltage winding between terminals X_1 and X_2.

   _____ volts.

6. Make certain that the incoming power leads are connected to terminals H_1 and H_4 as shown in Figure 19-9. Apply a voltage of 208 volts to the transformer and measure the voltage across terminals X_1 and X_2.

   _____ volts.

7. The measured voltage may be slightly higher than the computed voltage. The rated voltage of a transformer is based on full load. It is normal for the secondary voltage to be slightly higher when no load is connected to the transformer. Transformers are generally wound with a few extra turns of wire in the winding that is intended to be used as the load side. This helps overcome the voltage drop when load is added. The slight change in turns-ratio does not affect the operation of the transformer to a great extent.

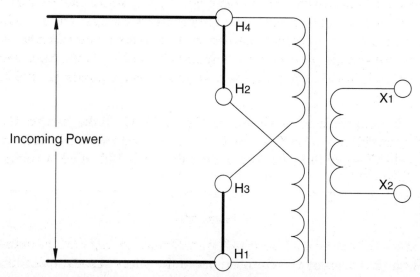

Figure 19-9 High-voltage windings connected in parallel.

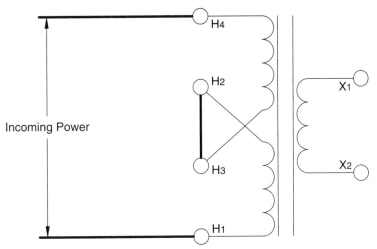

Figure 19-10 High-voltage windings connected in series.

8. **Turn off the power to the transformer.**

9. Disconnect the wires connected to the transformer and reconnect the transformer as shown in Figure 19-10. The two high-voltage windings are connected in series by connecting H_2 and H_3 together. This connection changes the turns-ratio of the transformer from 2:1 to 4:1. Make certain that the incoming power is connected to terminals H_1 and H_4.

10. Assume that a voltage of 208 volts is applied to the high-voltage windings. Compute the voltage across the low-voltage winding.

    _____ volts

11. Turn on the power and apply a voltage of 208 volts to the transformer. Measure the voltage across terminals X_1 and X_2.

    _____ volts

12. **Turn off the power.** Disconnect the power lines that are connected to terminals H_1 and H_4. Do not disconnect the wire between terminals H_2 and H_3.

13. In the next part of the exercise, the low-voltage winding will be used as the primary and the high-voltage windings will be used as the secondary. If the high-voltage windings are connected in series, the turns-ratio will be 1:4, which means that the secondary voltage will be four times greater than the primary voltage. The transformer has now become a step-up transformer instead of a step-down transformer. Assume that a voltage of 120 volts is connected to terminals X_1 and X_2. If the high-voltage windings are connected in series, compute the voltage across terminals H_1 and H_4.

    _____ volts

14. Connect the transformer as shown in Figure 19-11. Make certain that the voltage applied to terminals X_1 and X_2 is 120 volts and not 208 volts. Also make certain that the AC voltmeter is set for a higher range than the computed value of voltage in step 13.

Caution

The secondary voltage in this step will be 480 volts or higher. Use extreme caution when making this measurement. Be sure to wear safety glasses at all times.

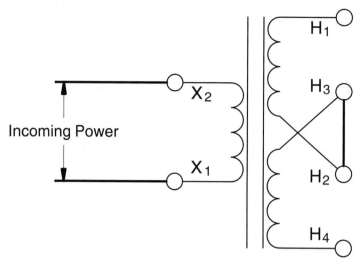

Figure 19-11 The incoming power is connected to terminals X_1 and X_2.

15. Turn on the power and measure the voltage across terminals H_1 and H_4.

    _____ volts

16. **Turn off the power supply.**

17. Disconnect the lead between terminals H_2 and H_3. Reconnect the transformer so that the high-voltage windings are connected in parallel by connecting H_1 and H_3 together and H_2 and H_4 together as shown in Figure 19-12. Do not disconnect the power leads to terminals X_1 and X_2. The transformer now has a turns-ratio of 1:2.

18. Assume that a voltage of 120 volts is connected to the low-voltage winding. Compute the voltage across the high-voltage winding.

    _____ volts

19. Make certain the power leads are still connected to terminals X_1 and X_2. Turn on the power and apply 120 volts to terminals X_1 and X_2. Measure the voltage across terminals H_1 and H_4.

    _____ volts

20. **Turn off the power supply** and disconnect all leads to the transformer. Return the components to their proper places.

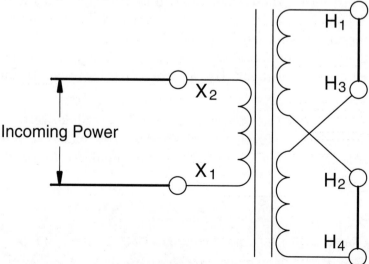

Figure 19-12 The transformer has a turns-ratio of 1:2.

Review Questions

1. What is a transformer?

_____

2. What are common efficiencies for transformers?

_____

3. What is an isolation transformer?

_____

4. All values of a transformer are proportional to its:

_____

5. A transformer has a primary voltage of 480 volts and a secondary voltage of 20 volts. What is the turns-ratio of the transformer?

_____

6. If the secondary of the transformer in question 5 supplies a current of 9.6 amperes to a load, what is the primary current (disregard excitation current)?

_____

7. Explain the difference between a step-up and a step-down transformer.

_____

8. A transformer has a primary voltage of 240 volts and a secondary voltage of 48 volts. What is the turns-ratio of this transformer?

_____

9. A transformer has an output of 750 volt-amps. The primary voltage is 120 volts. What is the primary current?

_____

10. A transformer has a turns-ratio of 1:6. The primary current is 18 amperes. What is the secondary current?

_____

Unit 20 Single-Phase Transformer Calculations

Objectives

After studying this unit, you should be able to:

- Discuss transformer excitation current.
- Compute values of primary current using the secondary current and the turns-ratio.
- Compute the turns-ratio of a transformer using measured values.
- Connect a step-down or step-up isolation transformer.

LABORATORY EXERCISE

Name _____ Date _____

Materials Required

480-240/120-volt, 0.5-kVA control transformer

4 150-ohm resistors

AC voltmeter

AC ammeter, in-line or clamp-on. (If a clamp-on type is used, the use of a 10:1 scale divider is recommended.)

In this experiment the excitation current of an isolation transformer will be measured. The transformer will then be connected as both a step-down and a step-up transformer. The turns-ratio will be determined from measured values and the primary current will be computed and then measured.

1. Connect the high-voltage windings of the transformer in parallel for 240 volt operation.
2. Connect the high-voltage winding to a 208 volt AC source with an AC ammeter connected in series with one of the lines, as shown in Figure 20-1.
3. Turn on the power source and measure the current. This is the *excitation* current of the transformer. The excitation current is the amount of current necessary to magnetize the iron in the transformer and will remain constant regardless of the load on the transformer.

   _____ amp(s)
4. Measure the voltage across the low-voltage winding at terminals X_1 - X_2.

   _____ volts

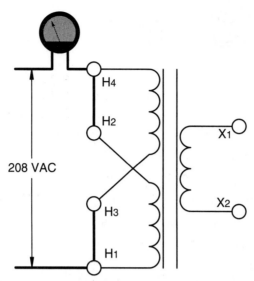

Figure 20-1 Connecting the high-voltage winding for parallel operation.

5. Compute the turns-ratio by dividing the primary voltage by the secondary voltage. Since the primary has the higher voltage, the larger number will be placed on the left side of the ratio, such as 3:1 or 4:1.

_____ ratio

6. **Turn off the power supply.**

7. Connect two 150-ohm resistors in parallel with the low-voltage winding of the transformer. Connect an AC ammeter in series with one of the lines, as shown in Figure 20-2.

8. Turn on the power and measure the current flow in the secondary circuit of the transformer.

_____ amp(s)

9. **Turn off the power supply.**

10. Compute the amount of primary current using the turns-ratio. Since the primary voltage is higher, the amount of primary current will be less. Divide the secondary current by the turns-ratio. Then add the excitation current to this value.

_____ $I_{(PRIMARY)}$

$$I_{(PRIMARY)} = \frac{I_{(SECONDARY)}}{Turns\text{-}ratio} + \text{Excitation current}$$

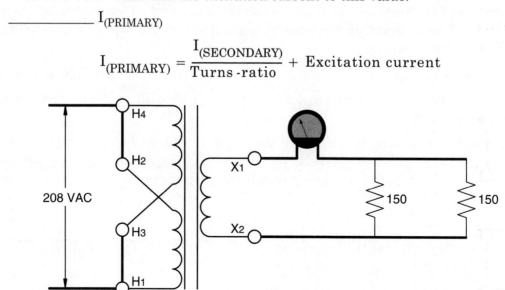

Figure 20-2 Two resistors are connected in parallel to the secondary winding.

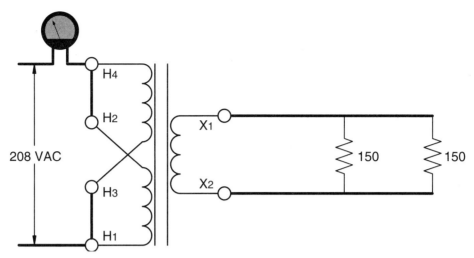

Figure 20-3 Measuring primary current.

11. Reconnect the AC ammeter in one of the primary lines, as shown in Figure 20-3.

12. Turn on the power supply and measure the primary current. Compare this value with the computed value.

_____ $I_{(PRIMARY)}$

13. **Turn off the power supply** and reconnect the AC ammeter in the secondary circuit and add two more 150-ohm resistors in parallel with the transformer secondary (Figure 20-4).

14. Turn on the power and measure the secondary current.

_____ amp(s)

15. **Turn off the power supply.**

16. Compute the amount of current flow that should be in the primary circuit using the turns-ratio. Be sure to add the excitation current.

_____ $I_{(PRIMARY)}$

17. Reconnect the AC ammeter in series with one of the lines of the primary winding of the transformer.

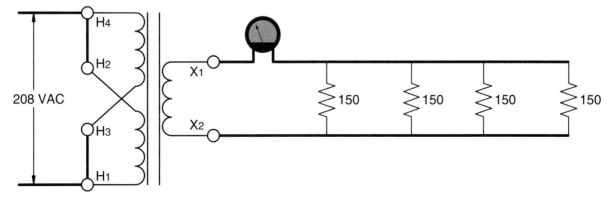

Figure 20-4 Adding load to the transformer secondary.

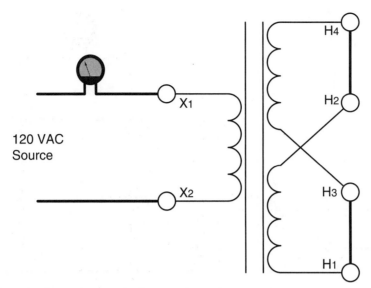

Figure 20-5 Using the low-voltage winding as the primary.

18. Turn on the power and measure the current flow. Compare the measured value with the computed value.

    _____ $I_{(PRIMARY)}$

19. **Turn off the power supply.**

20. Reconnect the transformer by connecting the low-voltage terminals, X_1 - X_2, to a 120 volt AC source. Connect an AC ammeter in series with one of the power lines, as shown in Figure 20-5.

21. Turn on the power and measure the excitation current of the transformer.

    _____ amp(s)

22. Measure the secondary voltage with an AC voltmeter.

    _____ volts

23. Determine the turns-ratio by dividing the secondary voltage by the primary voltage. Since the primary voltage is lower, the higher number will be placed on the right-hand side of the ratio: 1:3 or 1:4.

    _____ ratio

24. **Turn off the power supply.**

25. Connect two 150-ohm resistors in series. Connect these two resistors in parallel with the high-voltage winding. Connect an AC ammeter in series with one of the secondary leads, as shown in Figure 20-6.

26. Turn on the power supply and measure the secondary current.

    _____ amp(s)

27. Compute the primary current using the turns-ratio. Since the primary voltage is less than the secondary voltage, the primary current will be more than the secondary current. To determine the primary current, multiply the secondary current by the turns-ratio and add the excitation current.

    $$I_{(PRIMARY)} = I_{(SECONDARY)} \times \text{Turns-ratio} + \text{Excitation current}$$

    _____ $I_{(PRIMARY)}$

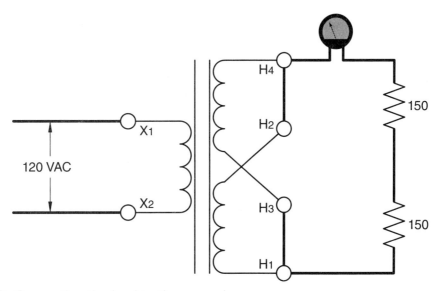

Figure 20-6 Connecting the load to the secondary.

28. **Turn off the power supply.**
29. Reconnect the AC ammeter in series with the primary side of the transformer.
30. Turn on the power supply and measure the primary current. Compare this value with the computed value.

       _____ $I_{(PRIMARY)}$

31. **Turn off the power supply.**
32. Reconnect the AC ammeter in series with the secondary winding. Add two more 150-ohm resistors that have been connected in series to the secondary circuit. These two resistors should be connected in parallel with the first two resistors, seen in Figure 20-7.
33. Turn on the power supply and measure the secondary current.

       _____ amp(s)

34. **Turn off the power supply.**

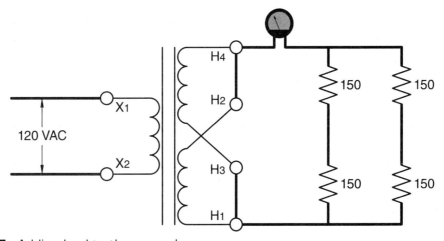

Figure 20-7 Adding load to the secondary.

35. Compute the amount of current that should flow in the primary circuit.

    _____ I$_{(PRIMARY)}$

36. Reconnect the AC ammeter in series with one of the primary lines.

37. Turn on the power supply and measure the primary current. Compare this value with the computed value.

    _____ I$_{(PRIMARY)}$

38. **Turn off the power supply** and disconnect the transformer and resistors.

39. Reconnect the transformer as shown in Figure 20-8 by connecting the two high-voltage windings in series. Connect four 150-ohm resistors in series and connect them to terminals H$_1$ and H$_4$ for the transformer. Connect an ammeter in series with the secondary winding.

Caution

The transformer now has a turns-ratio of 1:4. The output voltage will be approximately 480 volts when 120 volts are applied to terminals X$_1$ and X$_2$. Make sure that the power is turned off before making any adjustments to the circuit.

---

40. Turn on the power and measure the secondary current.

    _____ amp(s)

41. Make certain the voltmeter is set for a range greater than 480 volts. Measure the voltage across terminals H$_1$ and H$_4$. **Use caution when making this measurement.**

    _____ volts

42. **Turn off the power.**

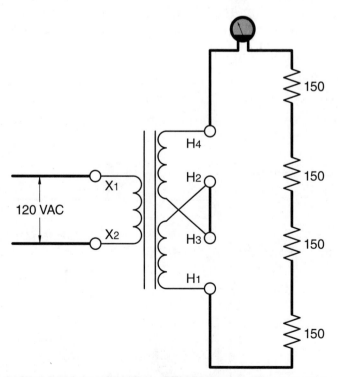

Figure 20-8 Connect the two high voltage windings in series. The transformer now has a turns-ratio of 1:4.

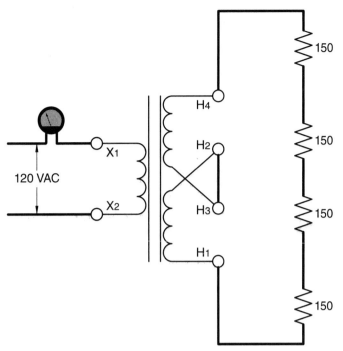

Figure 20-9   The ammeter is reconnected in the primary winding.

43. Use the turns-ratio to compute the primary current. Be sure to add the excitation current to the calculation.

    —————— I$_{(PRIMARY)}$

44. Reconnect the ammeter in series with the primary as shown in Figure 20-9.

45. Turn on the power and measure the amount of primary current. Compare this value with the computed value.

    —————— amp(s)

46. **Turn off the power.** Disconnect the circuit and return the components to their proper place.

Review Questions

1. A transformer has a primary voltage of 277 volts and a secondary voltage of 120 volts. What is the turns-ratio of this transformer?

   _____

2. A transformer has a turns-ratio of 1:6. Is this a step-up or a step-down transformer?

   _____

3. A transformer with a turns-ratio of 3.5:1 has a secondary current of 16 amperes. What is the primary current?

   _____

4. A transformer has a primary current of 18 amperes and a secondary current of 6 amperes. What is the turns-ratio of the transformer?

   _____

5. A transformer has a primary voltage of 240 volts and a secondary voltage of 60 volts. It has a power rating of 7.5 VA. What is the rated current of the secondary?

_____

6. A 75 kVA transformer has a secondary voltage of 480 volts and a current of 183 amperes. Is this transformer being operated within its power rating?

_____

7. A transformer has a primary voltage of 120 volts and a secondary voltage of 18 volts. The primary excitation current is 0.25 amp. The total primary current is 6.5 amperes. What is the secondary current?

_____

8. Would a 1 kVA transformer be large enough to supply the load of the transformer in question 7?

_____

9. A transformer has a primary voltage of 12,470 volts and a secondary voltage of 2,400 volts. If the secondary current is 22.6 amperes, what is the primary current (disregard excitation current)?

_____

10. Would a 75 kVA transformer supply the power needed by the load in question 7?

_____

Objectives

After studying this unit, you should be able to:

- Discuss buck and boost connections for a transformer.
- Connect a transformer for additive polarity.
- Connect a transformer for subtractive polarity.
- Determine the turns-ratio and calculate current values using measured values.

To understand what is meant by transformer polarity, the voltage produced across a winding must be considered during some point in time. In a 60 Hz AC circuit, the voltage changes polarity 120 times per second. When discussing transformer polarity, it is necessary to consider the relationship between the different windings at the same point in time. It will, therefore, be assumed that this point in time is when the peak positive voltage is being produced across the winding.

Polarity Markings on Schematics

When a transformer is shown on a schematic diagram, it is common practice to indicate the polarity of the transformer windings by placing a dot beside one end of each winding, as shown in Figure 21-1. These dots signify that the polarity is the same at that point in time for each winding. For example, assume the voltage applied to the primary winding is at its peak positive value at the terminal indicated by the dot. The voltage at the dotted lead of the secondary will be at its peak positive value at the same time.

This same type of polarity notation is used for transformers that have more than one primary or secondary winding. An example of a transformer with a multisecondary is shown in Figure 21-2.

Additive and Subtractive Polarities

The polarity of transformer windings can be determined by connecting one lead of the primary to one lead of the secondary and testing for an increase or decrease in voltage. This

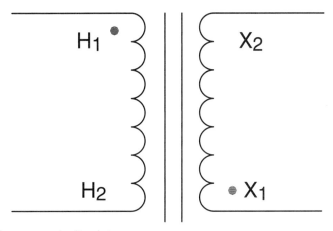

Figure 21-1 Transformer polarity dots.

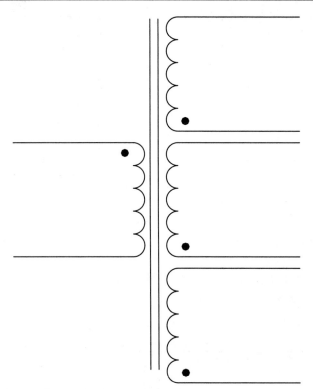

Figure 21-2 Polarity marks for multiple secondaries.

is often referred to as a *buck* or *boost* connection (Figure 21-3). The transformer shown in the example has a primary voltage rating of 120 volts and a secondary voltage rating of 24 volts. This same circuit has been redrawn in Figure 21-4 to show the connection more clearly. Notice that the secondary winding has been connected in series with the primary winding. When 120 volts is applied to the primary winding, the voltmeter connected across the secondary will indicate either the *SUM* of the two voltages or the *DIFFERENCE* between the two voltages. If this voltmeter indicates 144 volts ($120 + 24 = 144$), the windings are connected additive (boost) and polarity dots can be placed as shown in Figure 21-5. Notice in this connection that the secondary voltage is added to the primary voltage.

If the voltmeter connected to the secondary winding should indicate a voltage of 96 volts ($120 - 24 = 96$), the windings are connected subtractive (buck) and polarity dots would be placed as shown in Figure 21-6.

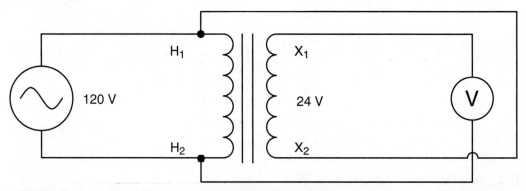

Figure 21-3 Connecting the secondary and primary windings forms an autotransformer.

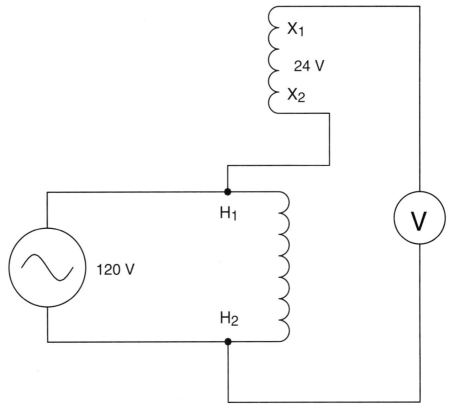

Figure 21-4 Redrawing the connection.

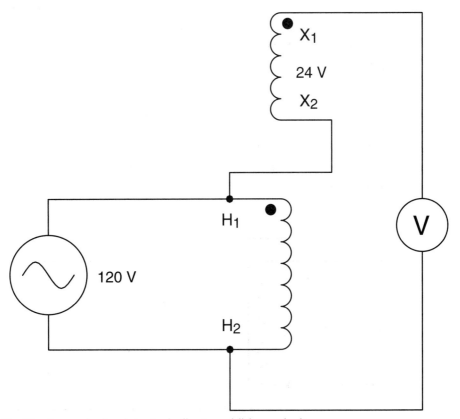

Figure 21-5 Placing polarity dots to indicate additive polarity.

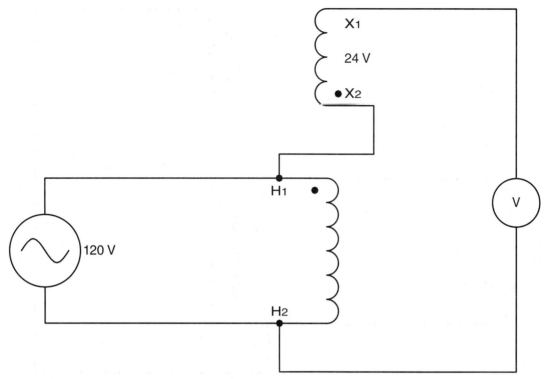

Figure 21-6 Polarity dots indicate subtractive polarity.

Using Arrows to Place Dots

To help in the understanding of additive and subtractive polarity, arrows can be used to indicate a direction of greater-than or less-than values. In Figure 21-7, arrows have been added to indicate the direction in which the dot is to be placed. In this example, the transformer is connected additive, or boost, and both of the arrows point in the same direction. Notice that the arrow points to the dot. In Figure 21-8, it is seen that the values of the two arrows add to produce 144 volts.

In Figure 21-9, arrows have been added to a subtractive, or buck, connection. In this instance, the arrows point in opposite directions and the voltage of one tries to cancel the voltage of the other. The result is that the smaller value is eliminated and the larger value is reduced as shown in Figure 21-10.

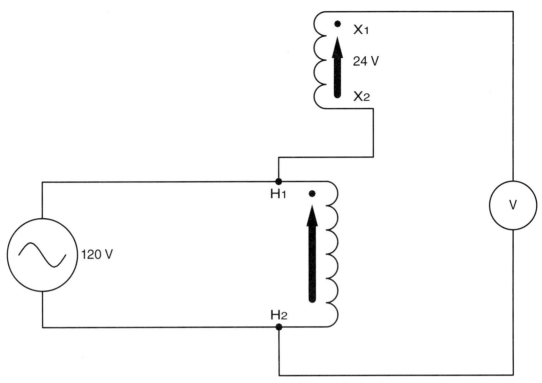

Figure 21-7 Arrows help indicate the placement of the polarity dots.

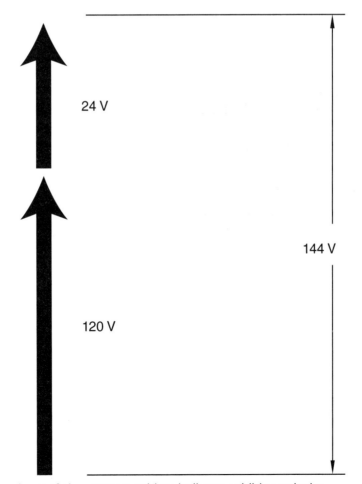

Figure 21-8 The values of the arrows add to indicate additive polarity.

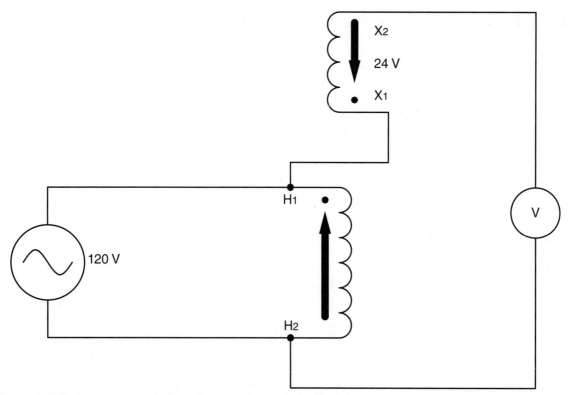

Figure 21-9   The arrows help indicate subtractive polarity.

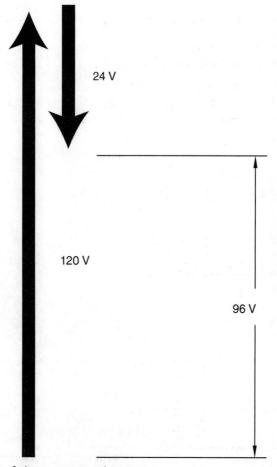

Figure 21-10   The value of the arrows subtract.

LABORATORY EXERCISE

Name _____ Date _____

Materials Required

480-240/120-volt 0.5-kVA control transformer

AC voltmeter

2 AC ammeters, in-line or clamp-on. (If a clamp-on type is used, a 10:1 scale divider is recommended.)

150 ohm resistors

4 100-ohm resistors

In this experiment a control transformer will be connected for both additive (boost) and subtractive (buck) polarity. Buck and boost connections are made by physically connecting the primary and secondary windings together. If they are connected in such a way that the primary and secondary voltages add, the transformer is connected additive, or boost. If the windings are connected in such a way that the primary and secondary voltages subtract, they are connected subtractive, or buck.

In this exercise only one of the high-voltage windings will be used. The other will not be connected.

1. Connect the circuit shown in Figure 21-11.

2. Turn on the power and measure the primary and secondary voltages.

   $E_{(PRIMARY)}$ _____ volts
   $E_{(SECONDARY)}$ _____ volts

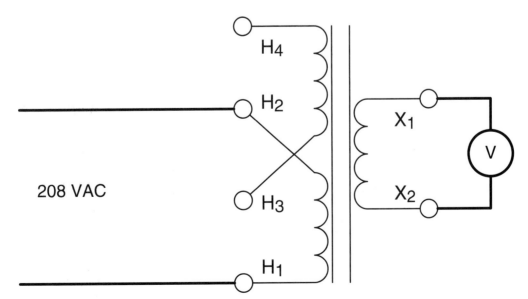

Figure 21-11 Measuring the secondary voltage.

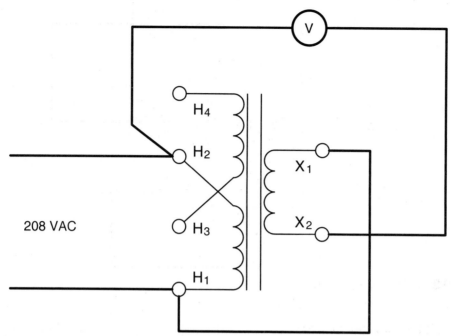

Figure 21-12 Connecting X_2 and H_2.

3. **Turn off the power supply.**

4. Determine the turns-ratio of this transformer connection by dividing the higher voltage by the lower voltage. Recall that if the primary winding has the higher voltage, the higher number will be placed on the left and 1 will be placed on the right. If the secondary has the higher voltage, a 1 will be placed on the left and the higher number will be placed on the right.

$$\text{Turns-ratio} = \frac{\text{Higher voltage}}{\text{Lower voltage}}$$

Ratio _____

5. Connect the circuit shown in Figure 21-12 by connecting X_1 to H_1. Connect a voltmeter across terminals X_2 and H_2.

6. Turn on the power supply and measure the voltage across X_2 and H_2.

_____ volts

7. **Turn off the power supply.**

8. Determine the turns-ratio of this transformer connection.

Ratio _____

9. If the measured voltage is the difference between the applied voltage and the secondary voltage, the transformer is connected subtractive polarity, or buck. If the measured voltage is the sum of the applied voltage and the secondary voltage, the transformer is connected additive, or boost. Is the transformer connected buck or boost?

_____

10. Connect an AC ammeter in series with one of the power supply lines.

11. Turn on the power supply and measure the excitation current of the transformer.

$I_{(EXC.)}$ _____ amp(s)

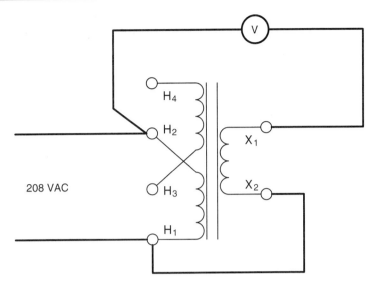

Figure 21-13 Connecting X_1 and H_2.

12. **Turn off the power supply.**

13. Reconnect the transformer as shown in Figure 21-13 by connecting X_2 to H_1. Connect an AC voltmeter across terminals X_1 and H_2.

14. Turn on the power supply and measure the voltage across terminals X_1 and H_2.

    _____ volts

15. **Turn off the power supply.**

16. Is the transformer connected buck or boost?

    _____

17. Determine the turns-ratio of this transformer connection.

    Ratio _____

18. Connect an AC ammeter in series with one of the primary leads.

19. Turn on the power supply and measure the excitation current of this connection.

    $I_{(EXC)}$ _____amp(s)

20. Compare the value of excitation current for the buck and boost connections. Is there any difference between these two values?

    _____

21. **Turn off the power supply.**

22. Figure 21-14 shows the proper location for the placement of polarity dots. Recall that polarity dots are used to indicate which windings of a transformer have the same polarity at the same time. To better understand how the dots are placed, redraw the two transformer windings in a series connection as shown in Figure 21-15. Place a dot beside one of the high-voltage terminals. In this example, a dot has been placed beside the H_1 terminal. Next, draw an arrow pointing to the dot. To place the second dot, draw an arrow in the same direction as the first arrow. This arrow should point to the dot that is to be placed beside the secondary terminal. Since terminal X_2 is connected to H_1, the arrow must point to terminal X_1.

23. Reconnect the transformer for subtractive polarity. If two ammeters are available, place one ammeter in series with one of the primary leads and the second ammeter

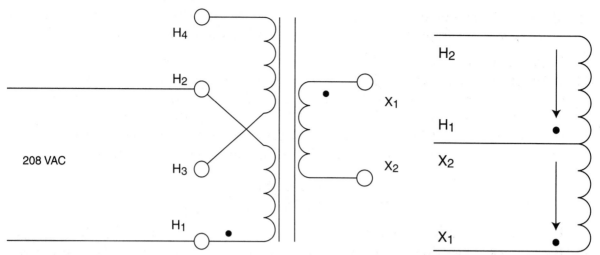

Figure 21-14 Placing polarity dots on the transformer windings.   **Figure 21-15** Determining the placement of polarity dots.

in series with the secondary lead that is not connected to the H_1 terminal. Connect a 100-ohm resistor in the secondary circuit, and connect a voltmeter in parallel with the resistor, as shown in Figure 21-16.

24. Turn on the power supply and measure the secondary current.

$I_{(SECONDARY)}$ ——————— amp(s)

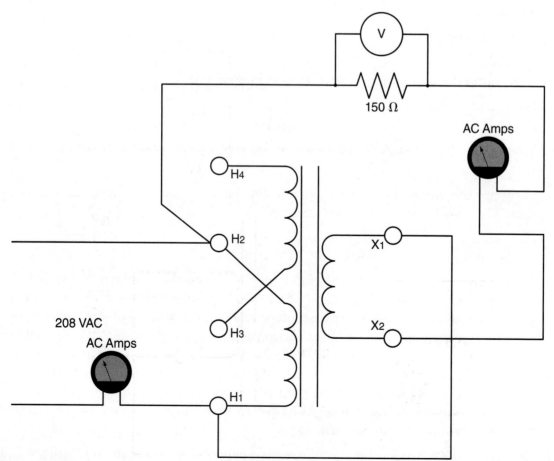

Figure 21-16 Connecting load to a subtractive polarity transformer.

25. Measure the secondary voltage. Since the resistor is the only load connected to the secondary, the voltage drop across the lamp will be the secondary voltage.

    E(SECONDARY) —————— volts

26. Calculate the amount of primary current using the measured value of secondary current and the turns-ratio. Be sure to use the turns-ratio for this connection as determined in step 8. Since the primary voltage is greater than the secondary voltage, the primary current should be less. Therefore, divide the secondary current by the turns-ratio and then add the excitation current measured in step 11.

$$I_{(PRIMARY)} = \frac{I_{(SECONDARY)}}{Turns\text{-}ratio} + I_{(EXC)}$$

    I(PRIMARY) —————— amp(s)

27. If necessary, **turn off the power supply** and connect an AC ammeter in series with one of the primary leads.

28. Turn on the power supply and measure the primary current. Compare this value with the calculated value.

    I(PRIMARY) —————— amp(s)

29. **Turn off the power supply.**

30. Connect a 150-ohm resistor in parallel with the 100-ohm resistor as shown in Figure 21-17. Reconnect the AC ammeter in series with the secondary winding if necessary.

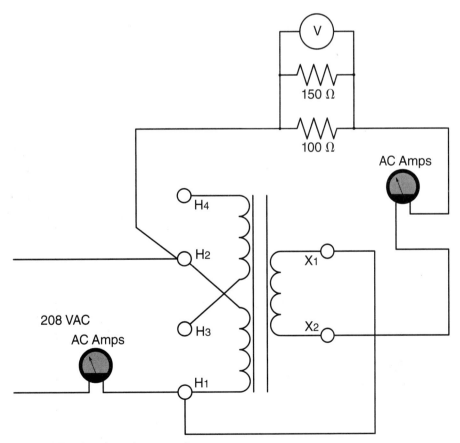

Figure 21-17 Adding load to the transformer.

31. Turn on the power supply and measure the amount of secondary current.

    $I_{(SECONDARY)}$ —————— amp(s)

32. Calculate the primary current.

    $I_{(PRIMARY)}$ —————— amp(s)

33. If necessary, **turn off the power supply** and connect the AC ammeter in series with one of the primary leads.

34. Turn on the power supply and measure the primary current. Compare this value with the computed value.

    $I_{(PRIMARY)}$ —————— amp(s)

35. **Turn off the power supply.**

36. Reconnect the transformer for the boost connection by connecting terminal X_2 to H_1. If two ammeters are available, connect one AC ammeter in series with one of the power supply leads and the second AC ammeter in series with the secondary. Connect four 150-ohm resistors in series with terminals X_1 and H_2 as shown in Figure 21-18. Connect an AC voltmeter across terminals X_2 and H_1.

37. Turn on the power and measure the secondary current.

    $I_{(SECONDARY)}$ —————— amp(s)

38. **Turn off the power supply.**

39. Compute the primary current using the turns-ratio. Be sure to use the turns-ratio for this connection as determined in step 17. Since the primary voltage in this connection

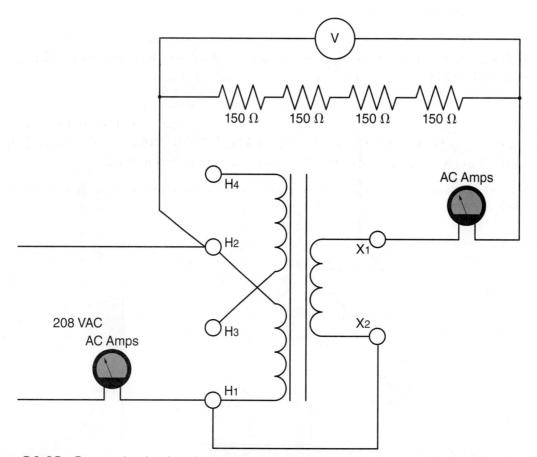

Figure 21-18 Connecting load to the boost connection.

is less than the secondary voltage, the primary current will be greater. To calculate the primary current, multiply the secondary current by the turns-ratio and then add the excitation current.

$$I_{(PRIMARY)} = (I_{(SECONDARY)} \times Turns\text{-}ratio) + I_{(EXC)}$$

$I_{(PRIMARY)}$ _____ amp(s)

40. If necessary, connect the AC ammeter in series with one of the power supply leads.

41. Turn on the power supply.

42. Measure the primary current. Compare this value with the calculated value.

$I_{(PRIMARY)}$ _____ amp(s)

43. **Turn off the power supply.**

44. Disconnect the circuit and return the components to their proper place.

Review Questions

1. What do the dots shown beside the terminal leads of a transformer represent on a schematic?

   _____

2. A transformer has a primary voltage rating of 240 volts and a secondary voltage rating of 80 volts. If the windings are connected subtractive, what voltage would appear across the entire connection?

   _____

3. If the windings of the transformer in question 2 were to be connected additive, what voltage would appear across the entire winding?

   _____

4. The primary leads of a transformer are labeled 1 and 2. The secondary leads are labeled 3 and 4. If polarity dots are placed beside leads 1 and 4, which secondary lead would be connected to terminal 2 to make the connection additive?

   _____

Unit 22 Autotransformers

Objectives

After studying this unit, you should be able to:

- Discuss the operation of an autotransformer.
- Connect a control transformer as an autotransformer.
- Calculate the turns-ratio from measured voltage values.
- Calculate primary current using the secondary current and the turns-ratio.
- Connect an autotransformer as a step-down transformer.
- Connect an autotransformer as a step-up transformer.

The word *auto* means self. An autotransformer is literally a *self-transformer*. It uses the same winding as both the primary and secondary. Recall that the definition of a primary winding is a winding that is connected to the source of power and the definition of a secondary winding is a winding that is connected to a load. Autotransformers have very high efficiencies, most in the range of 95% to 98%.

In Figure 22-1, the entire winding is connected to the power source, and part of the winding is connected to the load. In this illustration all the turns of wire form the primary and part of the turns form the secondary. Since the secondary part of the winding contains fewer turns than the primary section, the secondary will produce less voltage. This autotransformer is a step-down transformer.

In Figure 22-2, the primary section is connected across part of a winding and the secondary is connected across the entire winding. In this illustration the secondary section contains more windings than the primary. This autotransformer is a step-up transformer. Notice that autotransformers, like isolation transformers, can be used as step-up or step-down transformers.

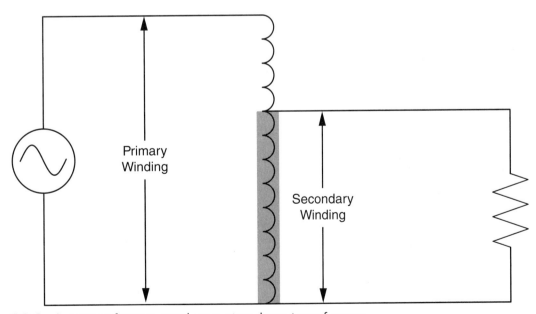

Figure 22-1 Autotransformer used as a step-down transformer.

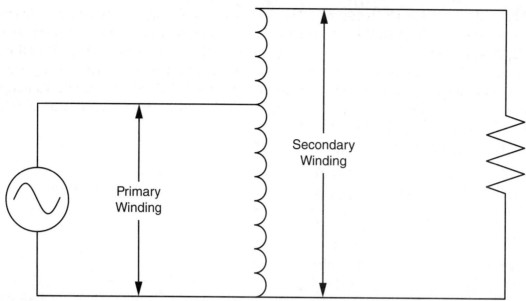

Figure 22-2 Autotransformer used as a step-up transformer.

Determining Voltage Values

Autotransformers are not limited to a single secondary winding. Many autotransformers have multiple taps to provide different voltages as shown in Figure 22-3. In this example there are 40 turns of wire between taps A and B, 80 turns of wire between taps B and C, 100 turns of wire between taps C and D, and 60 turns of wire between taps D and E. The primary section of the windings is connected between taps B and E. It will be assumed that the primary is connected to a source of 120 volts. The voltage across each set of taps will be determined.

There is generally more than one method that can be employed to determine values of a transformer. Since the number of turns between each tap is known, the volts-per-turn

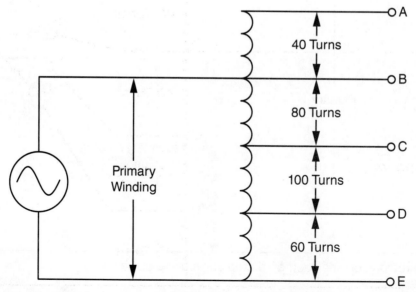

Figure 22-3 Autotransformer with multiple taps.

method will be used in this example. *The volts-per-turn for any transformer is determined by the primary winding.* In this illustration the primary winding is connected across taps B and E. The primary turns are, therefore, the sum of the turns between taps B and E (80 + 100 + 60 = 240 turns). Since 120 volts is connected across 240 turns, this transformer will have a volts-per-turn ratio of 0.5 (240 turns/120 volts = 0.5 volt-per-turn). To determine the amount of voltage between each set of taps, it becomes a simple matter of multiplying the number of turns by the volts-per-turn.

A-B (40 turns × 0.5 = 20 volts)

A-C (120 turns × 0.5 = 60 volts)

A-D (220 turns × 0.5 = 110 volts)

A-E (280 turns × 0.5 = 140 volts)

B-C (80 turns × 0.5 = 40 volts)

B-D (180 turns × 0.5 = 90 volts)

B-E (240 turns × 0.5 = 120 volts)

C-D (100 turns × 0.5 = 50 volts)

C-E (160 turns × 0.5 = 80 volts)

D-E (60 turns × 0.5 = 30 volts)

Using Transformer Formulas

The values of voltage and current for autotransformers can also be determined by using standard transformer formulas. The primary winding of the transformer shown in Figure 22-4 is between points B and N and has a voltage of 120 volts applied to it. If the turns of wire are counted between points B and N, it can be seen there are 120 turns of wire. Now assume that the selector switch is set to point D. The load is now connected between points

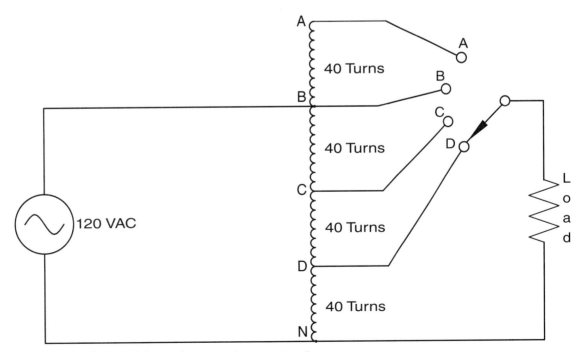

Figure 22-4 Determining voltage and current values.

D and N. The secondary of this transformer contains 40 turns of wire. If the amount of voltage applied to the load is to be computed, the following formula can be used:

$$\frac{E_P}{E_S} = \frac{N_P}{N_S}$$

$$\frac{120}{E_S} = \frac{120}{40}$$

$$120E_S = 4{,}800$$

$$E_S = 40 \text{ volts}$$

Assume that the load connected to the secondary has an impedance of 10 Ω. The amount of current flow in the secondary circuit can be computed using the following formula:

$$I = \frac{E}{Z}$$

$$I = \frac{40}{10}$$

$$I = 4 \text{ amps}$$

The primary current can be computed by using the same formula that was used to compute primary current for an isolation type of transformer.

$$\frac{E_P}{E_S} = \frac{I_S}{I_P}$$

$$\frac{120}{40} = \frac{4}{I_P}$$

$$I_P = 1.333 \text{ amps}$$

The amount of power input and output for the autotransformer must also be the same.

Primary	Secondary
120 × 1.333 = 160 volt-amps	40 × 4 = 160 volt-amps

Now assume that the rotary switch is connected to point A. The load is now connected to 160 turns of wire. The voltage applied to the load can be computed by:

$$\frac{E_P}{E_S} = \frac{N_P}{N_S}$$

$$\frac{120}{E_S} = \frac{120}{60}$$

$$120E_S = 19{,}200$$

$$E_S = 160 \text{ volts}$$

The amount of secondary current can be computed using the formula:

$$I = \frac{E}{Z}$$

$$I = \frac{160}{10}$$

$$I = 16 \text{ amps}$$

The primary current can be computed using the formula:

$$\frac{E_P}{E_S} = \frac{I_S}{I_P}$$

$$\frac{120}{160} = \frac{16}{I_P}$$

$$120I_P = 2,560$$

$$I_P = 21.333 \text{ amps}$$

The answers can be checked by determining if the power in and power out are the same.

Primary	Secondary
$120 \times 21.333 = 2,560$ volt-amps	$160 \times 16 = 2,560$ volt-amps

Current Relationships

An autotransformer with a 2:1 turns-ratio is shown in Figure 22-5. It is assumed that a voltage of 480 volts is connected across the entire winding. Since the transformer has a turns-ratio of 2:1, a voltage of 240 volts will be supplied to the load. Ammeters connected in series with each winding indicate the current flow in the circuit. It is assumed that the load produces a current flow of 4 amperes on the secondary. Note that a current flow of 2 amperes is supplied to the primary.

$$I_{\text{PRIMARY}} = \frac{I_{\text{SECONDARY}}}{\text{Ratio}}$$

$$I_P = \frac{4}{2}$$

$$I_P = 2 \text{ amperes}$$

If the rotary switch shown in Figure 22-4 were to be removed and replaced with a sliding tap that made contact directly to the transformer winding, the turns-ratio could be adjusted continuously. This type of transformer is commonly referred to as a Variac or Powerstat depending on the manufacturer. The windings are wrapped around a tape-wound torroid core inside a plastic case. The tops of the windings have been milled flat similar to a commutator. A carbon brush makes contact with the windings. When the

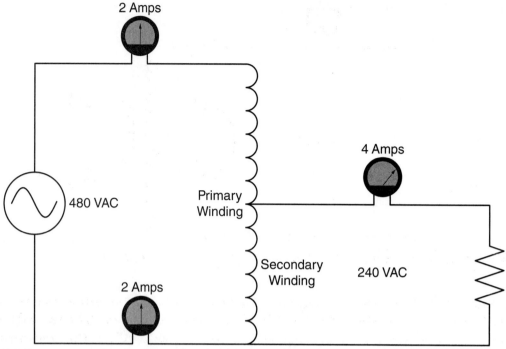

Figure 22-5   Current divides between primary and secondary.

brush is moved across the windings, the turns-ratio changes, which changes the output voltage. This type of autotransformer provides a very efficient means of controlling AC voltage. Autotransformers are often used by power companies to provide a small increase or decrease to the line voltage. They help provide voltage regulation to large power lines.

The autotransformer does have one disadvantage. Since the load is connected to one side of the power line, there is no line isolation between the incoming power and the load. This can cause problems with certain types of equipment and must be a consideration when designing a power system.

LABORATORY EXERCISE

Name _____ Date _____

Materials Required

480-240/120-volt, 0.5-kVA control transformer

AC voltmeter

2 AC ammeter, in-line or clamp-on. (If the clamp-on type is used, a 10:1 scale divider is recommended.)

4 150-ohm resistors

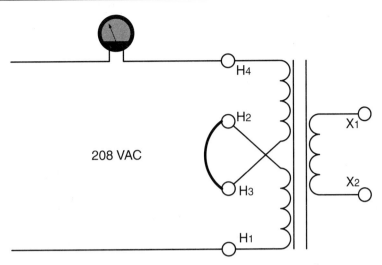

Figure 22-6 Connecting the high-voltage windings as an autotransformer.

In this experiment the control transformer will be connected for operation as an autotransformer. The low-voltage winding will not be used in this experiment. The two high-voltage windings will be connected in series to form one continuous winding. The transformer will be connected as both a step-down and a step-up transformer.

1. Series connect the two high-voltage windings by connecting terminals H_2 and H_3 together. The H_1 and H_4 terminals will be connected to a source of 208 VAC. Connect an ammeter in series with one of the power supply lines, as shown in Figure 22-6.

2. Turn on the power supply and measure the excitation current. The current will be small, and it may be difficult to determine this current value.

    $I_{(EXC)}$ _____ amp(s)

3. Measure the primary voltage across terminals H_1 and H_4.

    $E_{(PRIMARY)}$ _____ volts

4. Measure the secondary voltage across terminals H_1 and H_2. (Note: It is also possible to use terminals H_3 and H_4 as the secondary winding.)

    $E_{(SECONDARY)}$ _____ volts

5. Determine the turns-ratio of this transformer connection.

$$\text{Turns-ratio} = \frac{\text{Higher voltage}}{\text{Lower voltage}}$$

    Ratio _____

6. **Turn off the power supply.**

7. Connect an AC ammeter in series with the H_2 terminal and a 150-ohm resistor as shown in Figure 22-7. The secondary winding of the transformer will be between terminals H_2 and H_1.

8. Calculate the secondary current using Ohm's law.

    $I_{(SECONDERY)} =$ _____ A

9. Calculate the primary current using the turns-ratio.

$$I_{PRIMARY} = \frac{I_{SECONDARY}}{\text{Turns-Ratio}} + E_{EXC}$$

    $I_{(PRIMARY)} =$ _____ A

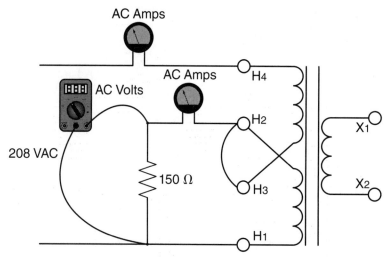

Figure 22-7   Connecting a load to the autotransformer.

10. Turn on the power and measure the secondary current, primary current, and voltage drop across the secondary. **Turn off the power.**

    $I_{(SECONDARY)}$ _____ A

    $I_{(PRIMARY)}$ _____ A

    $E_{(SECONDARY)}$ _____ volts

11. Compare the measured values with the calculated values in steps 8 and 9. Are they within 5% of each other?

    _____

12. Connect a second 150-ohm resistor in parallel with the first as shown in Figure 22-8. This should provide a total resistance of 75 ohms (150/2). Calculate the secondary current and primary current.

    $I_{(SECONDARY)}$ = _____ A

    $I_{(PRIMARY)}$ = _____ A

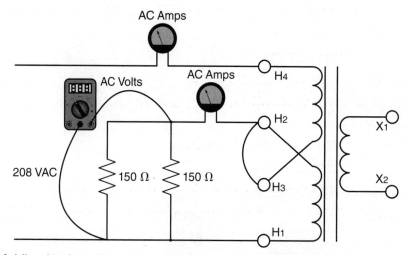

Figure 22-8   Adding load to the autotransformer.

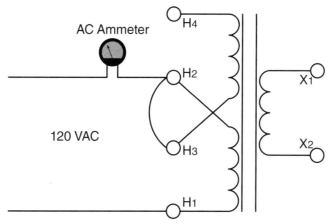

Figure 22-9 The autotransformer connected for high voltage.

13. Turn on the power and measure the secondary current, primary current, and voltage drop across the secondary. **Turn off the power**.

    $I_{(SECONDARY)}$ _____ A

    $I_{(PRIMARY)}$ _____ A

    $E_{(SECONDARY)}$ _____ volts

14. Compare the measured values with the calculated values in step 12. Are they within 5% of each other?

    _____

15. The transformer will now be connected as a step-up autotransformer. Connect the circuit shown in Figure 22-9. Turn on the power and measure the excitation current of the primary. **Turn off the power**.

    $I_{(EXC)}$ = _____ A

16. Connect an AC voltmeter across terminals H1 and H4 of the transformer. Turn on the power and measure the secondary voltage. **Turn off the power**.

    $E_{(SECONDARY)}$ = _____ volts

17. Determine the turns-ratio of the transformer.

    Ratio _____

18. Connect the circuit shown in Figure 22-10.

19. Calculate the secondary current and the primary current.

    $I_{(SECONDARY)}$ _____ A

    $I_{(PRIMARY)}$ _____ A

20. Turn on the power and measure the secondary current, primary current, and voltage drop across the transformer secondary winding. **Turn off the power**.

    $I_{(SECONDARY)}$ _____ A

    $I_{(PRIMARY)}$ _____ A

    $E_{(SECONDARY)}$ _____ volts

21. Compare the measured values with the calculated values in step 19. Are they within 5% of each other?

    _____

22. Disconnect the circuit and return the components to their proper place.

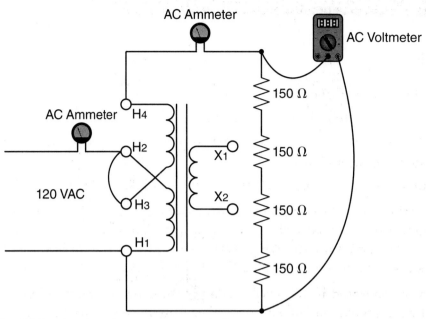

Figure 22-10 Adding load to the secondary winding.

Review Questions

1. An AC power source is connected across 325 turns of an autotransformer and the load is connected across 260 turns. What is the turns-ratio of this transformer?

_____

2. Is the transformer in question 1 a step-up or step-down transformer?

_____

3. An autotransformer has a turns-ratio of 3.2:1. A voltage of 208 volts is connected across the primary. What is the voltage of the secondary?

_____

4. A load impedance of 52 Ω is connected to the secondary winding of the transformer in question 3. How much current will flow in the secondary?

_____

5. How much current will flow in the primary of the transformer in question 4?

_____

6. The autotransformer shown in Figure 22-3 has the following number of turns between windings: A-B (120 turns), B-C (180 turns), C-D (250 turns), and D-E (300 turns). A voltage of 240 volts is connected across B and E. Find the voltages between each of the following points:

A-B _____   A-C _____   A-D _____   A-E _____   B-C _____   B-D _____

B-E _____   C-D _____   C-E _____   D-E _____

Unit 23 Three-Phase Circuits

Objectives

After studying this unit, you should be able to:

- Connect a wye connected, three-phase load.
- Calculate and measure voltage and current values for a wye connected load.
- Connect a delta connected load.
- Calculate and measure voltage and current values for a delta connected load.

Before beginning the study of three-phase transformers, it is appropriate to discuss three-phase power connections and basic circuit calculations. This unit may be review for some students and new ground for others. Whichever is the case, a working knowledge of three-phase circuits is essential before beginning the study of three-phase transformers.

Most of the power generated in the world today is three-phase. Three-phase power was first conceived by a man named Nikola Tesla. There are several reasons why three-phase power is superior to single-phase power.

1. The kVA rating of three-phase transformers is about 150% greater than for a single-phase transformer with a similar core size.

2. The power delivered by a single-phase system pulsates (Figure 23-1). The power falls to zero three times during each cycle. The power delivered by a three-phase circuit pulsates also, but the power never falls to zero (Figure 23-2). In a three-phase system, the power delivered to the load is the same at any instant.

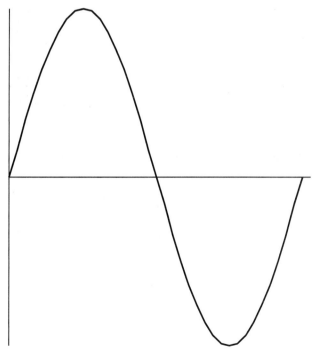

Figure 23-1 Single-phase power falls to zero three times each cycle.

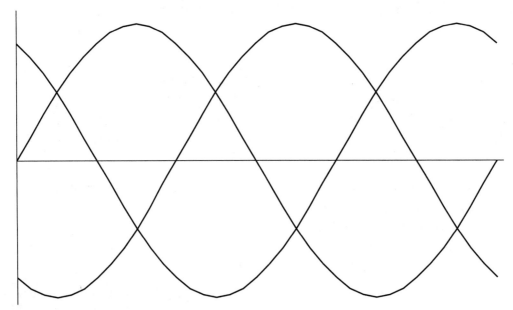

Figure 23-2 Three-phase power never falls to zero.

3. In a balanced three-phase system, the conductors need be only about 75% the size of conductors for a single-phase two-wire system of the same kVA rating. This helps offset the cost of supplying the third conductor required by three-phase systems.

A single-phase alternating voltage can be produced by rotating a magnetic field through the conductors of a stationary coil as shown in Figure 23-3. Since alternate polarities of the magnetic field cut through the conductors of the stationary coil, the induced voltage will change polarity at the same speed as the rotation of the magnetic field. The alternator shown in Figure 23-3 is single-phase because it produces only one AC voltage.

If three separate coils are spaced 120° apart as shown in Figure 23-4, three voltages 120° out of phase with each other will be produced when the magnetic field cuts through the coils. This is the manner in which a three-phase voltage is produced. There are two basic three-phase connections, the *wye* or *star*, and the *delta*.

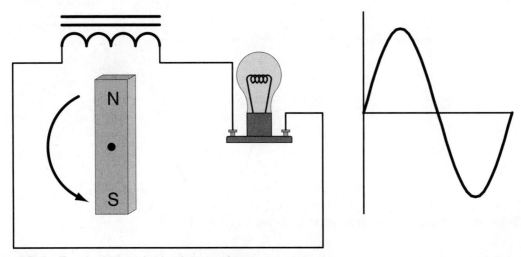

Figure 23-3 Producing a single-phase voltage.

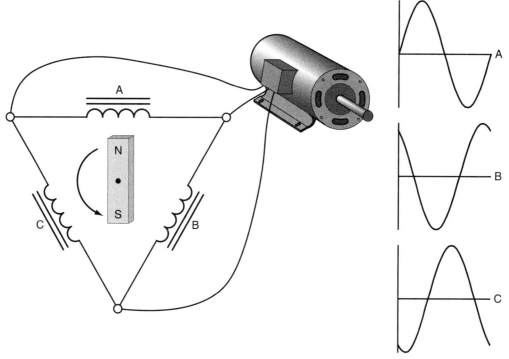

Figure 23-4 The voltages of a three-phase system are 120° out of phase with each other.

Wye Connection

The wye or star connection is made by connecting one end of each of the three-phase windings together, as shown is Figure 23-5. The voltage measured across a single winding or phase is known as the *phase* voltage as shown in Figure 23-6. The voltage measured between the lines is known as the line-to-line voltage or simply as the *line* voltage.

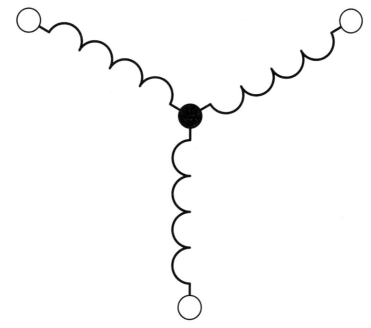

Figure 23-5 A wye connection is formed by joining one end of each winding together.

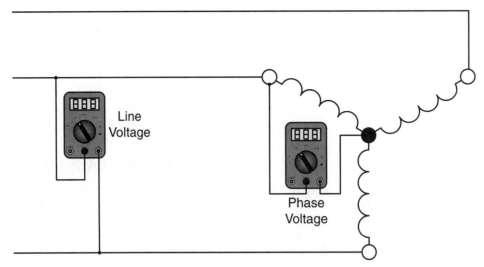

Figure 23-6 Line and phase voltages are different in a wye connection.

In Figure 23-7, ammeters have been placed in the phase winding of a wye connected load and in the line supplying power to the load. Voltmeters have been connected across the input to the load and across the phase. A line voltage of 208 volts has been applied to the load. Notice that the voltmeter connected across the lines indicates a value of 208 volts, but the voltmeter connected across the phase indicates a value of 120 volts.

In a wye connected system, the line voltage is higher than the phase voltage by a factor of $\sqrt{3}$ (1.732). Two formulas used to compute the voltage in a wye connected system are:

$$E_{PHASE} = \frac{E_{LINE}}{\sqrt{3}}$$

or

$$E_{LINE} = E_{PHASE} \times \sqrt{3}$$

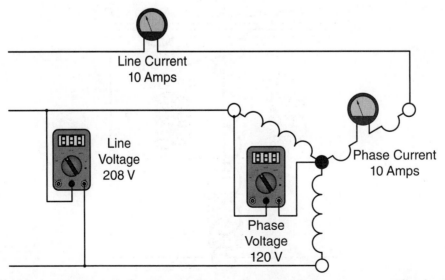

Figure 23-7 Line current and phase current are the same in a wye connection.

Also, notice in Figure 23-7 that there is 10 amps of current flow in both the phase and the line. In a wye connected system, phase current and line current are the same.

$$I_{LINE} = I_{PHASE}$$

Helpful Hint

In a wye connected system, the line voltage is higher than the phase voltage by a factor of $\sqrt{3}$ (1.732).

In a wye connected system, phase current and line current are the same.

$$I_{LINE} = I_{PHASE}$$

Voltage Relationships in a Wye Connection

Many students of electricity have difficulty at first understanding why the line voltage of the wye connection used in this illustration is 208 volts instead of 240 volts. Since line voltage is measured across two phases that have a voltage of 120 volts each, it would appear that the sum of the two voltages should be 240 volts. One cause of this misconception is that many students are familiar with the 240/120 volt connection supplied to most homes. If voltage is measured across the two incoming lines, a voltage of 240 volts will be seen. If voltage is measured from either of the two lines to the neutral, a voltage of 120 volts will be seen. The reason for this is that this connection is derived from the center tap of an isolation transformer, as shown in Figure 23-8. If the center tap is used as a common point, the two line voltages on either side of it will

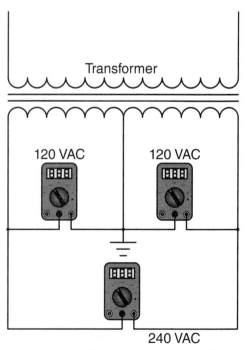

Figure 23-8 Single-phase transformer with grounded center tap.

be in phase with each other. Since the two voltages are in phase, they add similar to a boost connected transformer, as shown in Figure 23-9. The vector sum of these two voltages would be 240 volts.

Three-phase voltages are 120° apart, not in phase. If the three voltages are drawn 120° apart, it will be seen that the vector sum of these voltages is 208 volts, as shown in Figure 23-10. Another illustration of vector addition is shown in Figure 23-11. In this illustration, two-phase voltage vectors are added and the resultant is drawn from the starting point of one vector to the end point of the other. The parallelogram method of vector addition for the voltages in a wye connected three-phase system is shown in Figure 23-12.

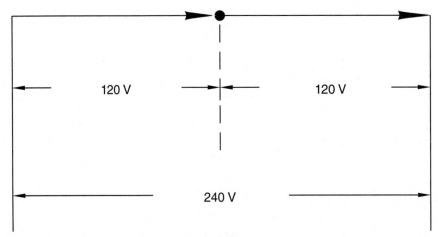

120 V   120 V

240 V

Figure 23-9 The two voltages are in phase with each other.

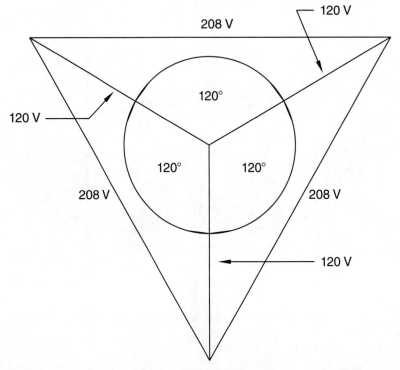

208 V   120 V

120°

120 V

120°   120°

208 V   208 V

120 V

Figure 23-10 Vector sum of the voltages in a three-phase wye connection.

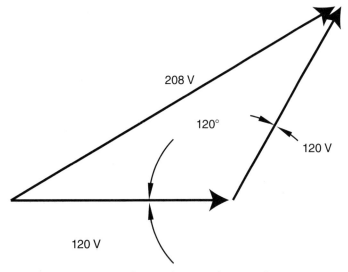

Figure 23-11 Adding voltage vectors of two-phase voltage values.

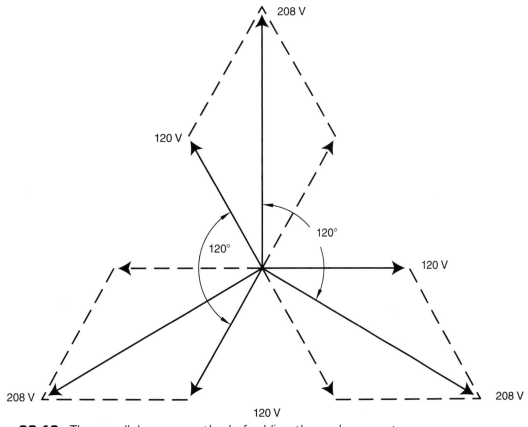

Figure 23-12 The parallelogram method of adding three-phase vectors.

Delta Connection

In Figure 23-13, three separate inductive loads have been connected to form a delta connection. This connection receives its name from the fact that a schematic diagram of this connection resembles the Greek letter delta (Δ). In Figure 23-14, voltmeters have been connected across the lines and across the phase. Ammeters have been connected in the

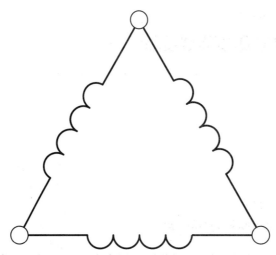

Figure 23-13 Three-phase delta connection.

line and in the phase. In the delta connection, line voltage and phase voltage are the same. Notice that both voltmeters indicated a value of 480 volts.

$$E_{LINE} = E_{PHASE}$$

Helpful Hint

In the delta connection, line voltage and phase voltage are the same.

Notice that the line current and phase current are different, however. The line current of a delta connection is higher than the phase current by a factor of $\sqrt{3}$ (1.732). In the example shown, it is assumed that each of the phase windings has a current flow of 10 amperes. The current in each of the lines, however, is 17.32 amperes. The reason for this

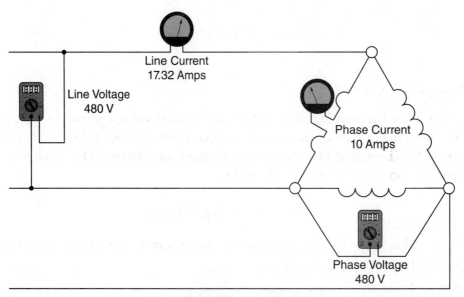

Figure 23-14 Voltage and current relationships in a delta connection.

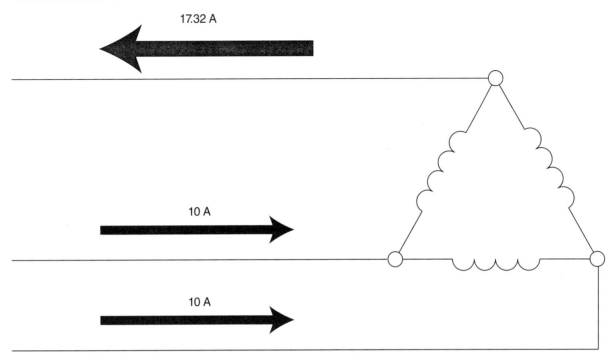

Figure 23-15 Division of currents in a delta connection.

difference in current is that current flows through different windings at different times in a three-phase circuit. During some periods of time, current will flow between two lines only. At other times, current will flow from two lines to the third, seen in Figure 23-15. The delta connection is similar to a parallel connection because there is always more than one path for current flow. Since these currents are 120° out of phase with each other, vector addition must be used when finding the sum of the currents (Figure 23-16). Formulas for determining the current in a delta connection are:

$$I_{PHASE} = \frac{I_{LINE}}{\sqrt{3}}$$

or

$$I_{LINE} = I_{PHASE} \times \sqrt{3}$$

Three-Phase Power

Students sometimes become confused when computing values of power in three-phase circuits. One reason for this confusion is because there are actually two formulas that can be used. If LINE values of voltage and current are known, the apparent power of the circuit can be computed using the formula:

$$VA = \sqrt{3} \times E_{LINE} \times I_{LINE}$$

If the PHASE values of voltage and current are known, the apparent power can be computed using the formula:

$$VA = 3 \times E_{PHASE} \times I_{PHASE}$$

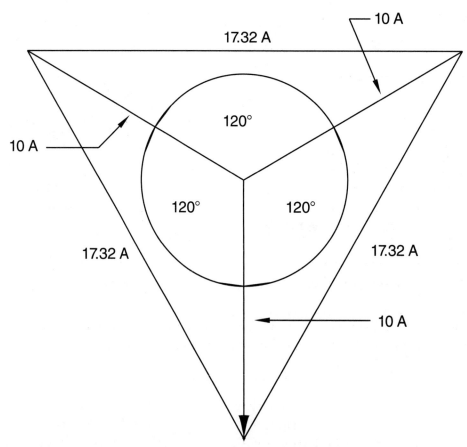

Figure 23-16 Vector addition is used to compute the sum of the currents in a delta connection.

Notice that in the first formula, the line values of voltage and current are multiplied by the square root of 3. In the second formula, the phase values of voltage and current are multiplied by 3. The first formula is the most used because it is generally more convenient to obtain line values of voltage and current because they can be measured with a voltmeter and clamp-on ammeter.

Watts and VARs

Watts and VARs can be computed in a similar manner. Watts can be computed by multiplying the apparent power by the power factor:

$$P = \sqrt{3} \times E_{LINE} \times I_{LINE} \times PF$$

or

$$P = 3 \times E_{PHASE} \times I_{PHASE} \times PF$$

Note: When computing the power of a pure resistive load, the voltage and current are in phase with each other and the power factor is 1.

VARs can be computed in a similar manner, except that voltage and current values of a pure reactive load are used. For example, a pure capacitive load is shown in Figure 23-17.

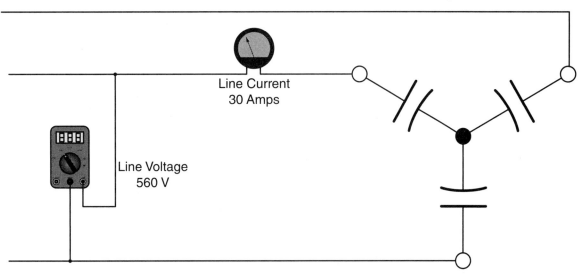

Figure 23-17 Pure capacitive three-phase load.

In this example, it is assumed that the line voltage is 480 volts and the line current is 30 amperes. Capacitive VARs can be computed using the formula:

$$VARs_C = \sqrt{3} \times E_{LINE(CAPACITIVE)} \times I_{LINE(CAPACITIVE)}$$

$$VARs_C = 1.732 \times 560 \times 30$$

$$VARs_C = 29,097.6$$

Three-Phase Circuit Calculations

In the following examples, values of line and phase voltage, line and phase current, and power will be computed for different types of three-phase connections.

Example #1. A wye connected, three-phase alternator supplies power to a delta connected resistive load, as shown in Figure 23-18. The alternator has a line voltage of 480 volts. Each resistor of the delta load has 8 Ω of resistance. Find the following values:

$E_{L(LOAD)}$ - Line voltage of the load

$E_{P(LOAD)}$ - Phase voltage of the load

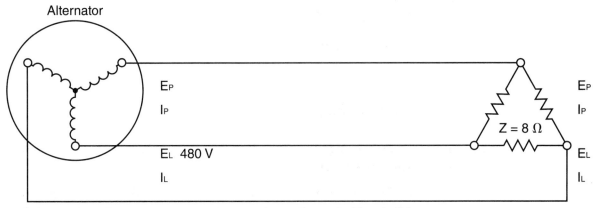

Figure 23-18 Computing three-phase values: Example Circuit #1.

$I_{P(LOAD)}$ - Phase current of the load

$I_{L(LOAD)}$ - Line current to the load

$I_{L(ALT)}$ - Line current delivered by the alternator

$E_{P(ALT)}$ - Phase voltage of the alternator

P - True power

Solution: The load is connected directly to the alternator. Therefore, the line voltage supplied by the alternator is the line voltage of the load.

$$E_{L(LOAD)} = 480 \text{ volts}$$

The three resistors of the load are connected in a delta connection. In a delta connection, the phase voltage is the same as the line voltage.

$$E_{P(LOAD)} = E_{I(LOAD)}$$

$$E_{P(LOAD)} = 480 \text{ volts}$$

Each of the three resistors in the load comprises one phase of the load. Now that the phase voltage is known (480 volts), the amount of phase current can be computed using Ohm's law.

$$I_{P(LOAD)} = \frac{E_{P(LOAD)}}{Z}$$

$$I_{P(LOAD)} = \frac{480}{8}$$

$$I_{P(LOAD)} = 60 \text{ amps}$$

In this example the three load resistors are connected as a delta with 60 amperes of current flow in each phase. The line current supplying a delta connection must be 1.732 times greater than the phase current.

$$I_{L(LOAD)} = I_{P(LOAD)} \times 1.732$$

$$I_{L(LOAD)} = 60 \times 1.732$$

$$I_{L(LOAD)} = 103.92 \text{ amps}$$

The alternator must supply the line current to the load or loads to which it is connected. In this example, there is only one load connected to the alternator. Therefore, the line current of the load will be the same as the line current of the alternator.

$$I_{L(ALT)} = 103.92 \text{ amps}$$

The phase windings of the alternator are connected in a wye connection. In a wye connection, the phase current and line current are equal. The phase current of the alternator will, therefore, be the same as the alternator line current.

$$I_{P(ALT)} = 103.92 \text{ amps}$$

The phase voltage of a wye connection is less than the line voltage by a factor of the square root of 3 ($\sqrt{3}$). The phase voltage of the alternator will be:

$$E_{P(ALT)} = \frac{E_{L(ALT)}}{\sqrt{3}}$$

$$E_{P(ALT)} = \frac{480}{1.732}$$

$$E_{P(ALT)} = 277.13 \text{ volts}$$

In this circuit the load is pure resistive. The voltage and current are in phase with each other, which produces a unity power factor of 1. The true power in this circuit will be computed using the formula:

$$P = 1.732 \times E_{L(ALT)} \times I_{L(ALT)} \times PF$$

$$P = 1.732 \times 480 \times 103.92 \times 1$$

$$P = 86,394.93 \text{ watts}$$

Example #2. In the next example, a delta connected alternator is connected to a wye connected resistive load, as shown in Figure 23-19. The alternator produces a line voltage of 240 volts and the resistors have a value of 6 Ω each. The following values will be found:

$E_{L(LOAD)}$ - Line voltage of the load
$E_{P(LOAD)}$ - Phase voltage of the load
$I_{P(LOAD)}$ - Phase current of the load
$I_{L(LOAD)}$ - Line current to the load
$I_{L(ALT)}$ - Line current delivered by the alternator
$E_{P(ALT)}$ - Phase voltage of the alternator
P - True power

As in the first example, the load is connected directly to the output of the alternator. The line voltage of the load must, therefore, be the same as the line voltage of the alternator.

$$E_{L(LOAD)} = 240 \text{ volts}$$

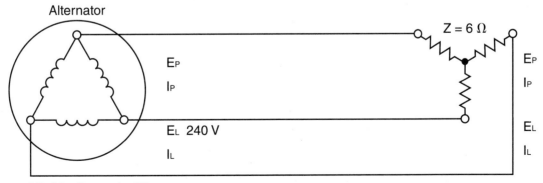

Figure 23-19 Example #2.

The phase voltage of a wye connection is less than the line voltage by a factor of 1.732.

$$E_{P(LOAD)} = \frac{240}{1.732}$$

$$E_{P(LOAD)} = 138.57 \text{ volts}$$

Each of the three 6 Ω resistors comprises one phase of the wye connected load. Since the phase voltage is 138.57 volts, this voltage is applied to each of the three resistors. The amount of phase current can now be determined using Ohm's law.

$$I_{P(LOAD)} = \frac{E_{P(LOAD)}}{Z}$$

$$I_{P(LOAD)} = \frac{138.57}{6}$$

$$I_{P(LOAD)} = 23.1 \text{ amps}$$

The amount of line current needed to supply a wye connected load is the same as the phase current of the load.

$$I_{L(LOAD)} = 23.1 \text{ amps}$$

In this example there is only one load connected to the alternator. The line current supplied to the load is the same as the line current of the alternator.

$$I_{L(ALT)} = 23.1 \text{ amps}$$

The phase windings of the alternator are connected in delta. In a delta connection the phase current is less than the line current by a factor of 1.732.

$$I_{P(ALT)} = \frac{I_{P(ALT)}}{1.732}$$

$$I_{P(ALT)} = \frac{23.1}{1.732}$$

$$I_{P(ALT)} = 13.34 \text{ amps}$$

The phase voltage of a delta is the same as the line voltage.

$$E_{P(ALT)} = 240 \text{ volts}$$

Since the load in this example is pure resistive, the power factor has a value of unity or 1. Power will be computed by using the line values of voltage and current.

$$P = 1.732 \times E_L \times I_L \times PF$$

$$P = 1.732 \times 240 \times 23.1 \times 1$$

$$P = 9,602.21 \text{ watts}$$

LABORATORY EXERCISE

Name _____ Date _____

Materials Required

2 AC voltmeters

AC ammeter, in-line or clamp-on. (If a clamp-on type is used, it is recommended to use a 10:1 scale divider.)

6 150-ohm resistors

In this experiment six 150-ohm resistors will be connected to form different three-phase loads. Two lamps will be connected in series to form three separate loads. These loads will be connected to form wye or delta connections.

1. Connect the two 150-ohm resistors in series to form three separated load banks. Connect the load banks in wye by connecting one end of each bank together to form a center point, as shown in Figure 23-20. It is assumed that this load is to be connected to a 208 VAC three-phase line. Connect an AC ammeter in series with the line supplying power to the load.

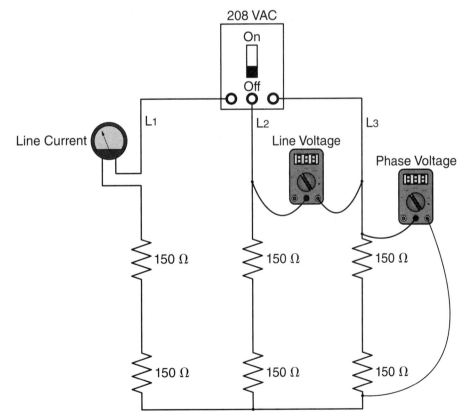

Figure 23-20 Measuring the line current in a wye connected load.

2. Turn on the power and measure the line voltage supplied to the load.

   $E_{(LINE)}$ —————— volts

3. Calculate the value of phase voltage for a wye connected load.

$$E_{PHASE} = \frac{E_{LINE}}{\sqrt{3}}$$

   $E_{(PHASE)}$ —————— volts

4. Measure the phase voltage and compare this value with the computed value.

   $E_{(PHASE)}$ —————— volts

5. Measure the line current.

   $I_{(LINE)}$ —————— amp(s)

6. **Turn off the power supply.**

7. In a wye connected system, the line current and phase current are the same. Reconnect the circuit as shown in Figure 23-21.

8. Turn on the power and measure the phase current.

   $I_{(PHASE)}$ —————— amp(s)

9. **Turn off the power supply.**

10. Reconnect the three banks of lamps to form a delta connected load, as shown in Figure 23-22.

11. Turn on the power and measure the line voltage supplied to the load.

   $E_{(LINE)}$ —————— volts

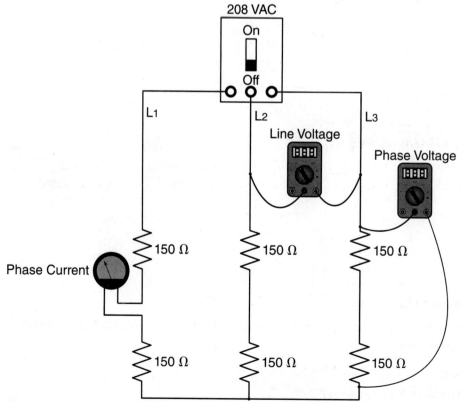

Figure 23-21 Measuring the phase current in a wye connected load.

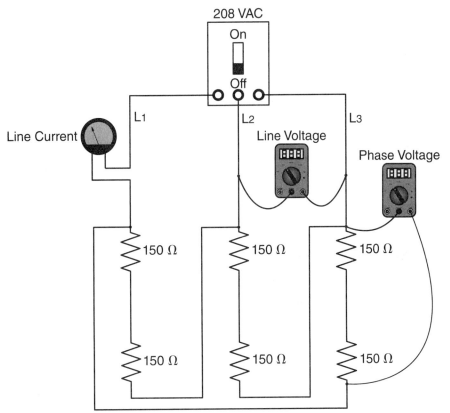

Figure 23-22 Measuring the voltage and line current values of a delta connected load.

12. Measure the phase value of voltage.

    $E_{(PHASE)}$ ——————— volts

13. Are the line and phase voltage values the same or different?

    ———————

14. Measure the line current.

    $I_{(LINE)}$ ——————— amp(s)

15. **Turn off the power supply.**

16. In a delta connected system, the phase current will be less than the line current by a factor of 1.732. Calculate the phase current value for this connection.

$$I_{PHASE} = \frac{I_{LINE}}{1.732}$$

    $I_{(PHASE)}$ ——————— amp(s)

17. Reconnect the circuit as shown in Figure 23-23.

18. Turn on the power supply and measure the phase current. Compare this value with the computed value.

    $I_{(PHASE)}$ ——————— amp(s)

19. **Turn off the power supply.**

20. Disconnect the circuit and return the components to their proper place.

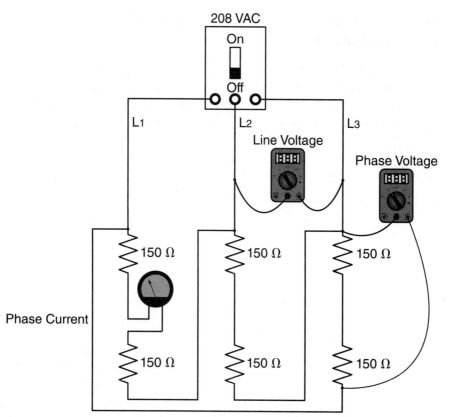

Figure 23-23 Measuring the voltage and phase current values of a delta connected load.

Review Questions

1. How many degrees out of phase with each other are the voltages of a three-phase system?

   _____

2. What are the two main types of three-phase connections?

   _____

3. A wye connected load has a voltage of 480 volts applied to it. What is the voltage dropped across each phase?

   _____

4. A wye connected load has a phase current of 25 amps. How much current is flowing through the lines supplying the load?

   _____

5. A delta connection has a voltage of 560 volts connected to it. How much voltage is dropped across each phase?

   _____

6. A delta connection has 30 amps of current flowing through each phase winding. How much current is flowing through each of the lines supplying power to the load?

   _____

7. A three-phase load has a phase voltage of 240 volts and a phase current of 18 amperes. What is the apparent power of this load?

   _____

8. If the load in question 7 is connected in a wye, what would be the line voltage and line current supplying the load?

9. An alternator with a line voltage of 2,400 volts supplies a delta connected load. The line current supplied to the load is 40 amperes. Assuming the load is a balanced three-phase load, what is the impedance of each phase?

10. What is the apparent power of the circuit in question 9?

Unit 24 Three-Phase Transformers

Objectives

After studying this unit, you should be able to:

- Connect three single-phase transformers to form a three-phase bank.
- Connect transformer windings in a delta configuration.
- Connect transformer windings in a wye configuration.
- Compute values of voltage, current, and turns-ratio for different three-phase connections.
- Compute the values for an open delta connected transformer bank.

Three-phase transformers are used throughout industry to change values of three-phase voltage and current. Since three-phase power is the major way in which power is produced, transmitted, and used, an understanding of how three-phase transformer connections are made is essential. This unit discusses different types of three-phase transformer connections and presents examples of how values of voltage and current for these connections are computed.

A three-phase transformer is constructed by winding three single-phase transformers on a single core, as shown in Figure 24-1. The transformer is enclosed in a case and may be dry or mounted in an enclosure that will be filled with a dielectric oil. The dielectric oil performs several functions. Since it is a dielectric, it provides electrical insulation between the windings and the case. It is also used to help provide cooling and to prevent the formation of moisture, which can deteriorate the winding insulation.

Three-Phase Transformer Connections

Three-phase transformers are connected in delta or wye configurations. A wye-delta transformer, for example, has its primary winding connected in a wye and its secondary winding

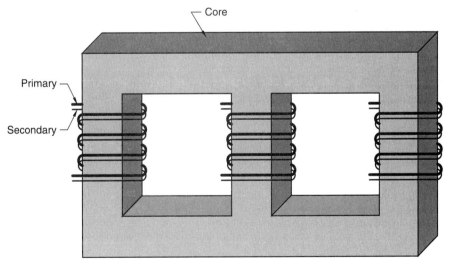

Figure 24-1 Basic construction of a three-phase transformer.

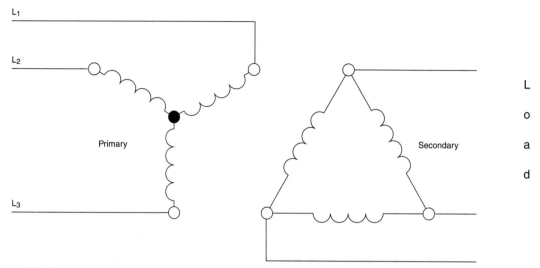

Figure 24-2 Wye-delta connected three-phase transformer.

connected in a delta, as shown in Figure 24-2. A delta-wye transformer would have its primary winding connected in delta and its secondary connected in wye, as shown in Figure 24-3.

Connecting Single-Phase Transformers into a Three-Phase Bank

If three-phase transformer is needed, and a three-phase transformer of the proper size and turns-ratio is not available, three single-phase transformers can be connected to form a three-phase bank. When three single-phase transformers are used to make a three-phase transformer bank, their primary and secondary windings are connected in a wye or delta connection. The three transformer windings in Figure 24-4 have been labeled A, B, and C. One end of each primary lead has been labeled H_1 and the other end has been labeled H_2. One end of each secondary lead has been labeled X_1 and the other end has been labeled X_2.

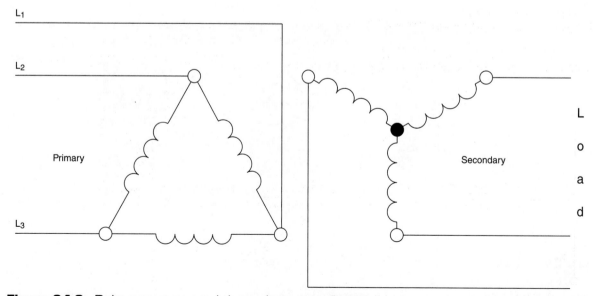

Figure 24-3 Delta-wye connected three-phase transformer.

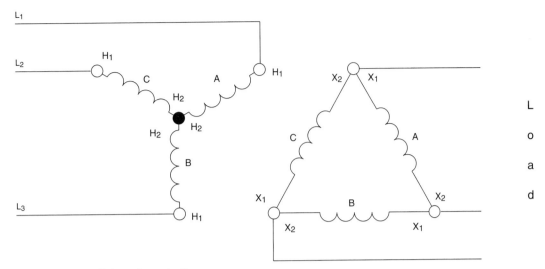

Figure 24-4 Identifying the windings.

Figure 24-5 shows three single-phase transformers labeled A, B, and C. The primary leads of each transformer have been labeled H_1 and H_2, and the secondary leads have been labeled X_1 and X_2. The schematic diagram of Figure 24-4 will be used to connect the three single-phase transformers into a three-phase wye-delta connection, as shown in Figure 24-6.

The primary winding will be tied into a wye connection first. The schematic in Figure 24-4 shows that the H_2 lead of each primary winding is connected together, and the H_1 lead of each winding is open for connection to the incoming power line. Notice in Figure 24-6 that the H_2 lead of each primary winding has been connected together, and the H_1 lead of each winding has been connected to the incoming power line.

Figure 24-4 also shows the X_1 lead of transformer A is connected to the X_2 lead of transformer C. Notice that this same connection has been made in Figure 24-6. The X_1 lead of transformer B is connected to the X_2 lead of transformer A, and the X_1 lead to transformer C is connected to the X_2 lead of transformer B. The load is connected to the points of the delta connection.

Although Figure 24-4 illustrates the proper schematic symbology for a three-phase transformer connection, some electrical schematics and wiring diagrams do not illustrate three-phase transformer connections in this manner. One type of diagram, called the one line diagram, would illustrate a delta-wye connection, as shown in Figure 24-7. These diagrams are generally used to show the main power distribution system of a large industrial plant.

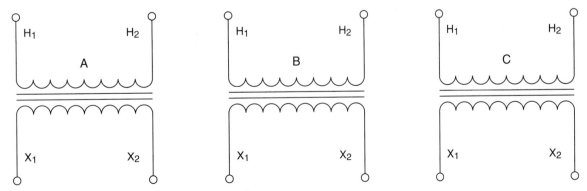

Figure 24-5 Three single-phase transformers.

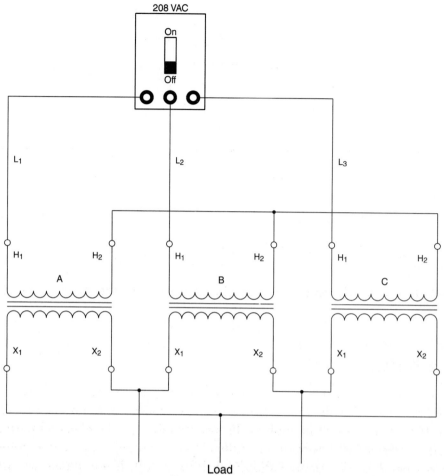

Figure 24-6 Connecting three single-phase transformers to form a wye-delta three-phase bank.

The one line diagram in Figure 24-8 shows the main power to the plant and the transformation of voltages to different subfeeders. Notice that each transformer shows whether the primary and secondary are connected as a wye or delta, and the secondary voltage of the subfeeder.

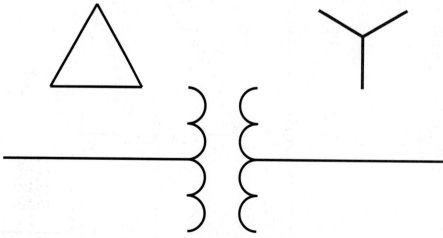

Figure 24-7 One line diagram symbol used to represent a delta-wye three-phase transformer connection.

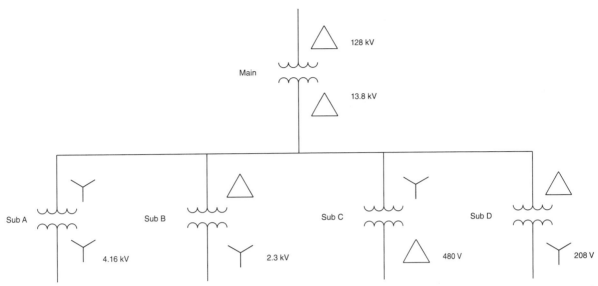

Figure 24-8 One line diagrams are generally used to show the main power distribution of a plant.

Closing a Delta

Delta connections should be checked for proper polarity before making the final connection and applying power. If the phase winding of one transformer is reversed, an extremely high current will flow when power is applied. Proper phasing can be checked with a voltmeter, as shown in Figure 24-9. If power is applied to the transformer bank before the delta connection is closed, the voltmeter should indicate 0 volt. If one phase winding has been reversed, however, the voltmeter will indicate double the amount of voltage. For example, assume the output voltage of a delta secondary is 240 volts. If the voltage is checked

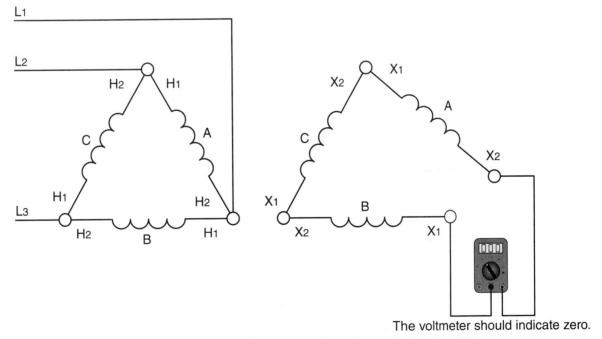

The voltmeter should indicate zero.

Figure 24-9 Testing for proper transformer polarity before closing the delta.

before the delta is closed, the voltmeter should indicate a voltage of 0 volt if all windings have been phased properly. If one winding has been reversed, however, the voltmeter will indicate a voltage of 480 volts (240 + 240). This test will confirm whether a phase winding has been reversed, but it will not indicate if the reversed winding is located in the primary or secondary. If either primary or secondary windings have been reversed, the voltmeter will indicate double the output voltage.

It should be noted, however, that a voltmeter is a high impedance device. It is not unusual for a voltmeter to indicate some amount of voltage before the delta is closed, especially if the primary has been connected as a wye and the secondary as a delta. When this is the case, however, the voltmeter will generally indicate close to the normal output voltage if the connection is correct and double the output voltage if the connection is incorrect. Regardless of whether the primary is connected as a delta or wye, the voltmeter will indicate twice the normal output voltage of the secondary if the connection is incorrect.

Three-Phase Transformer Calculations

When computing the values of voltage and current for three-phase transformers, the formulas used for making transformer calculations and three-phase calculations must be followed. Another very important rule that must be understood is that only phase values of voltage and current can be used when computing transformer values. When three-phase transformers are connected as a wye or delta, the primary and secondary windings themselves become the phases of a three-phase connection. This is true whether a three-phase transformer is used or whether three single-phase transformers are employed to form a three-phase bank. Refer to transformer A in Figure 24-5. All transformation of voltage and current takes place between the primary and secondary windings. Since these windings form the phase values of the three-phase connection, only phase, not line, values can be used when calculating transformed voltages and currents.

> ### Helpful Hint
>
> Only phase values of voltage and current can be used when computing transformer values.

Example #1: A three-phase transformer connection is shown in Figure 24-10. Three single-phase transformers have been connected to form a wye-delta bank. The primary is connected to a three-phase line of 13,800 volts, and the secondary voltage is 480. A three-phase resistive load with an impedance of 2.77 Ω per phase is connected to the secondary of the transformer. The following values will be computed for this circuit.

$E_{P(PRIMARY)}$ = Phase voltage of the primary

$E_{P(SECONDARY)}$ = Phase voltage of the secondary

Ratio = Turns-ratio of the transformer

$E_{P(LOAD)}$ = Phase voltage of the load bank

$I_{P(LOAD)}$ = Phase current of the load bank

$I_{L(SECONDARY)}$ = Secondary line current

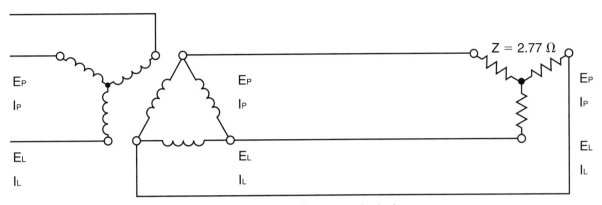

Figure 24-10 Example #1: Three-phase transformer calculations.

$I_{P(SECONDARY)}$ = Phase current of the secondary
$I_{P(PRIMARY)}$ = Phase current of the primary
$I_{L(PRIMARY)}$ = Line current of the primary

The primary windings of the three single-phase transformers have been connected to form a wye connection. In a wye connection, the phase voltage is less than the line voltage by a factor of 1.732 ($\sqrt{3}$). Therefore, the phase value of voltage can be computed using the following formula:

$$E_{P(PRIMARY)} = \frac{E_L}{1.732}$$

$$E_{P(PRIMARY)} = \frac{13,800}{1.732}$$

$$E_{P(PRIMARY)} = 7,967.67 \text{ volts}$$

The secondary windings are connected as a delta. In a delta connection, the phase voltage and line voltage are the same.

$$E_{P(SECONDARY)} = E_{L(SECONDARY)}$$

$$E_{P(SECONDARY)} = 480 \text{ volts}$$

The turns-ratio can be computed by comparing the phase voltage of the primary to the phase voltage of the secondary.

$$\text{Ratio} = \frac{\text{Primary phase voltage}}{\text{Secondary phase voltage}}$$

$$\text{Ratio} = \frac{7,967.67}{480}$$

$$\text{Ratio} = 16.6:1$$

The load bank is connected in a wye connection. The voltage across the phase of the load bank will be less than the line voltage by a factor of 1.732.

$$E_{P(LOAD)} = \frac{E_{L(LOAD)}}{1.732}$$

$$E_{P(LOAD)} = \frac{480}{1.732}$$

$$E_{P(LOAD)} = 277 \text{ volts}$$

Now that the voltage across each of the load resistors is known, the current flow through the phase of the load can be computed using Ohm's law.

$$I_{P(LOAD)} = \frac{E_{P(LOAD)}}{R}$$

$$I_{P(LOAD)} = \frac{277}{2.77}$$

$$I_{P(LOAD)} = 100 \text{ amperes}$$

Since the load is connected as a wye connection, the line current will be the same as the phase current. Therefore, the line current supplied by the secondary of the transformer is equal to the phase current of the load.

$$I_{L(SECONDARY)} = 100 \text{ amperes}$$

The secondary of the transformer bank is connected as a delta. The phase current of the delta is less than the line current by a factor of 1.732.

$$I_{P(SECONDARY)} = \frac{I_{L(SECONDARY)}}{1.732}$$

$$I_{P(SECONDARY)} = \frac{100}{1.732}$$

$$I_{P(SECONDARY)} = 57.74 \text{ amps}$$

The amount of current flow through the primary can be computed using the turns-ratio. Since the primary has a higher voltage, it will have a lower current. (Volts × Amps input must equal Volts × Amps output.)

$$I_{P(PRIMARY)} = \frac{I_{P(SECONDARY)}}{\text{Ratio}}$$

$$I_{P(PRIMARY)} = \frac{57.74}{16.6}$$

$$I_{P(PRIMARY)} = 3.48 \text{ amperes}$$

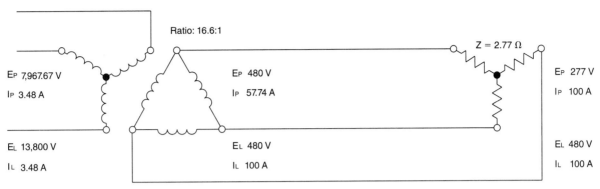

Figure 24-11 Example #1 with all missing values.

Recall that all transformed values of voltage and current take place across the phases; the primary has a phase current of 3.48 amps.

In a wye connection, the phase current is the same as the line current.

$$I_{L(PRIMARY)} = 3.48 \text{ amps}$$

The transformer connection with all computed values is shown in Figure 24-11.

Example #2: In the next example, a three-phase transformer is connected in a delta-delta configuration (Figure 24-12). The load is connected as a wye and each phase has an impedance of 7 Ω. The primary is connected to a line voltage of 4,160 volts and the secondary line voltage is 440 volts. The following values will be found:

$E_{P(PRIMARY)}$ = Phase voltage of the primary

$E_{P(SECONDARY)}$ = Phase voltage of the secondary

Ratio = Turns-ratio of the transformer

$E_{P(LOAD)}$ = Phase voltage of the load bank

$I_{P(LOAD)}$ = Phase current of the load bank

$I_{L(SECONDARY)}$ = Secondary line current

$I_{P(SECONDARY)}$ = Phase current of the secondary

$I_{P(PRIMARY)}$ = Phase current of the primary

$I_{L(PRIMARY)}$ = Line current of the primary

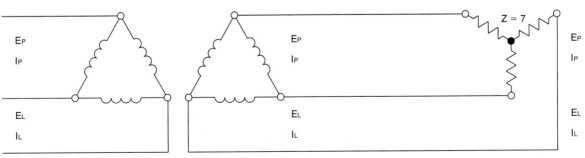

Figure 24-12 Example #2: Three-phase transformer calculations.

The primary is connected as a delta. The phase voltage will be the same as the applied line voltage.

$$E_{P(PRIMARY)} = E_{L(PRIMARY)}$$

$$E_{P(PRIMARY)} = 4{,}160 \text{ volts}$$

The secondary of the transformer is connected as a delta, also. Therefore, the phase voltage of the secondary will be the same as the line voltage of the secondary.

$$E_{P(SECONDARY)} = 440 \text{ volts}$$

All transformer values must be computed using phase values of voltage and current. The turns-ratio can be found by dividing the phase voltage of the primary by the phase voltage of the secondary.

$$\text{Ratio} = \frac{\text{Primary phase voltage}}{\text{Secondary phase voltage}}$$

$$\text{Ratio} = \frac{4{,}160}{440}$$

$$\text{Ratio} = 9.45{:}1$$

The load is connected directly to the output of the secondary. The line voltage applied to the load must, therefore, be the same as the line voltage of the secondary.

$$E_{L(LOAD)} = 440 \text{ volts}$$

The load is connected in a wye. The voltage applied across each phase will be less than the line voltage by a factor of 1.732.

$$E_{P(LOAD)} = \frac{E_{L(LOAD)}}{1.732}$$

$$E_{P(LOAD)} = \frac{440}{1.732}$$

$$E_{P(LOAD)} = 254 \text{ volts}$$

The phase current of the load can be computed using Ohm's law.

$$I_{P(LOAD)} = \frac{E_{P(LOAD)}}{Z}$$

$$I_{P(LOAD)} = \frac{254}{7}$$

$$I_{P(LOAD)} = 36.29 \text{ amps}$$

The amount of line current supplying a wye connected load will be the same as the phase current of the load.

$$I_{L(LOAD)} = 36.39 \text{ amps}$$

Since the secondary of the transformer is supplying current to only one load, the line current of the secondary will be the same as the line current of the load.

$$I_{L(SECONDARY)} = 36.29 \text{ amps}$$

The phase current in a delta connection is less than the line current by a factor of 1.732.

$$I_{P(SECONDARY)} = \frac{I_{L(SECONDARY)}}{1.732}$$

$$I_{P(SECONDARY)} = \frac{36.29}{1.732}$$

$$I_{P(SECONDARY)} = 20.95 \text{ amps}$$

The phase current of the transformer primary can now be computed using the phase current of the secondary and the turns-ratio.

$$I_{P(PRIMARY)} = \frac{I_{P(SECONDARY)}}{Ratio}$$

$$I_{P(PRIMARY)} = \frac{20.95}{9.45}$$

$$I_{P(PRIMARY)} = 2.27 \text{ amps}$$

In this example, the primary of the transformer is connected as a delta. The line current supplying the transformer will be higher than the phase current by a factor of 1.732.

$$I_{L(PRIMARY)} = I_{P(PRIMARY)} \times 1.732$$

$$I_{L(PRIMARY)} = 2.27 \times 1.732$$

$$I_{L(PRIMARY)} = 3.93 \text{ amps}$$

The circuit with all computed values is shown in Figure 24-13.

Open Delta Connections

The open delta transformer connection can be made with only two transformers instead of three (Figure 24-14). This connection is often used when the amount of three-phase power needed is not excessive, such as in a small business. It should be noted that the

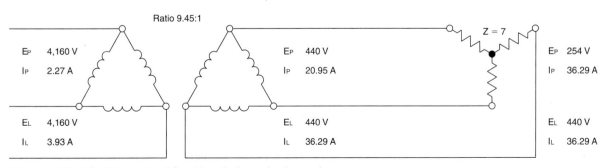

Figure 24-13 Example #2 with all the missing values.

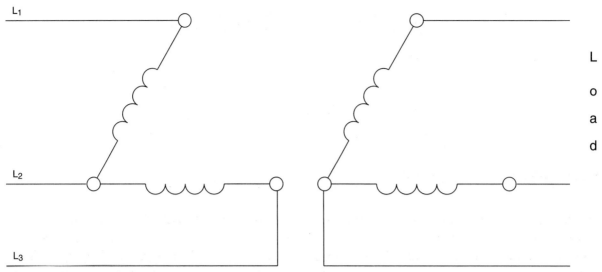

Figure 24-14   Open delta connection.

output power of an open delta connection is only 86.6% of the rated power of the two transformers. For example, assume two transformers, each having a capacity of 25 kVA (kilovolt-amperes), are connected in an open delta connection. The total output power of this connection is 43.3 kVA (50 kVA × 0.866 = 43.3 kVA).

Another figure given for this calculation is 57.7%. This percentage assumes a closed delta bank containing three transformers. If three 25 kVA transformers were connected to form a closed delta connection, the total output power would be 75 kVA (3 × 25 kVA = 75 kVA). If one of these transformers were to be removed, and the transformer bank operated as an open delta connection, the output power would be reduced to 57.7% of its original capacity of 75 kVA. The output capacity of the open delta bank is 43.3 kVA (75 kVA × 0.577 = 43.3 kVA).

The voltage and current values of an open delta connection are computed in the same manner as a standard delta-delta connection when three transformers are employed. The voltage and current rules for a delta connection must be used when determining line and phase values of voltage and current.

LABORATORY EXERCISE

Name _____ Date _____

Note: Due to the length of this laboratory exercise it has been divided into two parts.

Materials Required

3 480-240/120-volt, 0.5-kVA control transformers

AC voltmeter

2 AC ammeter, in-line or clamp-on. (If the clamp-on type is used, it is recommended to use a 10:1 scale divider.)

6 150-ohm resistors

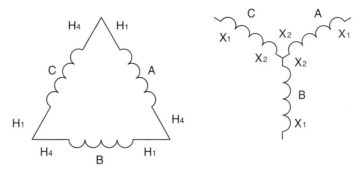

Figure 24-15 A delta-wye transformer connection.

In this experiment, three single-phase control transformers will be connected to form different three-phase transformer banks. Values of voltage, current, and turns-ratios will be computed and then measured. The three transformers will be operated with their high-voltage windings connected in parallel for low-voltage operation. The high-voltage windings are used as the primary for each connection.

The Delta-Wye Connection (PART 1)

A delta-wye connected three-phase transformer bank has its primary windings connected in a delta configuration and its secondary windings connected in a wye configuration (Figure 24-15). Notice that the three primary windings have been labeled A, B, and C. The H_1 terminal of transformer A is connected to the H_4 terminal of transformer C. The H_4 terminal of transformer A is connected to the H_1 terminal of transformer B, and the H_4 terminal of transformer B is connected to the H_1 terminal of transformer C. The secondary windings form a wye by connecting all the X_2 terminals together.

1. Connect the circuit shown in Figure 24-16. Notice that the three transformers have been labeled A, B, and C. The H_1 terminal of transformer A is connected to the H_4 terminal of transformer C, the H_4 terminal of transformer A is connected to the H_1 terminal of transformer B, and the H_4 terminal of transformer B is connected to the H_1 terminal of transformer C. This is the same connection shown in the schematic drawing of Figure 24-15. Also, notice that the X_2 terminal of each transformer is connected together to form a wye connected secondary.

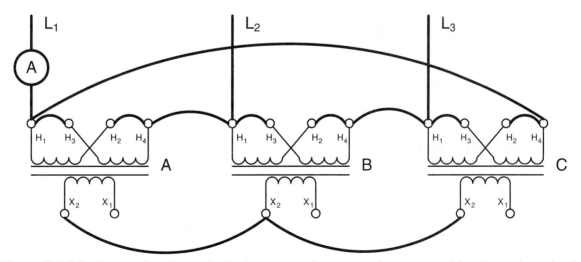

Figure 24-16 Connecting three single-phase transformers to form a wye-delta three-phase bank.

> **Helpful Hint**
>
> When calculating three-phase transformer values, draw the circuit as illustrated in Figures 24-10, 24-11, 24-12, and 24-13. Insert the known quantities in the drawing. This method helps avoid confusion when dealing with wye and delta connections.

2. The transformers in this exercise have a turns ratio of 2:1. Assuming that 208 volts is applied to the primary windings, calculate the phase voltage of the secondary windings.

   $E_{(SECONDARY\ PHASE)}$ = _____ volts.

3. Calculate the line voltage of the secondary.

   $E_{(SECONDARY\ LINE)}$ = _____ volts.

4. Turn on the power and measure the excitation current of the transformers. The excitation current should remain constant as long as the transformers are connected in this configuration.

   $E_{(EXC)}$ = _____ A

5. Measure the phase voltage of the primary winding with an AC voltmeter. This can be accomplished by measuring the voltage across H_1 and H_4 of any transformer.

   $E_{(PHASE\ PRIMARY)}$ = _____ volts

6. Measure the phase voltage of the secondary winding. This can be accomplished by measuring the voltage across X_1 and X_2 of any transformer.

   $E_{(PHASE\ SECONDARY)}$ = _____ volts

7. Measure the line voltage of the secondary winding. This can be accomplished by measuring the voltage across any two X_1 terminals of the three transformers. **Turn off the power.**

   $E_{(LINE\ SECONDARY)}$ = _____ volts

8. Compare the measured values with the computed values above. Are they within 5% of each other?

   _____

9. Connect the circuit shown in Figure 24-17. In this circuit, three 150-ohm resistors have been connected to form a wye load. A wye connection can be readily identified because one end of each load is connected together to form a center point.

10. Calculate the phase voltage of the wye connected load.

    $E_{(PHASE\ LOAD)}$ _____ volts

11. Using Ohm's law, calculate the phase current in the load.

    $I_{(PHASE\ LOAD)}$ = _____ A

12. Calculate the line current supplied by the transformer secondary to the load.

    $I_{(LINE\ SECONDARY)}$ = _____ A

13. Calculate the current in the phase winding of the transformer secondary.

    $I_{(PHASE\ SECONDARY)}$ = _____ A

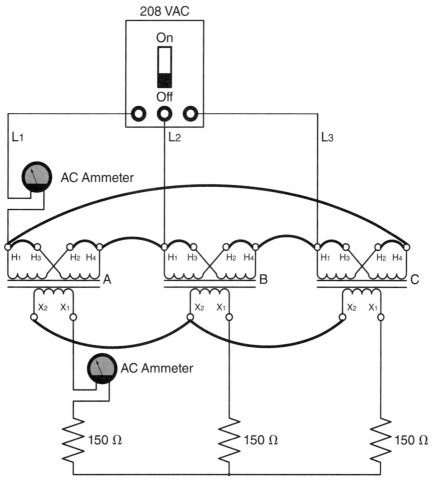

Figure 24-17 Adding load to the connection.

14. Calculate the current in the phase winding of the primary using the turns ratio. Because this is a step-down transformer, the primary current will be less than the secondary current by a factor of the turns-ratio.

    $I_{\text{(PHASE PRIMARY)}} = $ _____ A

15. Calculate the line current of the primary. Make sure to add the excitation current to the calculation.

    $I_{\text{(LINE PRIMARY)}} = $ _____ A

16. Turn on the power and measure the line current of the secondary and primary. **Turn off the power**.

    $I_{\text{(LINE SECONDARY)}} = $ _____ A

    $I_{\text{(LINE PRIMARY)}} = $ _____ A

    Compare these measured values with the computed values in step 12 and step 15. Are the values within 5% of each other?

    _____

17. Connect a second 150-ohm resistor in parallel with each of the load resistors, as shown in Figure 24-18. This will provide a resistance of 75 ohms for each phase of the load.

$$R_T = \frac{R}{N} \qquad R_T = \frac{150}{2} \qquad R_T = 75 \ \Omega$$

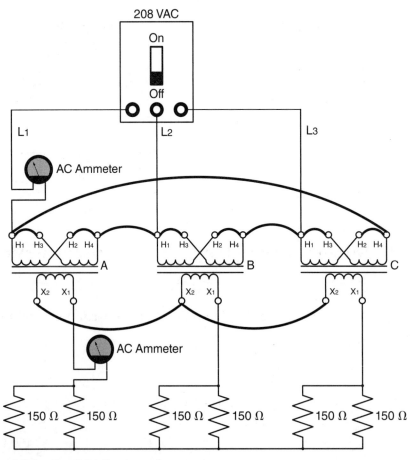

Figure 24-18   Adding load to the transformers.

18. Using the value of phase voltage for the load determined in step 10, calculate the phase current of the load using Ohm's law.

$I_{(PHASE\ LOAD)}$ = _____ A

19. Calculate the line current supplied by the transformer secondary to the load.

$I_{(LINE\ SECONDARY)}$ = _____ A

20. Calculate the current in the phase winding of the transformer secondary.

$I_{(PHASE\ SECONDARY)}$ = _____ A

21. Calculate the current in the phase winding of the primary using the turns-ratio. Because this is a step-down transformer, the primary current will be less than the secondary current by a factor of the turns-ratio.

$$I_{(PHASE\ PRIMARY)} = \frac{I_{(PHASE\ SECONDARY)}}{Turns\text{-}ratio}$$

$I_{(PHASE\ PRIMARY)}$ = _____ A

Helpful Hint

When calculating primary current using the secondary current and the turns-ratio, the phase value of secondary current must be used to determine the phase value of the primary current.

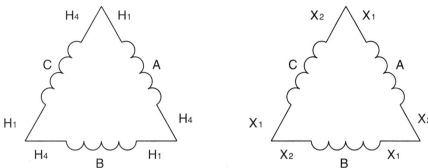

Figure 24-19 Delta-wye transformer connection.

22. Calculate the line current of the primary. Make sure to add the excitation current to the calculation.

$I_{(LINE\ PRIMARY)}$ = _____ A

23. Turn on the power and measure the line current of the secondary and primary. **Turn off the power**.

$I_{(LINE\ SECONDARY)}$ = _____ A

$I_{(LINE\ PRIMARY)}$ = _____ A

Compare these measured values with the computed values in step 12 and step 15. Are the values within 5% of each other?

_____

Delta-Delta Connection

The three transformers will now be reconnected to form a delta-delta connection. The schematic diagram for a delta-delta connection is shown in Figure 24-19.

24. Reconnect the transformers as shown in Figure 24-20. In this connection, the primary windings remain connected in a delta configuration, but the secondary windings have been reconnected from a wye to a delta. Because the primary windings have not been changed, the excitation current will remain the same.

25. Assuming a primary voltage of 208 volts, calculate the phase voltage of the secondary. The transformers have a ratio of 2:1.

$E_{(PHASE\ SECONDARY)}$ _____ volts

26. Calculate the line voltage of the secondary.

$E_{(LINE\ SECONDARY)}$ _____ volts

27. Calculate the phase voltage of the wye connected load resistors.

$E_{(PHASE\ LOAD)}$ _____ volts

28. Calculate the phase current of the load using Ohm's law.

$I_{(PHASE\ LOAD)}$ _____ A

29. Calculate the line current supplied to the load by the secondary of the transformer.

$I_{(LINE\ SECONDARY)}$ _____ A

30. Calculate the phase current of the secondary winding.

$I_{(PHASE\ SECONDARY)}$ _____ A

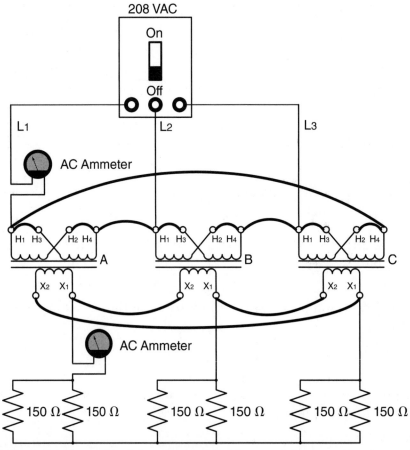

Figure 24-20 Transformers with delta connected primary, delta connected secondary, and wye connected load.

31. Using the turns-ratio of the transformer, calculate the phase current of the primary winding.

    $I_{(PHASE\ PRIMARY)}$ —————— A

32. Calculate the line current supplied to the primary. Make sure to add the excitation current to the calculation.

    $I_{(LINE\ PRIMARY)}$ —————— A

33. Turn on the power and measure the line current of the secondary and the line current of the primary. **Turn off the power**.

    $I_{(LINE\ SECONDARY)}$ —————— A

    $I_{(LINE\ PRIMARY)}$ —————— A

    Compare the measured values with the calculated values in step 29 and step 32. Are the values within 5% of each other?

    ——————

34. Reconnect the load resistors to form a delta connection instead of a wye connection as shown in Figure 24-21.

35. Assuming a primary voltage of 208 volts, calculate the phase voltage of the secondary. The transformers have a ratio of 2:1.

    $E_{(PHASE\ SECONDARY)}$ —————— volts

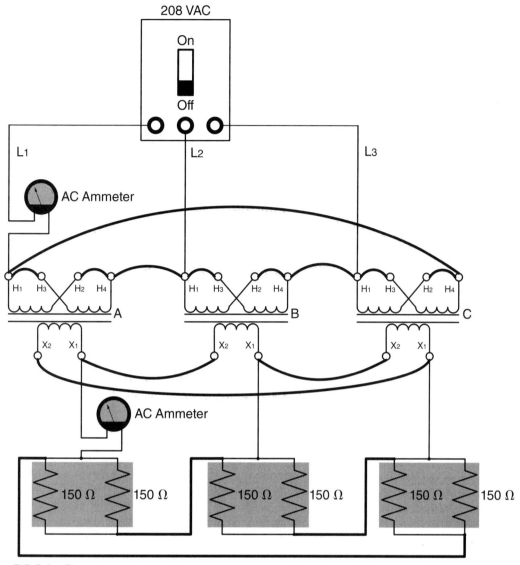

Figure 24-21 Changing the load from a wye connection to a delta connection.

36. Calculate the line voltage of the secondary.

    $E_{\text{(LINE SECONDARY)}}$ _____ volts

37. Calculate the phase voltage of the delta connected load resistors.

    $E_{\text{(PHASE LOAD)}}$ _____ volts

38. Calculate the phase current of the load using Ohm's law.

    $I_{\text{(PHASE LOAD)}}$ _____ A

39. Calculate the line current supplied to the load by the secondary of the transformer.

    $I_{\text{(LINE SECONDARY)}}$ _____ A

40. Calculate the phase current of the secondary winding.

    $I_{\text{(PHASE SECONDARY)}}$ _____ A

41. Using the turns-ratio of the transformer, calculate the phase current of the primary winding.

    $I_{\text{(PHASE PRIMARY)}}$ _____ A

42. Calculate the line current supplied to the primary. Make sure to add the excitation current to the calculation.

$I_{(LINE\ PRIMARY)}$ _____ A

43. Turn on the power and measure the line current of the secondary and the line current of the primary. **Turn off the power**.

$I_{(LINE\ SECONDARY)}$ _____ A

$I_{(LINE\ PRIMARY)}$ _____ A

Compare the measured values with the calculated values in step 39 and step 42. Are the values within 5% of each other?

_____

Wye-Delta Connection (Part 2)

In the next part of the exercise, the three transformers will be connected to form a wye-delta bank. The schematic drawing of the connection is shown in Figure 24-22. Notice that all the H_4 terminals have been joined to form the wye connection. Power will be applied to the H_1 terminals. The secondary windings will remain in a delta connection.

44. Reconnect the transformers as shown in Figure 24-23. For the first part of the exercise, no load will be connected to the secondary.

45. Assuming a line voltage of 208 volts, calculate the phase voltage of the wye connected primary.

$E_{(PHASE\ PRIMARY)}$ = _____ volts

46. The transformers have a turns-ratio of 2:1. Calculate the phase voltage of the secondary.

$E_{(PHASE\ SECONDARY)}$ = _____ volts

47. Turn on the power and measure the phase voltage of the primary by measuring the voltage across terminals H_1 and H_4 of any transformer.

$E_{(PHASE\ PRIMARY)}$ = _____ volts

48. Measure the phase voltage of the secondary by measuring the voltage across the X_1 and X_2 terminals of any transformer.

$E_{(PHASE\ SECONDARY)}$ = _____ volts

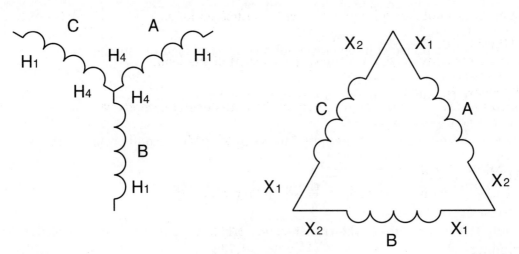

Figure 24-22   A wye-delta transformer connection.

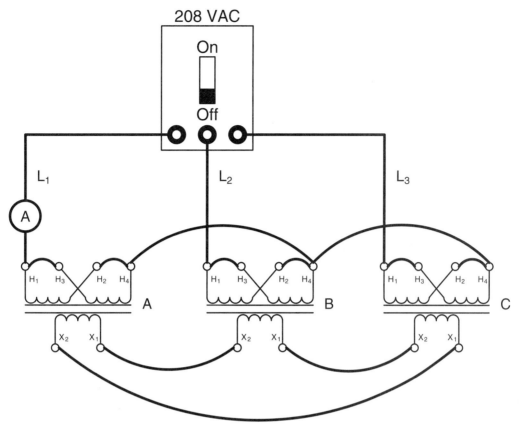

Figure 24-23 Transformer bank with a wye connected primary and delta connected secondary.

49. Measure the excitation current of the primary winding. **Turn off the power**.

$I_{(EXC)}$ = _____ A

50. Compare the measured values of primary and secondary phase voltage with the calculated values in step 46 and step 47. Are the two values within 5% of each other?

_____

51. Reconnect the delta connected resistor bank to the transformer secondary as shown in Figure 24-24.

52. Calculate the phase voltage of the delta connected load.

$E_{(PHASE\ LOAD)}$ = _____ volts

53. Using Ohm's law, calculate the phase current of the load.

$I_{(PHASE\ LOAD)}$ = _____ A

54. Calculate the line current supplied by the transformer secondary to operate the load.

$$I_{LINE} = I_{PHASE} \times 1.732$$

$I_{(LINE\ SECONDARY)}$ = _____ A

55. Calculate the phase current of the secondary winding.

$$I_{PHASE} = \frac{I_{LINE}}{1.732}$$

$I_{(PHASE\ SECONDARY)}$ = _____ A

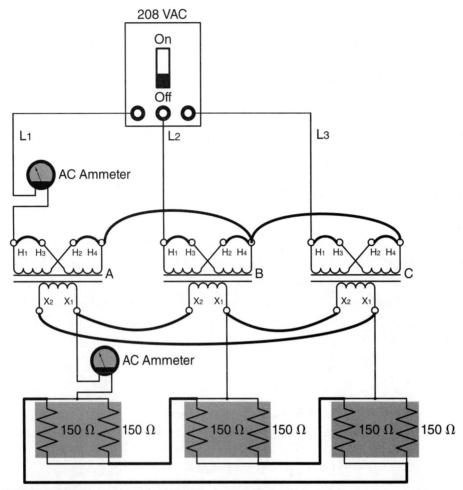

Figure 24-24 Adding load to the transformer connection.

56. Using the turns-ratio of the transformer and the phase current of the secondary winding, calculate the phase current of the primary winding.

$I_{\text{(PHASE PRIMARY)}}$ = _____ A

57. Calculate the line current supplied to the primary windings. Make certain to add the excitation current value.

$I_{\text{(LINE PRIMARY)}}$ = _____ A

58. Turn on the power and measure the secondary line current and the primary line current. **Turn off the power**.

$I_{\text{(LINE SECONDARY)}}$ = _____ A

$I_{\text{(LINE PRIMARY)}}$ = _____ A

Compare the measured values with the calculated values in step 54 and step 57. Are the values within 5% of each other?

_____

Wye-Wye Connection

The next section of this exercise deals with transformers connected in a wye-wye configuration. The schematic diagram for this connection is shown in Figure 24-25.

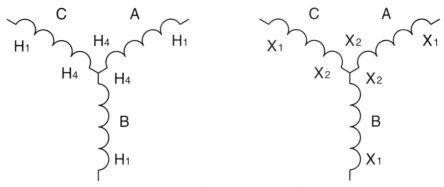

Figure 24-25 A wye-wye three-phase transformer connection.

59. Connect the circuit shown in Figure 24-26. Note that only the secondary windings of the transformers must be reconnected. The primary windings remain connected in a wye configuration and the load resistors remain connected in a delta configuration.

60. Assuming a line voltage of 208 volts, calculate the phase voltage of the primary winding. (In a wye connection, the phase voltage is less than the line voltage by a factor of 1.732.)

$E_{\text{(PHASE PRIMARY)}}$ = _____ volts

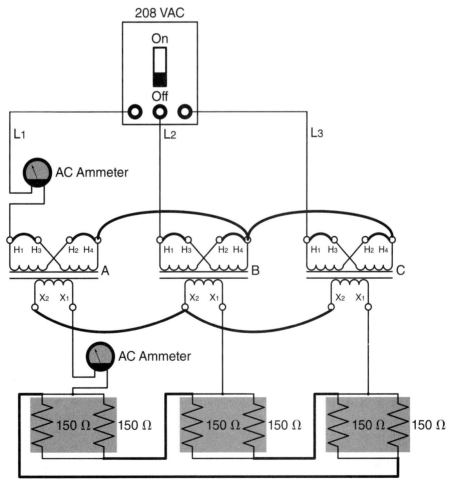

Figure 24-26 Transformers with a wye connected primary, wye connected secondary, and delta connected load.

61. Calculate the phase voltage of the secondary. The transformers have a ratio of 2:1.

   $E_{(PHASE\ SECONDARY)}$ _____ volts

62. Calculate the line voltage of the secondary.

   $E_{(LINE\ SECONDARY)}$ = _____ volts

63. Calculate the phase voltage of the delta connected load resistors.

   $E_{(PHASE\ LOAD)}$ = _____ volts

64. Calculate the phase current of the load using Ohm's law.

   $I_{(PHASE\ LOAD)}$ = _____ A

65. Calculate the line current supplied to the load by the secondary of the transformer.

   $I_{(LINE\ SECONDARY)}$ = _____ A

66. Calculate the phase current of the secondary winding.

   $I_{(PHASE\ SECONDARY)}$ = _____ A

67. Using the turns-ratio of the transformer, calculate the phase current of the primary winding.

   $I_{(PHASE\ PRIMARY)}$ = _____ A

68. Calculate the line current supplied to the primary. Make sure to add the excitation current to the calculation.

   $I_{(LINE\ PRIMARY)}$ _____ A

69. Turn on the power and measure the line current of the secondary and the line current of the primary. **Turn off the power**.

   $I_{(LINE\ SECONDARY)}$ _____ A

   $I_{(LINE\ PRIMARY)}$ _____ A

   Compare the measured values with the calculated values in step 65 and step 68. Are the values within 5% of each other?

   _____

Open Delta Connection

The last connection to be made is the open delta. The open delta connection requires the use of only two transformers to supply three-phase power to a load. The schematic diagram for an open delta connection is shown in Figure 24-27. It should be noted that

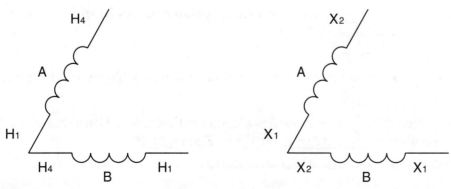

Figure 24-27 Open delta connection.

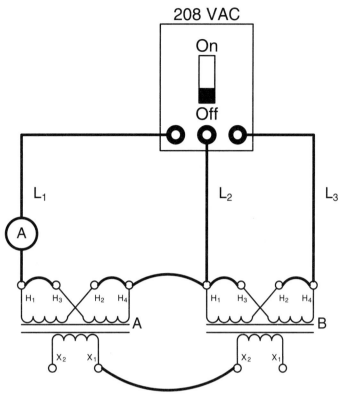

Figure 24-28 Two transformers connected in an open delta.

the open delta connection can supply on 86.6% of the combined kVA capacity of the two transformers.

70. Connect the circuit shown in Figure 24-28. Do not connect a load to the transformer at this time.

71. Assuming an applied voltage of 208 volts, calculate the phase voltage of the primary winding.

   $E_{(PHASE\ PRIMARY)}$ = _____ volts

72. Calculate the phase voltage of the secondary windings.

   $E_{(PHASE\ SECONDARY)}$ = _____ volts

73. Turn on the power and measure the primary phase voltage and secondary phase voltage.

   $E_{(PHASE\ PRIMARY)}$ = _____ volts

   $E_{(PHASE\ SECONDARY)}$ = _____ volts

   Compare the measured values with the calculated values in step 71 and step 72. Are the values within 5% of each other?

   _____

74. Measure the excitation current of the primary winding. **Turn off the power**.

   $I_{(EXC)}$ = _____ A

75. Connect three 150-ohm resistors in a delta configuration. Connect the load resistors to the transformer secondary as shown in Figure 24-29.

76. Calculate the line voltage of the secondary.

   $E_{(LINE\ SECONDARY)}$ =_____ volts

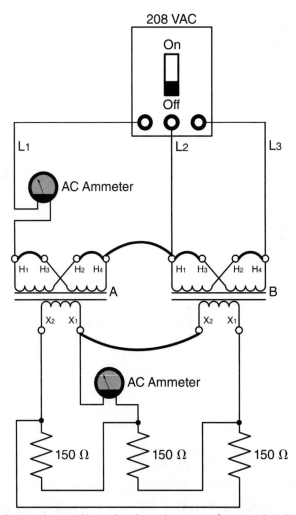

Figure 24-29 Connecting a three-phase load to the transformer bank.

77. Calculate the phase voltage of the delta connected load resistors.

$E_{(PHASE\ LOAD)}$ =_____ volts

78. Calculate the phase current of the load using Ohm's law.

$I_{(PHASE\ LOAD)}$ =_____ A

79. Calculate the line current supplied to the load by the secondary of the transformer.

$I_{(LINE\ SECONDARY)}$ =_____ A

80. Calculate the phase current of the secondary winding.

$I_{(PHASE\ SECONDARY)}$ =_____ A

81. Using the turns-ratio of the transformer, calculate the phase current of the primary winding.

$I_{(PHASE\ PRIMARY)}$ =_____ A

82. Calculate the line current supplied to the primary. Make sure to add the excitation current to the calculation.

$I_{(LINE\ PRIMARY)}$ _____ A

83. Turn on the power and measure the line current of the secondary and the line current of the primary. **Turn off the power**.

$I_{\text{(LINE SECONDARY)}}$ ——————— A

$I_{\text{(LINE PRIMARY)}}$ ——————— A

Compare the measured values with the calculated values in step 79 and step 82. Are the values within 5% of each other?

———————

84. Disconnect the circuit and return the components to their proper place.

Review Questions

1. How many transformers are needed to make an open delta connection?

_____

2. Two transformers rated at 100 kVA each are connected in an open delta connection. What is the total output power that can be supplied by this bank?

_____

3. When computing values of voltage and current for a three-phase transformer, should the line values of voltage and current be used or the phase values?

_____

Refer to Figure 24-30 to answer the following questions:

4. Assume a line voltage of 2,400 volts is connected to the primary of the three-phase transformer and the line voltage of the secondary is 240 volts. What is the turns-ratio of the transformer?

_____

5. Assume the load has an impedance of 3.5 Ω per phase. What is the line current provided by the transformer secondary?

_____

6. How much current is flowing through the secondary winding?

_____

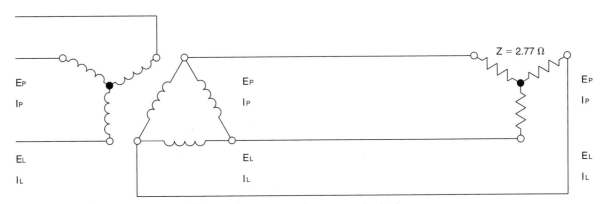

Figure 24-30 Example #1: Three-phase transformer calculation.

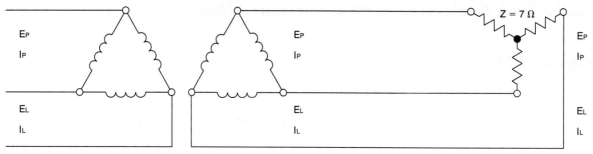

Figure 24-31 Example #2: Three-phase transformer calculation.

7. How much current is flowing through the primary winding?

_____

Refer to Figure 24-31 to answer the following questions:

8. Assume a line voltage of 12,470 volts is connected to the primary of the transformer and the line voltage of the secondary is 480 volts. What is the turns-ratio of the transformer?

_____

9. Assume the load has an impedance of 6 Ω per phase. What is the secondary line current?

_____

10. How much current is flowing in the secondary winding?

_____

11. How much current is flowing in the primary winding?

_____

12. What is the line current of the primary?

_____

Unit 25 Three-Phase Motors

Objectives

After studying this unit, you should be able to:

- Determine if a 9 lead dual voltage three-phase motor is connected wye or delta.
- Test continuity of motor windings.
- Test motor windings for poor insulation.
- Connect a dual voltage 9 lead three-phase motor.

Many three-phase squirrel cage motors are designed in such a manner that they can be connected to 240 or 480 volts. The majority of these motors have 9 "T" leads at the terminal connection box. The manner in which these T leads are connected determines if the motor will operate on 240 or 480 volts.

Wye and Delta Connections

There are two basic ways of connecting the stator windings for three-phase dual voltage motors: wye and delta. The standard numbering for stator windings of both wye and delta connected motors is shown in Figure 25-1. Stator windings are numbered by starting at an outer point with #1 and proceeding around in an ever-decreasing spiral. Notice that the opposite end of the winding that begins with #1 is #4. The opposite end of the winding that begins with #2 is #5. These numbers have been standardized and are used by most manufacturers.

High- and Low-Voltage Connections

When a motor is to be connected for high-voltage operation (generally 480 to 575 volts), the windings are connected in series, as shown in Figure 25-2. In a series circuit the sum of

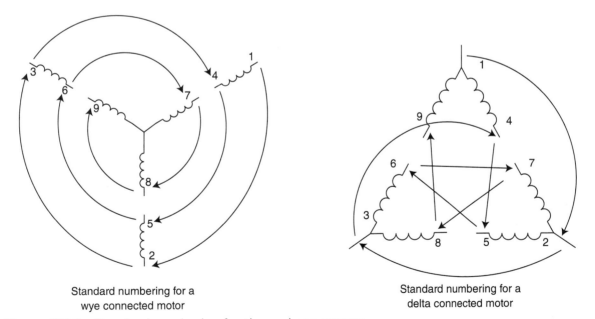

Standard numbering for a
wye connected motor

Standard numbering for a
delta connected motor

Figure 25-1 Standard numbering for three-phase motors.

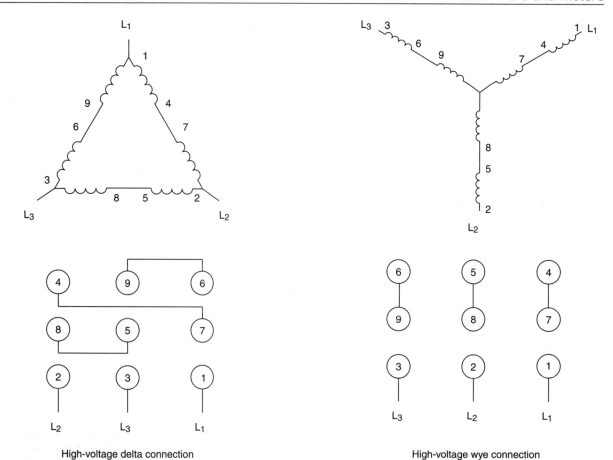

High-voltage delta connection   High-voltage wye connection

Figure 25-2 High-voltage connections.

the voltage drops across the components must equal the applied voltage. If 480 volts were to be connected across the delta connected motor, each of the windings of that phase would have voltage drop of 240 volts.

When a motor is to be connected to operate on low voltage (generally 240 to 208 volts), the stator windings are connected in parallel, as shown in Figure 25-3. Components connected in parallel have the same voltage applied to them. Therefore, if 208 volts are connected to the motor, all stator windings would have an applied voltage of 208 volts.

Determining If the Windings Are Connected Wye or Delta

An ohmmeter can be used to determine if a 9 lead motor is connected in wye or delta. In a wye connected motor, the opposite ends of leads 7, 8, and 9 are connected together, as shown in Figure 25-1. If an ohmmeter indicates continuity between these three leads, the motor is wye connected. In a delta connected motor, the ohmmeter should indicate continuity between 1, 4, and 9; 2, 5, and 7; and 3, 6, and 8.

Testing Stator Windings

To test the condition of the insulation of stator windings, it is necessary to use an ohmmeter that can supply a high voltage and measure high resistance. This device is a megohmmeter or megger, as shown in Figure 25-4. The megger measures the resistance

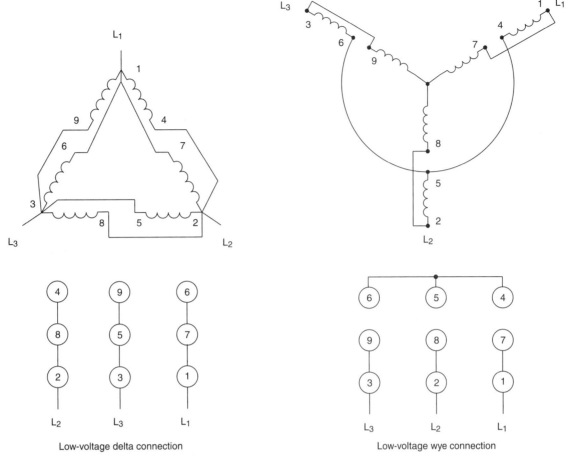

Figure 25-3 Low-voltage connections.

Figure 25-4 Hand-crank megger.

of the insulation. To test the motor, connect one megger lead to the case of the motor and the other to one of the motor line leads (it is assumed that the motor is connected for high or low voltage). The meter should indicate several million ohms of resistance. A low resistance reading indicates a motor that may fail in the near future. Low readings can also be caused by moisture in the stator winding. If a motor is left standing in a high humidity climate for long periods without being operated, moisture can form in the stator. If this is the case, it may be necessary to place the motor in a warm, dry environment until the moisture evaporates.

LABORATORY EXERCISE

Name _____ Date _____

Materials Required

1 9 lead dual voltage three-phase motor

1 208-volt three-phase power supply

1 ohmmeter

1 megohmmeter

1. Using the 9 lead three-phase motor provided, separate all nine leads.

2. Use an ohmmeter to test for continuity between T7, T8, and T9. Does the meter indicate continuity between these leads?

    Yes/no _____

3. If the answer to step 2 is yes, proceed to step 4. If the answer is no, skip step 4.

4. Use the ohmmeter to test for continuity between the following T leads. Write yes beside the leads that indicate continuity and no beside the ones that do not.

T1-T2 _____	T1-T3 _____	T1-T4 _____	T1-T5 _____
T1-T6 _____	T1-T7 _____	T1-T8 _____	T1-T9 _____
T2-T3 _____	T2-T4 _____	T2-T5 _____	T2-T6 _____
T2-T7 _____	T2-T8 _____	T2-T9 _____	T3-T4 _____
T3-T5 _____	T3-T6 _____	T3-T7 _____	T3-T8 _____
T3-T9 _____	T4-T5 _____	T4-T6 _____	T4-T7 _____
T4-T8 _____	T4-T9 _____	T5-T6 _____	T5-T7 _____
T5-T8 _____	T5-T9 _____	T6-T7 _____	T6-T8 _____
T6-T9 _____	T7-T8 _____	T7-T9 _____	T8-T9 _____

5. Compare the continuity readings with the schematic drawing of a 9 lead wye connected motor shown in Figure 25-1. Do the readings confirm the connection diagram?

    Yes/no _____

6. Use a megohmmeter to test the insulation of the stator windings. Set the megger output voltage for a value close to the rated voltage of the motor. Connect one lead of the megger to the case of the motor. Measure the resistance between the motor case and each of the following:

   T1-case _____ Ω T2-case _____ Ω

   T3-case _____ Ω T7-case _____ Ω

7. Connect the motor for low-voltage operation. Refer to the connection diagram on the motor nameplate or the diagram shown in Figure 25-3 for a low-voltage wye connection.

8. Connect the motor to a 208 volt, three-phase power source.

9. Turn on the power and notice the direction of rotation. Viewing the motor from the rear, does the motor turn in a clockwise or counterclockwise direction?

   _____

10. **Turn off the power supply.**

11. Reverse two of the line leads connected to the motor.

12. Turn on the power and notice the direction of rotation. Did the motor reverse its direction of rotation?

    Yes/no _____

13. **Turn off the power supply** and disconnect the motor. Return the components to their proper place.

Review Questions

1. What are the two ways that the stator windings of three-phase motors can be connected?

   _____

2. An ohmmeter reveals that a 9 lead dual voltage motor has continuity between T7, T8, and T9. Is the motor wye or delta connected?

   _____

3. An ohmmeter reveals that there is continuity between T1 and T4. Does this indicate a wye connected motor?

   _____

4. An ohmmeter reveals continuity between T5 and T7. Does this indicate a delta connected motor?

   _____

5. A 9 lead dual voltage motor is to be connected for high-voltage operation. An ohmmeter test indicates that the stator windings are delta connected. To which T lead(s) should T5 be connected?

   _____

6. A 9 lead dual voltage motor is to be connected for low-voltage operation. An ohmmeter test indicates that the motor is delta connected. To which T lead(s) should T6 be connected?

   _____

7. A 9 lead dual voltage motor is to be connected for low-voltage operation. An ohmmeter test reveals that the motor is delta connected. To which T lead(s) should T2 be connected?

   _____

8. A 9 lead dual voltage motor is to be connected for low-voltage operation. An ohmmeter test reveals that the motor is wye connected. To which T lead(s) should T4 be connected?

   _____

9. What piece of test equipment should be used to test the insulation of the stator winding?

   _____

10. How is the direction of rotation of a three-phase motor reversed?

   _____

Motor Controls

Unit 26 Start-Stop Push-Button Control

Objectives

After studying this unit, you should be able to:

- Place wire numbers on a schematic diagram.
- Place corresponding numbers on control components.
- Draw a wiring diagram from a schematic diagram.
- Define the difference between a schematic or ladder diagram and a wiring diagram.
- Connect a start-stop push-button control circuit.

In this experiment a schematic diagram of a start-stop push-button control will be converted to a wiring diagram and then connected in the laboratory. A schematic diagram shows components in their electrical sequence without regard for the physical location of any component (Figure 26-1). A wiring diagram is a pictorial representation of components with connecting wires. The pictorial representation of the components is shown in Figure 26-2.

Helpful Hint

A schematic diagram shows components in their electrical sequence without regard for the physical location of any component (Figure 26-1). A wiring diagram is a pictorial representation of components with connecting wires.

To simplify the task of converting the schematic diagram into a wiring diagram, wire numbers will be added to the schematic diagram. These numbers will then be transferred to the control components, as shown in Figure 26-2. The rules for numbering a schematic diagram are as follows:

1. A set of numbers can be used only once.
2. Each time you go through a component the number set must change.
3. All components that are connected together will have the same number.

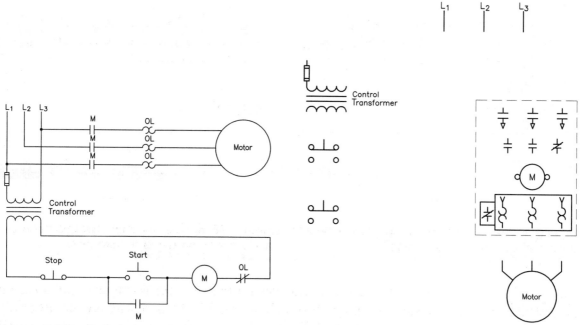

Figure 26-1 Schematic diagram of a basic start-stop push-button control circuit.

Figure 26-2 Components of the basic start-stop control circuit.

To begin the numbering procedure, begin at Line 1 (L_1) with the number 1 and place a number 1 beside each component that is connected to L_1 (Figure 26-3). The number 2 is placed beside each component connected to L_2 (Figure 26-4), and a 3 is placed beside each component connected to L_3 (Figure 26-5). The number 4 will be placed on the other side of the M load contact that already has a number 1 on one side and on one side of the overload heater (Figure 26-6). Number 5 is placed on the other side of the M load contact, which has one side numbered with a 2, and a 5 will be placed beside the second overload heater. The other side of the M load contact, which has been numbered with a 3, will be numbered with a 6, and one side of the third overload heater will be labeled with a 6. Numbers 7, 8, and 9 are placed between the other side of the overload heaters and the motor T leads.

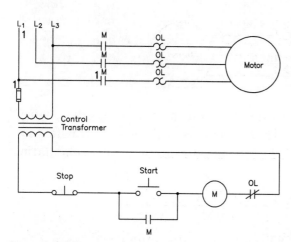

Figure 26-3 The number 1 is placed beside each component connected to L_1.

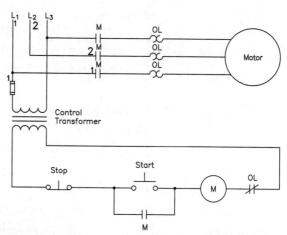

Figure 26-4 A number 2 is placed beside each component connected to L_2.

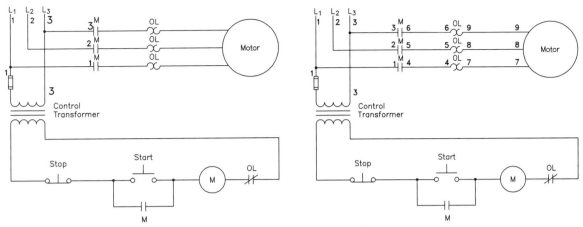

Figure 26-5 A number 3 is placed beside each component connected to L₃.

Figure 26-6 The number changes each time you proceed across a component.

The number 10 will begin at one side of the control transformer secondary and go to one side of the normally closed stop push button. The number 11 is placed on the other side of the stop button and on one side of the normally open start push button and normally open M auxiliary contact. A number 12 is placed on the other side of the start button and M auxiliary contact and on one side of M coil. Number 13 is placed on the other side of the coil to one side of the normally closed overload contact. Number 14 is placed on the other side of the normally closed overload contact and on the other side of the control transformer secondary winding. See Figure 26-7.

Numbering the Components

Now that the components on the schematic have been numbered, the next step is to place the same numbers on the corresponding components of the wiring diagram. The schematic diagram in Figure 26-7 shows that the number 1 has been placed beside L₁, the fuse on the control transformer, and one side of a load contact on M starter (Figure 26-8). The number 2 is placed beside L₂ and the second load contact on M starter (Figure 26-9). The number 3 is placed beside L₃, the third load contact on M starter, and the other side of the primary

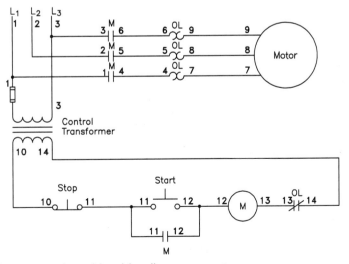

Figure 26-7 Numbers are placed beside all components.

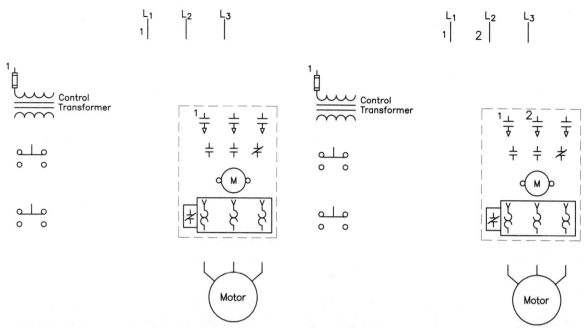

Figure 26-8 A 1 is placed beside L_1, the control transformer fuse, and M load contact.

Figure 26-9 The number 2 is placed beside L_2 and the second load contact on M starter.

winding on the control transformer. Numbers 4, 5, 6, 7, 8, and 9 are placed beside the components that correspond to those on the schematic diagram (Figure 26-10). Note on connection points 4, 5, and 6 from the output of the load contacts to the overload heaters, that these connections are factory made on a motor starter and do not have to be made in the field. These connections are not shown in the diagram for the sake of simplicity. If a separate contactor and overload relay are being used, however, these connections will have to

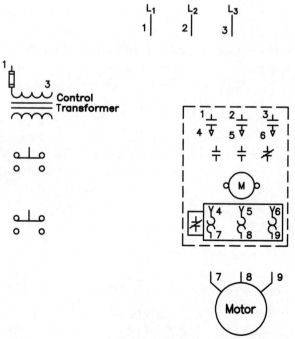

Figure 26-10 Placing numbers 3, 4, 5, 6, 7, 8, and 9 beside the proper components.

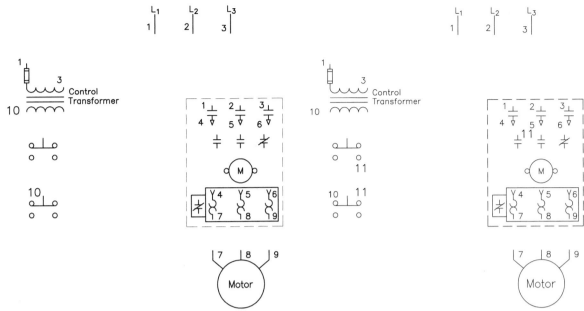

Figure 26-11 Wire number 10 connects from the transformer secondary to the stop button.

Figure 26-12 Number 11 connects to the stop button, start button, and holding contact.

be made. Recall that a contactor is a relay that contains load contacts and may or may not contain auxiliary contacts. A motor starter is a contactor and overload relay combined.

The number 10 starts at the secondary winding of the control transformer and goes to one side of the normally closed stop push button. When making this connection, care must be taken to make certain that connection is made to the normally closed side of the push button. Since this is a double-acting push button, it contains both normally closed and normally open contacts (Figure 26-11).

The number 11 starts at the other side of the normally closed stop button and goes to one side of the normally open start push button and to one side of a normally open M auxiliary contact (Figure 26-12). The starter in this example shows three auxiliary contacts: two normally open and one normally closed. It makes no difference which normally open contact is used.

This same procedure is followed until all circuit components have been numbered with the number that corresponds to the same component on the schematic diagram (Figure 26-13).

Connecting the Wires

Now that numbers have been placed beside the components, wiring the circuit becomes a matter of connecting numbers. Connect all components labeled with a number 1 together (Figure 26-14). All components numbered with a 2 are connected together (Figure 26-15). All components numbered with a 3 are connected together (Figure 26-16). This procedure is followed until all the numbered components are connected together, with the exception of 4, 5, and 6, which are assumed to be factory connected (Figure 26-17).

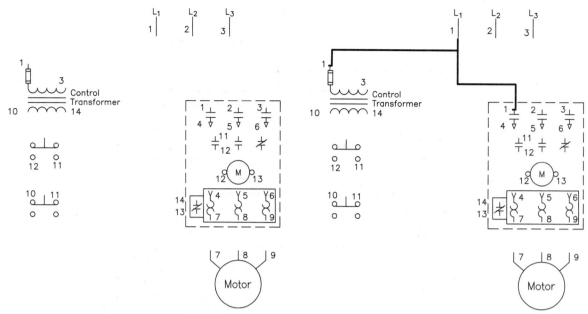

Figure 26-13 All components have been numbered.

Figure 26-14 Connecting all components numbered with a 1 together.

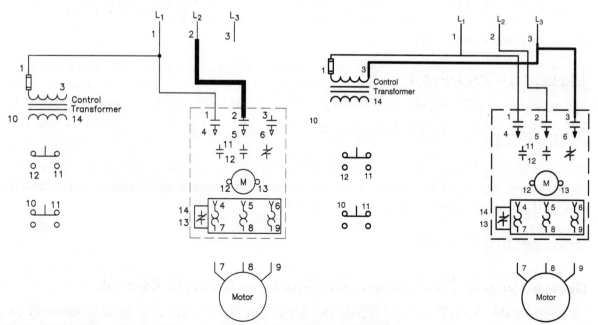

Figure 26-15 Connecting all components numbered with a 2 together.

Figure 26-16 Connecting all components numbered with a 3 together.

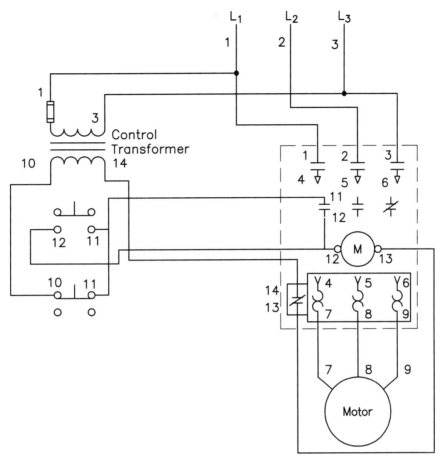

Figure 26-17 Completing the wiring diagram.

LABORATORY EXERCISE

Name _____ Date _____

Materials Required

Three-phase power supply

Three-phase squirrel cage induction motor or simulated load

2 double-acting push buttons (N.O./N.C. on same button)

Three-phase motor starter or contactor with overload relay containing three load contacts and at least one normally open auxiliary contact

Control transformer

Connecting a Start-Stop Push-Button Control Circuit

To connect the control circuit, follow the same procedure that was used to develop the wiring diagram. Use the schematic diagram shown in Figure 26-7. It is sometimes helpful to use a highlighter to mark the diagram as connections are made.

1. Connect all components that are labeled with a number 1. Make certain to connect to a load contact on the starter or contactor.

2. Connect all components labeled with a number 2. Again make sure to connect to a load contact on the starter or contactor.

3. Connect all components labeled with a 3.

4. Wire connections 4, 5, and 6 may or may not have to be made depending on whether you are using a starter or a contactor and separate overload relay.

5. Wires 7, 8, and 9 connect from the output of the heaters on the overload relay(s) to the motor T leads. Your circuit may contain a single three-phase overload relay or three separate overload relays if you are using a contactor and separate overload relay(s).

6. Wire number 10 connects from the secondary winding of the control transformer to one side of the normally closed push button used for the stop button. If using a double-acting push button, make certain to connect to the closed side.

7. Wire number 11 connects from the other side of the normally closed push button to the normally open push button used for the start button. If a double-acting push button is being used, make certain to connect to the open side. Wire number 11 also connects to a normally open auxiliary contact on M starter. Auxiliary contacts are smaller than the load contacts and are used as part of the control circuit. Make certain to connect to one side of an open contact.

8. Wire number 12 connects from the other side of the normally open start button to the other side of the normally open auxiliary contact and to one side of the coil on M starter.

9. Wire number 13 connects from the other side of the coil on M starter to one side of the normally open contact located on the overload relay. If a three-phase motor starter is being used, or if a separate three-phase overload relay is being used, there will be only one overload contact. Note the number of contacts on the overload relay. Some overload relays contain both normally open and normally closed contacts, and some do not. Make certain that connection is made to the normally closed contact if the relay contains more than one contact. If three separate single-phase overload relays are being used, each overload relay contains an overload contact. These three contacts will have to be connected in series so that if one opens, the circuit will be broken.

10. Wire number 14 connects from the other side of the normally closed overload contact to the other side of the secondary winding on the control transformer.

11. Check with your instructor before turning on the power.

12. Test the circuit for proper operation.

13. If the circuit works properly, **turn off the power** and disconnect the circuit. Return the wires and components to their proper place.

Review Questions

1. Refer to the circuit shown in Figure 26-7. If wire number 11 were disconnected at the normally open auxiliary M contact, how would the circuit operate?

_____

_____

_____

2. Assume that when the start button is pressed, M starter does not energize. List seven possible causes for this problem:

   a. _____

   b. _____

   c. _____

   d. _____

   e. _____

   f. _____

   g. _____

3. Explain the difference between a motor starter and a contactor.

   _____

   _____

   _____

4. Refer to the schematic in Figure 26-7. Assume that when the start button is pressed, the control transformer fuse blows. What is the most likely cause of this trouble?

   _____

   _____

   _____

5. Explain the difference between load and auxiliary contacts.

   _____

   _____

   _____

Objectives

After studying this unit, you should be able to:

- Place wire numbers on a schematic diagram.
- Place corresponding numbers on control components.
- Draw a wiring diagram from a schematic diagram.
- Connect a control circuit using two stop and two start push buttons.

There may be times when it is desirable to have more than one start-stop push-button station to control a motor. In this experiment the basic start-stop push-button control circuit discussed previously will be modified to include a second stop and start push button.

When a component is used to perform the function of stop in a control circuit, it will generally be a normally closed component and be connected in series with the motor starter coil. In this example a second stop push button is to be added to an existing start-stop control circuit. The second push button will be added to the control circuit by connecting it in series with the existing stop push button (Figure 27-1).

When a component is used to perform the function of start, it is generally normally open and connected in parallel with the existing start button (Figure 27-2). If either start button is pressed, a circuit will be completed to M coil. When M coil energizes, all M contacts change position. The three load contacts connected between the three-phase power line and the motor close to connect the motor to the line. The normally open auxiliary contact connected in parallel with the two start buttons close to maintain the circuit to M coil when the start button is released.

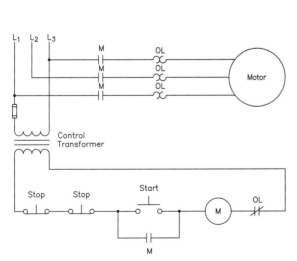

Figure 27-1 Adding a stop button to the circuit.

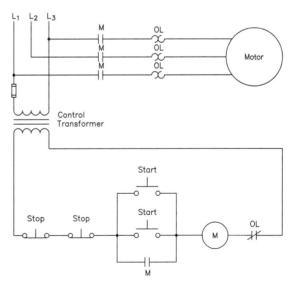

Figure 27-2 A second start button is added to the circuit.

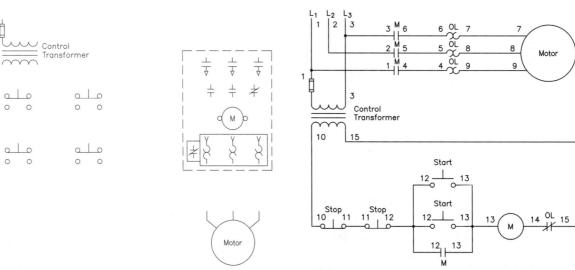

Figure 27-3 Components needed to produce a wiring diagram.

Figure 27-4 Numbering the schematic diagram.

Developing the Wiring Diagram

Now that the circuit logic has been developed in the form of a schematic diagram, a wiring diagram will be drawn from the schematic. The components needed to connect this circuit are shown in Figure 27-3. Following the same procedure discussed in Experiment 1, wire numbers will be placed on the schematic diagram (Figure 27-4). After wire numbers are placed on the schematic, corresponding numbers will be placed on the control components (Figure 27-5).

LABORATORY EXERCISE

Name _____ Date _____

Materials Required

Three-phase power supply

Three-phase squirrel cage induction motor or simulated load

4 double-acting push buttons (N.O./N.C. on same button)

Three-phase motor starter or contactor with overload relay containing three load contacts and at least one normally open auxiliary contact

Control transformer

Connecting the Circuit

1. Using the schematic in Figure 27-4 or the diagram with numbered components in Figure 27-5, connect the circuit in the laboratory by connecting all like numbers together.

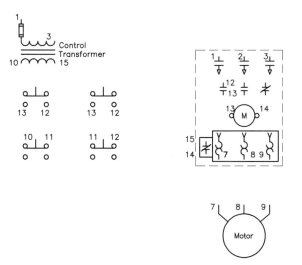

Figure 27-5 Numbering the components.

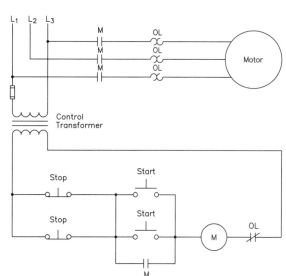

Figure 27-6 The stop buttons are connected in parallel.

2. After the circuit has been connected, check with your instructor before turning on the power.

3. Turn on the power and test the circuit for proper operation.

4. **Turn off the power** and disconnect the circuit. Return all components to their proper place.

Review Questions

1. When a component is to be used for the function of start, is the component generally normally open or normally closed?

2. When a component is to be used for the function of stop, is the component generally normally open or normally closed?

3. The two stop push buttons in Figure 27-2 are connected in series with each other. What would be the action of the circuit if they were to be connected in parallel, as shown in Figure 27-6?

4. What would be the action of the circuit if both start buttons were to be connected in series, as shown in Figure 27-7?

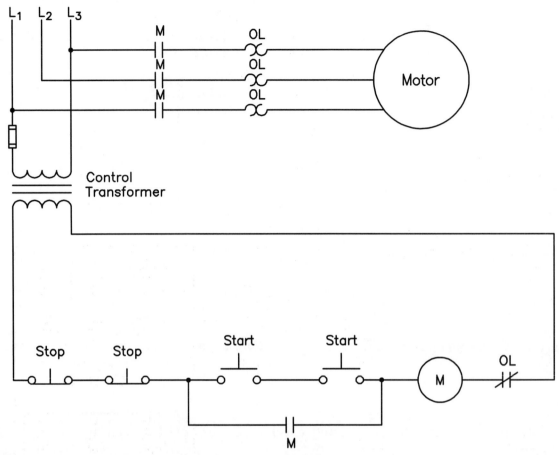

Figure 27-7 The start buttons are connected in series.

5. Following the procedure discussed previously, place wire numbers on the schematic in Figure 27-7. Place corresponding wire numbers on the components shown in Figure 27-8.

_____

_____

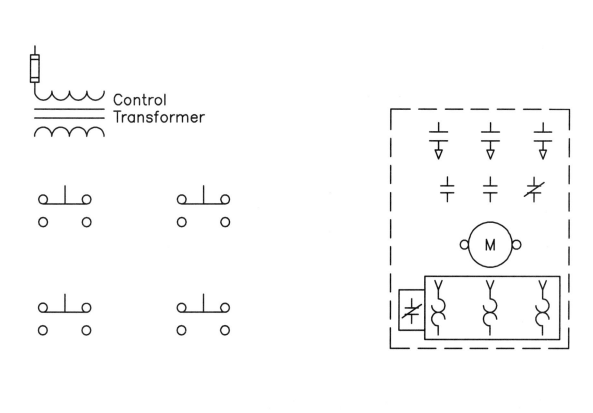

Figure 27-8 Add wire numbers to these components.

Unit 28 Forward-Reverse Control

Objectives

After studying this unit, you should be able to:

- Discuss cautions that must be observed in reversing circuits.
- Explain how to reverse a three-phase motor.
- Discuss interlocking methods.
- Connect a forward-reverse motor control circuit.

The direction of rotation of any three-phase motor can be reversed by changing any two motor T leads. Since the motor is connected to the power line regardless of which direction it operates, a separate contactor is needed for each direction. Since only one motor is in operation, however, only one overload relay is needed to protect the motor. True reversing controllers contain two separate contactors and one overload relay built into one unit.

Interlocking

Interlocking prevents some action from taking place until some other action has been performed. In the case of reversing starters, interlocking is used to prevent both contactors from being energized at the same time. This would result in two of the three-phase lines being shorted together. Interlocking forces one contactor to be de-energized before the other one can be energized.

Most reversing controllers contain mechanical interlocks as well as electrical interlocks. Mechanical interlocking is accomplished by using the contactors to operate a mechanical lever that prevents the other contactor from closing, while the other is energized.

Electrical interlocking is accomplished by connecting the normally closed auxiliary contacts on one contactor in series with the coil of the other contactor (Figure 28-1). Assume that the forward push button is pressed and F coil energizes. This causes all F contacts to change position. The three F load contacts close and connect the motor to the line. The normally open F auxiliary contact closes to maintain the circuit when the forward push button is released, and the normally closed F auxiliary contact connected in series with R coil opens (Figure 28-2). (*Note*: Figure 28-2 illustrates the circuit as it is when the forward starter has been energized. Schematics of this type are used throughout this laboratory manual to help students understand how relay logic operates. This can lead to confusion, however, because contacts that are connected normally open will be shown closed and normally closed contacts will be shown open. To help avoid confusion, normally open contacts that are closed during the stage the circuit is in at that moment will use double lines to indicate the contact is now closed. Contacts that are normally closed, but open at that stage of circuit operation, will show a line at the edges of the contact, but the contact will be open in the middle, as shown in Figure 28-3.)

If the opposite direction of rotation is desired, the stop button must be pressed first. If the reverse push button were to be pressed first, the now open F auxiliary contact connected in series with R coil would prevent a complete circuit from being established. Once the stop button has been pressed, however, F coil de-energizes and all F contacts return to their normal position. The reverse push button can now be pressed to energize R coil (Figure 28-4).

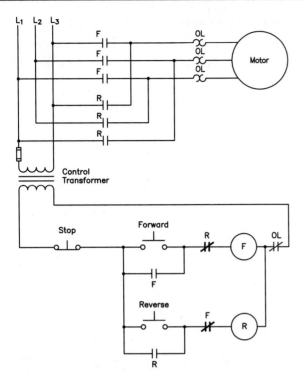

Figure 28-1   Forward-reverse control with interlock.

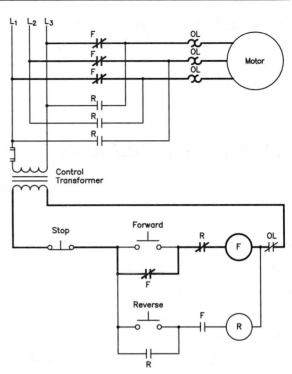

Figure 28-2   Motor operating in the forward direction.

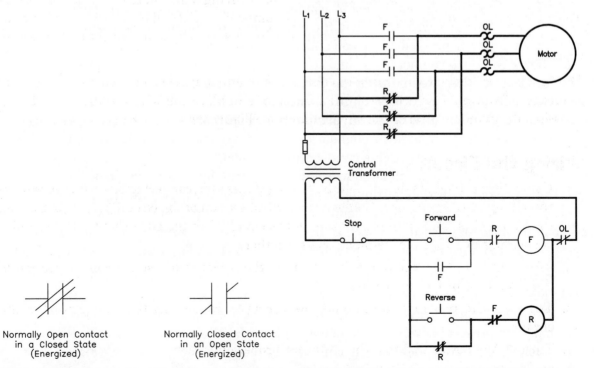

Figure 28-3   Contacts to illustrate circuit logic.

Figure 28-4   Motor operating in the reverse direction.

When R coil energizes, all R contacts change position. The three R load contacts close and connect the motor to the line. Notice, however, that two of the motor T leads are connected to different lines. The normally closed R auxiliary contact opens to prevent the possibility of F coil being energized until R coil is de-energized.

LABORATORY EXERCISE

Name _____ Date _____

Materials Required

Three-phase power supply

Control transformer

One of the following:

1. A three-phase reversing starter

2. Two three-phase contactors with at least one normally open and one normally closed auxiliary contact on each contactor; one three-phase overload relay or three single-phase overload relays

Three-phase squirrel cage motor or simulated motor load

3 double-acting push buttons (N.O./N.C. on each button)

Developing a Wiring Diagram

The same basic procedure will be used to develop a wiring diagram from the schematic as was followed in the previous experiments. The components needed to construct this circuit are shown in Figure 28-5. In this example it will be assumed that two contactors and a separate three-phase overload relay will be used.

The first step is to place wire numbers on the schematic diagram. A suggested numbering sequence is shown in Figure 28-6. The next step is to place the wire numbers beside the corresponding components of the wiring diagram (Figure 28-7).

Wiring the Circuit

1. Using the components listed at the beginning of this unit, connect a forward-reverse control circuit with interlocks. Connect the control section of the circuit before connecting the load section. This will permit the control circuit to be tested without the possibility of shorting two of the three-phase lines together.

2. After checking with the instructor, turn on the power and test the control section of the circuit for proper operation.

3. **Turn off the power** and complete the wiring by connecting the load portion of the circuit.

4. Turn on the power and test the motor for proper operation.

5. **Turn off the power** and disconnect the circuit. Return the components to their proper place.

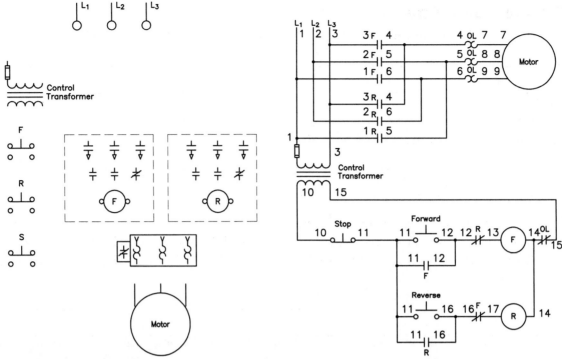

Figure 28-5 Components needed to construct a reversing circuit.

Figure 28-6 Placing wire numbers on the schematic.

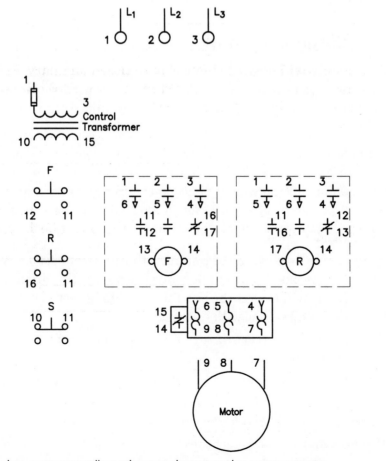

Figure 28-7 Placing corresponding wire numbers on the components.

Review Questions

1. How can the direction of rotation of a three-phase motor be changed?

   _____

   _____

   _____

2. What is interlocking?

   _____

   _____

   _____

3. Referring to the schematic shown in Figure 28-1, how would the circuit operate if the normally closed R contact connected in series with F coil were to be connected normally open?

   _____

   _____

   _____

4. What would be the danger, if any, if the circuit were to be wired as stated in review question 3?

   _____

   _____

   _____

5. How would the circuit operate if the normally closed auxiliary contacts were to be connected so that F contact was connected in series with F coil and R contact was connected in series with R coil (Figure 28-8)?

   _____

   _____

   _____

6. Assume that the circuit shown in Figure 28-1 were to be connected as shown in Figure 28-9. In what way would the operation of the circuit be different, if at all?

   _____

   _____

   _____

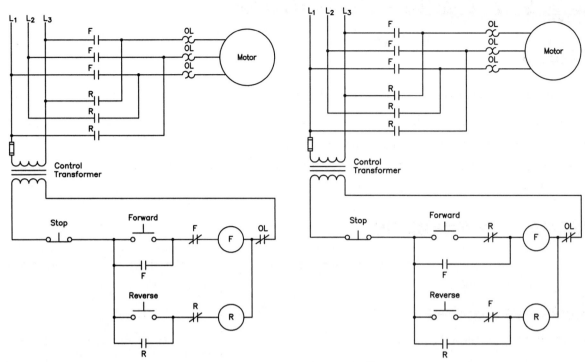

Figure 28-8 F and R normally open auxiliary contacts are connected incorrectly.

Figure 28-9 The position of the holding contacts has been changed.

Unit 29 Sequence Control

Objectives

After studying this unit, you should be able to:

- Define sequence control.
- Discuss methods of obtaining sequence control.
- Connect a control circuit for three motors that must be started in a predetermined sequence.

LABORATORY EXERCISE

Name _____ Date _____

Materials Required

Three-phase power supply

Control transformer

3 motor starters containing at least three load contacts and two normally open auxiliary contacts

3 squirrel cage motors or three simulated motor loads

4 double-acting push buttons (N.O./N.C. on each button)

Sequence control forces a circuit to operate in a predetermined manner. In this experiment three motors are to be started in sequence from 1 to 3. The requirements for the circuit are as follows:

1. The motors must start in sequence from #1 to #3. For example, motor #1 must be started before motor #2 can be started, and motor #2 must start before motor #3 can be started. Motor #2 cannot start before motor #1, and motor #3 cannot start before motor #2.
2. Each motor is started by a separate push button.
3. One stop button will stop all motors.
4. An overload on any motor will stop all three motors.

As a general rule, there is more than one way to design a circuit that will meet the specified requirements, just as there is generally more than one road that can be taken to reach a destination. One design that will meet the requirements is shown in Figure 29-1. Since the logic of the circuit is of primary interest, the load contacts and motors are not shown. In this circuit, push button #1 must be pressed before power can be provided to push button #2. When motor starter #1 energizes, the normally open auxiliary contact 1M closes, providing power to coil 1M and to push button #2. Motor starter #2 can now be started by pressing push button #2. Once motor starter #2 energizes, auxiliary contact 2M closes and provides power to coil 2M and push button #3. If the stop button should be pressed or any overload contact open, power will be interrupted to all starters.

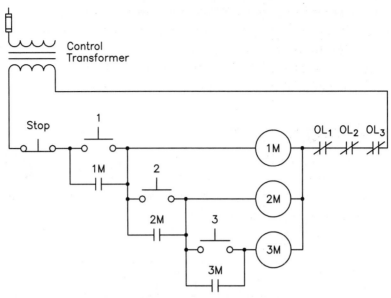

Figure 29-1 First example of starting the motors in sequence.

A Second Circuit for Sequence Control

A second method of providing sequence control is shown in Figure 29-2. In this circuit, normally open auxiliary contacts located on motor starters 1M and 2M are used to ensure

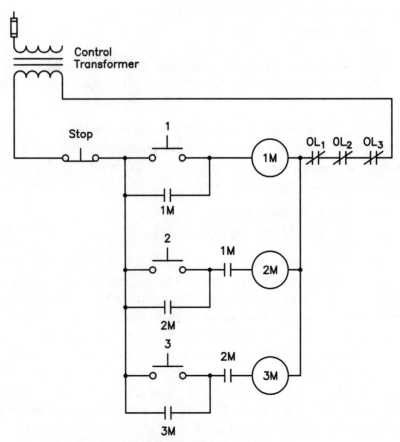

Figure 29-2 A second circuit for sequence control.

that the three motors start in the proper sequence. A normally open 1M auxiliary contact connected in series with starter coil 2M prevents motor #2 from starting before motor #1, and a normally open 2M auxiliary contact connected in series with coil 3M prevents motor #3 from starting before motor #2. If the stop button should be pressed or if any overload contact should open, power will be interrupted to all starters.

Developing a Wiring Diagram

The schematic shown in Figure 29-2 is shown with the motors in Figure 29-3. A drawing of the components needed to connect this circuit is shown in Figure 29-4. The schematic diagram, Figure 29-3, is shown with wire numbers in Figure 29-5. The components with corresponding wire numbers are shown in Figure 29-6.

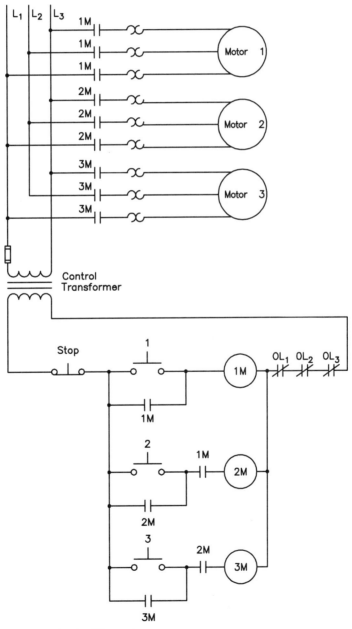

Figure 29-3 Sequence control with motors.

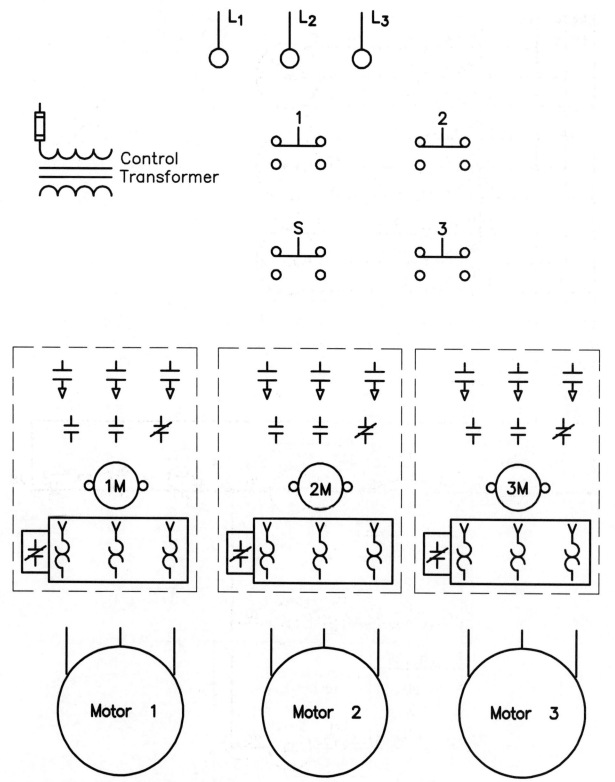

Figure 29-4 Components needed to connect the circuit.

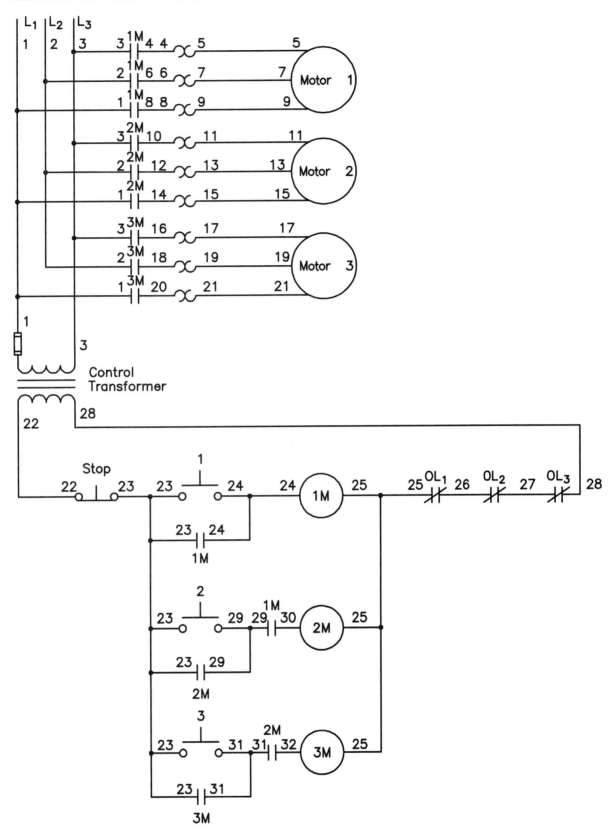

Figure 29-5 Numbering the schematic.

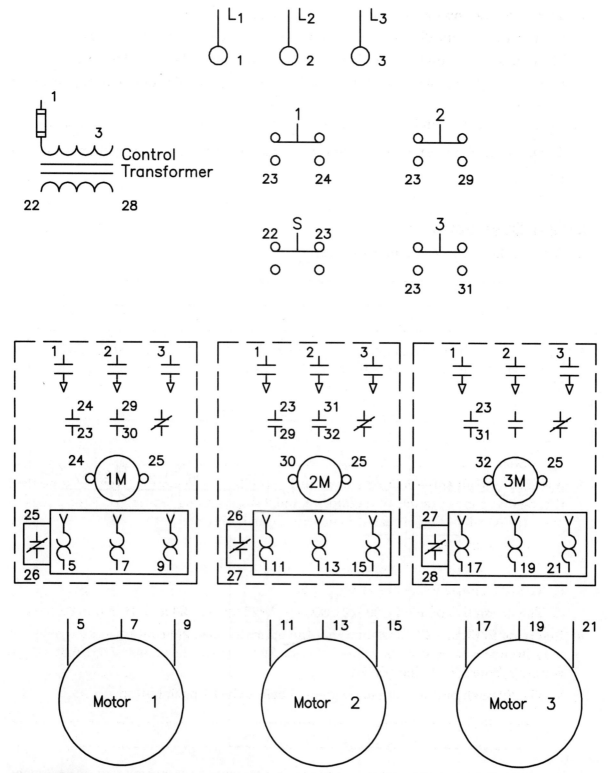

Figure 29-6 Numbering the components.

Connecting the Circuit

1. Using the materials listed at the beginning of this experiment, connect the circuit shown in Figure 29-5. Follow the number sequence shown.

2. After checking with the instructor, turn on the power and test the circuit for proper operation.

3. **Turn off the power** and disconnect the circuit.

4. Using the schematic diagram shown in Figure 29-1, add wire numbers to the schematic.

5. Place these wire numbers beside the proper components shown in Figure 29-4.

6. Connect the circuit shown in Figure 29-1 by following the wire numbers placed on the schematic.

7. Turn on the power and test the circuit for proper operation.

8. **Turn off the power** and disconnect the circuit. Return the components to their proper places.

Review Questions

1. What is the purpose of sequence control?

   _____

   _____

   _____

2. Refer to the schematic diagram in Figure 29-5. Assume that the 1M contact located between wire numbers 29 and 30 had been connected normally closed instead of normally open. How would this circuit operate?

   _____

   _____

   _____

3. Assume that all three motors shown in Figure 29-5 are running. Now assume that the stop button is pressed and motors 1 and 2 stop running, but motor 3 continues to operate. Which of the following could cause this problem?

   a. Stop button is shorted.

   b. 2M contact between wire numbers 31 and 32 is hung closed.

   c. The 3M load contacts are welded shut.

   d. The normally open 3M contact between wire numbers 23 and 31 is hung closed.

4. Referring to Figure 29-5, assume that the normally open 2M contact located between wire numbers 23 and 29 is welded closed. Also assume that none of the motors are running. What would happen if:

   a. The #2 push button were to be pressed before the #1 push button?

   _____

   _____

   _____

   b. The #1 push button were to be pressed first?

   _____

   _____

   _____

5. In the control circuit shown in Figure 29-2, if an overload occurs on any motor, all

three motors will stop running. In the space provided in Figure 29-7, redesign the circuit so that the motors must still start in sequence from 1 to 3, but an overload on any motor will stop only that motor. If an overload should occur on motor 1, for example, motors 2 and 3 would continue to operate.

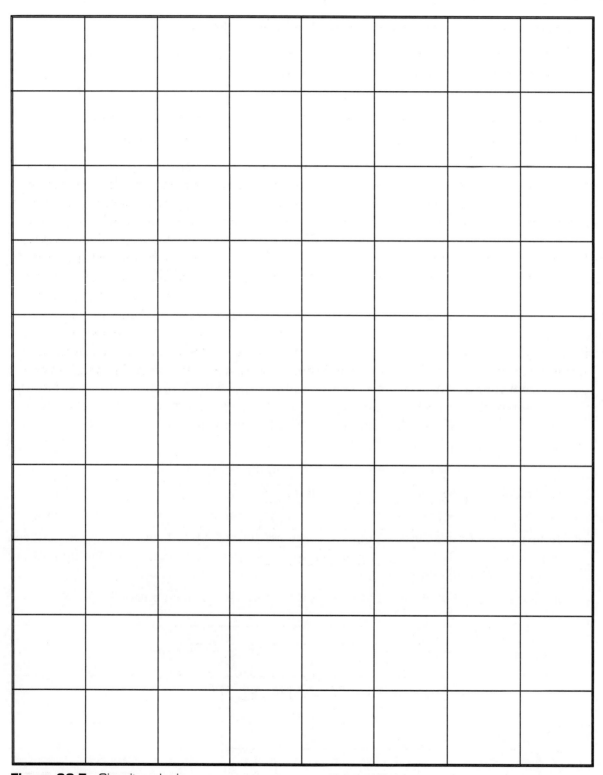

Figure 29-7   Circuit redesign.

Unit 30 Jogging Controls

Objectives

After studying this unit, you should be able to:

- Describe the difference between inching and jogging circuits.
- Discuss different jogging control circuits.
- Draw a schematic diagram of a jogging circuit.
- Discuss the connection of an 8-pin control relay.
- Connect a jogging circuit in the laboratory using double-acting push buttons.
- Connect a jogging circuit in the laboratory using an 8-pin control relay.

Jogging or inching control is used to help position objects by permitting the motor to be momentarily connected to power. Jogging and inching are very similar and these terms are often used synonymously. Both involve starting a motor with short jabs of power. The difference between jogging and inching is that when a motor is jogged, it is started with short jabs of power at full voltage. When a motor is inched, it is started with short jabs at reduced power. Inching circuits require the use of two contactors, one to run the motor at full power and the other to start the motor at reduced power (Figure 30-1). The run contactor is generally a motor starter that contains an overload relay while the inching contactor does not. In the circuit shown in Figure 30-1, if the inch push button is pressed, a circuit is completed to S contactor coil causing all S contacts to close. This connects the motor to the line through a set of series resistors used to reduce power to the motor. Note that there is no S holding contact in parallel with the inch push button. When the push button is released, S contactor de-energizes and

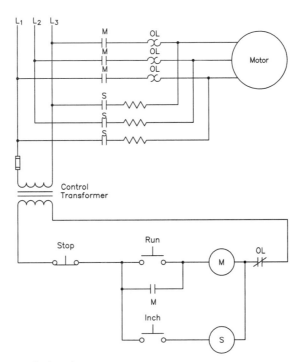

Figure 30-1 Inching control circuit.

all S contacts reopen and disconnect the motor from the power line. If the run push button is pressed, M contactor energizes and connects the motor directly to the power line. Note the normally open M auxiliary contact connected in parallel with the run push button to maintain the circuit when the button is released.

Other Jogging Circuits

Like most control circuits, jog circuits can be connected in different ways. One method is shown in Figure 30-2. In this circuit a simple single-pole switch is inserted in series with the normally open M auxiliary contact connected in parallel with the start button. When the switch is open, it is in the *jog* position and prevents M holding contact from providing a complete path to M coil. When the start button is pushed, M coil will energize and connect the motor to the power line. When the start button is released, M coil will de-energize and disconnect the motor from the line. If the switch is closed, it is in the *run* position and permits the holding contact to complete a circuit around the start button.

Another method of constructing a run-jog control is shown in Figure 30-3. This circuit employs a double-acting push button as the jog button. The normally closed section of the jog push button is connected in series with the normally open M auxiliary holding contact. If the jog button is pressed, the normally closed section of the button opens to disconnect the

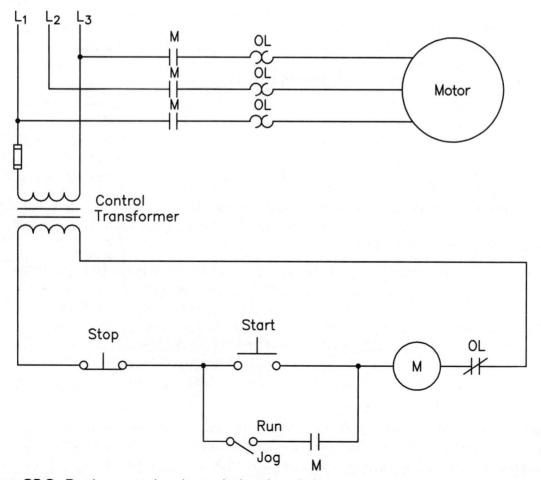

Figure 30-2 Run-jog controls using a single-pole switch.

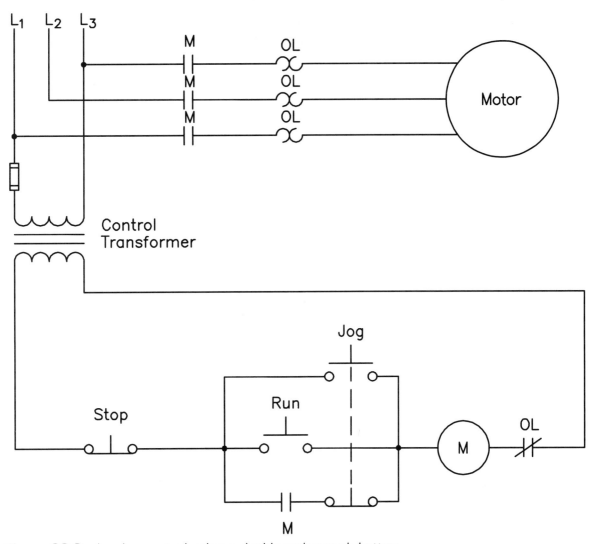

Figure 30-3 Jogging control using a double-acting push button.

holding contacts before the normally open section of the button closes. Although M auxiliary contact closes when M coil energizes, the now open jog button prevents it from completing a circuit to the coil. When the jog button is released, the normally open section reopens and breaks contact before the normally closed section can reclose.

Although a double-acting push button can be used to construct a run-jog circuit, it is not generally done because there is a possibility that the normally closed section of the jog button could reclose before the normally open section reopens. This could cause the holding contacts to lock the circuit in the run position causing an accident. To prevent this possibility, a control relay is often employed (Figure 30-4). In the circuit shown in Figure 30-4, if the jog push button is pressed, M contactor energizes and connects the motor to the line. When the jog button is released, M coil de-energizes and disconnects the motor from the line.

When the run push button is pressed, CR relay energizes and closes both CR contacts. The CR contacts connected in parallel with the run button close to maintain the circuit to CR coil, and the CR contacts connected in parallel with the jog button close and complete a circuit to M coil.

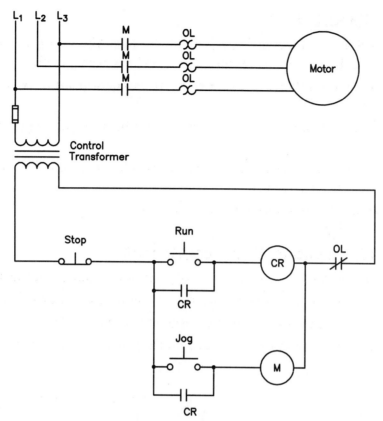

Figure 30-4 Run-jog control using a control relay.

LABORATORY EXERCISE

Name _____ Date _____

Materials Required

Three-phase power supply

Three-phase motor starter

1 three-phase motor or equivalent motor load

3 double-acting push buttons (N.O./N.C. on each button)

1 8-pin tube socket

1 8-pin control relay

1 single-pole switch

Control transformer

Connecting Jogging Circuits

In this experiment four different jog circuits will be connected in the laboratory. Three of these circuits are illustrated in Figures 30-2, 30-3, and 30-4. The fourth circuit will be designed by the student in accord with given circuit parameters.

Connecting Circuit 1

1. Refer to the schematic diagram in Figure 30-2. Place wire numbers beside the components following the procedure discussed in previous experiments.

2. Using the components shown in Figure 30-5, place corresponding wire numbers beside the components.

3. Connect the circuit by following the wire numbers in the schematic diagram in Figure 30-2.

4. Turn on the power and test the circuit for proper operation. The motor should jog when the switch is open and run when the switch is closed.

5. **Turn off the power** and disconnect the circuit.

Connecting the Second Run-Jog Circuit

1. Using the schematic shown in Figure 30-3, place wire numbers beside the components.

2. Place corresponding wire numbers beside the components shown in Figure 30-6.

3. Connect the circuit using the schematic diagram in Figure 30-3.

4. After checking with the instructor, turn on the power and test the circuit for proper operation.

5. **Turn off the power** and disconnect the circuit.

Connecting the Third Run-Jog Circuit

The third run-jog circuit involves the use of a control relay. In this circuit, an 8-pin control relay will be used. Eight-pin relays are designed to fit into an 8-pin tube socket. Therefore, the socket is the device to which connection is made, not the relay itself. Eight-pin relays commonly have coils with different voltage ratings such as 12 VDC, 24 VDC, 24 VAC, and 120 VAC, so make certain that the coil of the relay you use is rated for the circuit control voltage. Most 8-pin relays contain two single-pole, double-throw contacts. A diagram showing the standard pin connection for 8-pin relays with two sets of contacts is shown in Figure 30-7.

Connecting the Tube Socket

When making connections to tube sockets, it is generally helpful to place the proper relay pin numbers beside the component on the schematic diagram. To distinguish pin numbers from wire numbers, pin numbers will be circled. The schematic in Figure 30-4 is shown in Figure 30-8 with the addition of relay pin numbers. The connection diagram in Figure 30-7 shows that the relay coil is connected to pins 2 and 7. Note that CR relay coil in Figure 30-8 has a circled 2 and 7 placed beside it.

The connection diagram also indicates that the relay contains two sets of normally open contacts. One set is connected to pins 1 and 3, and the other set is connected to pins 8 and 6. Note in the schematic of Figure 30-8 that one of the normally open CR contacts has the circled numbers 1 and 3 beside it and the other normally open CR contact has the circled numbers 8 and 6 beside it.

1. Using the drawing in Figure 30-8, place wire numbers on the schematic.

2. Using the wire numbers placed on the schematic diagram in Figure 30-8, place corresponding wire numbers beside the proper components shown in Figure 30-9.

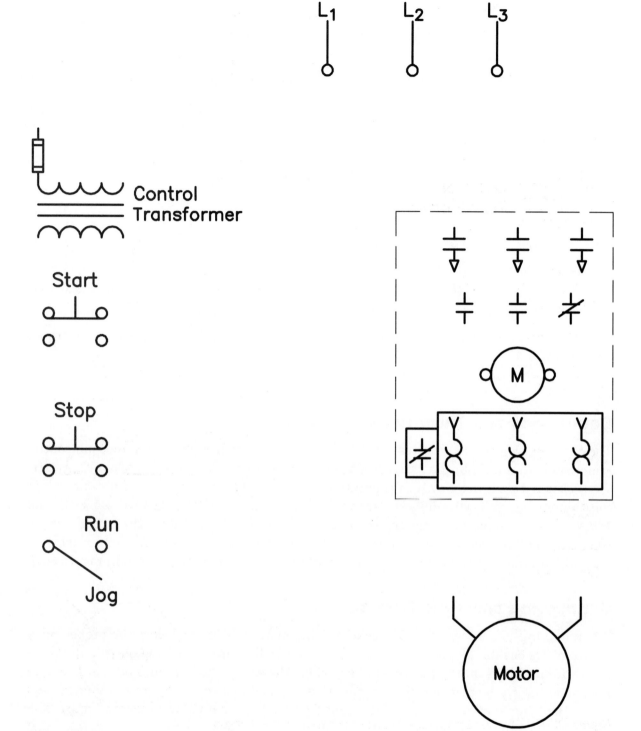

Figure 30-5 Components needed to connect circuit 1.

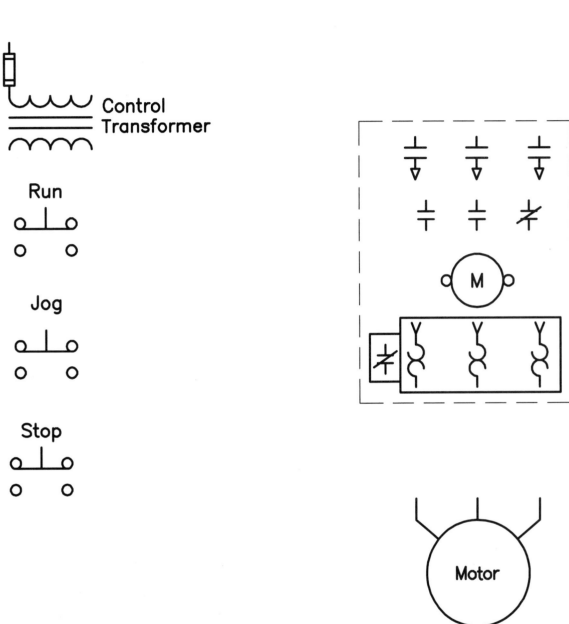

Figure 30-6 Components needed to connect the second run-jog circuit.

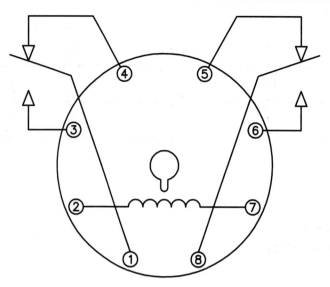

Typical in Connection for an 8-pin Pelay

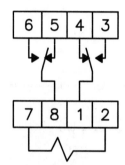

Typical 8-pin Socket Connection

Figure 30-7   Standard diagram for an 8-pin control relay.

3. Connect the circuit shown in Figure 30-8.

4. After checking with the instructor, turn on the power and test the circuit for proper operation.

5. **Turn off the power** and disconnect the circuit.

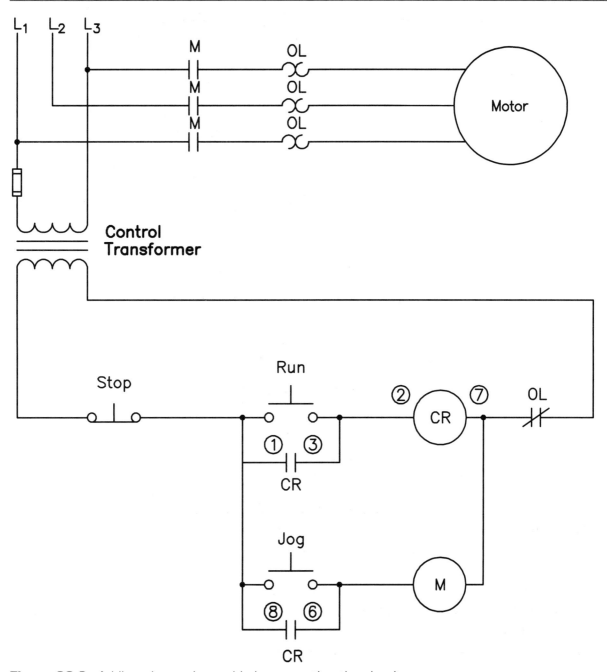

Figure 30-8 Adding pin numbers aids in connecting the circuit.

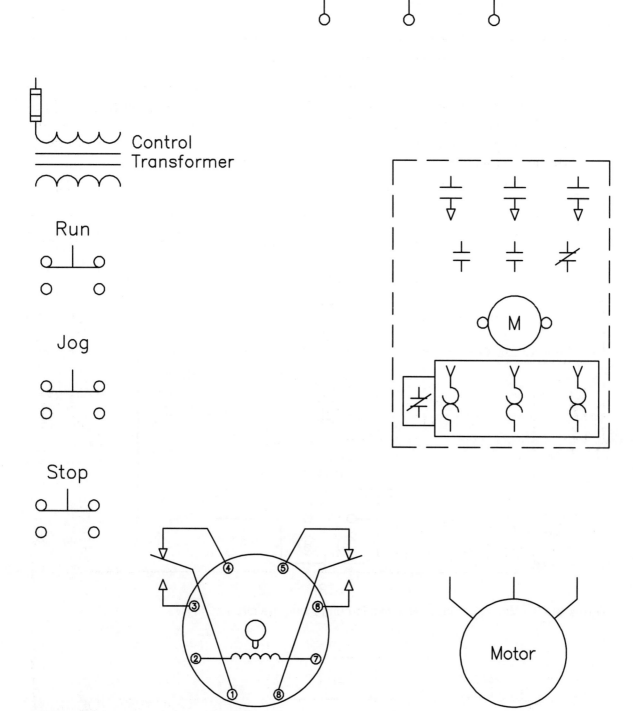

Figure 30-9 Components for circuit.

Review Questions

1. Explain the difference between inching and jogging.

   _____

   _____

2. What is the main purpose of jogging?

   _____

3. Refer to the circuit shown in Figure 30-10. In this circuit, the jog button has been connected incorrectly. The normally closed section has been connected in parallel with the run push button and the normally open section has been connected in series with the holding contacts. Explain how this circuit operates.

   _____

   _____

   _____

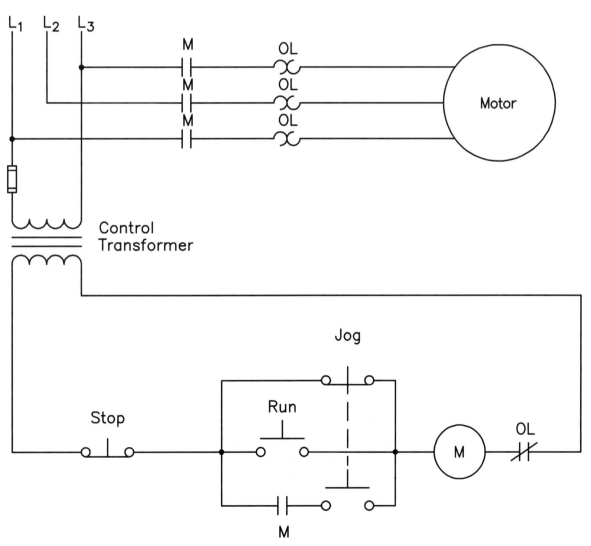

Figure 30-10 Jog button has been connected incorrectly.

4. Refer to the circuit shown in Figure 30-11. In this circuit the jog push button has again been connected incorrectly. The normally closed section of the button has been connected in series with the normally open run push button and the normally open section of the jog button is connecting in parallel with the holding contacts. Explain how this circuit operates.

_____

_____

_____

_____

_____

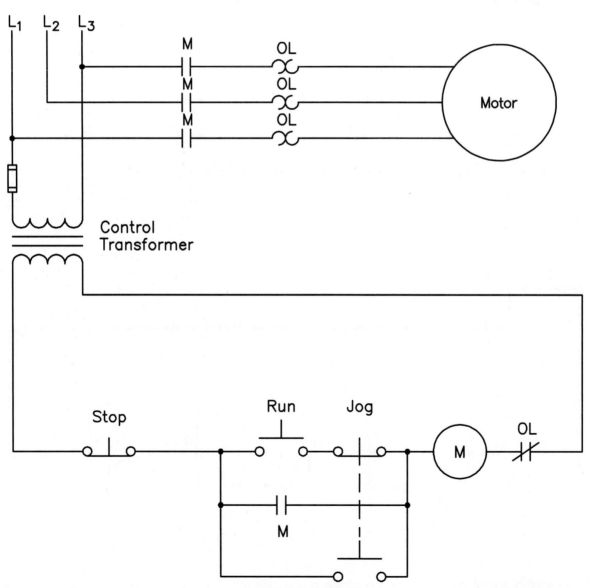

Figure 30-11 Push button has been connected incorrectly.

5. In the space provided in Figure 30-12, design a run-jog circuit to the following specfications:

   a. The circuit contains two push buttons, a normally closed stop button and a normally open start button.

   b. When the start button is pressed, the motor will run normally. When the stop button is pressed, the motor will stop.

   c. If the stop button is manually held in, the motor can be jogged by pressing the start button.

   d. The circuit contains a control transformer, motor, and three-phase motor starter with at least one normally open auxiliary contact.

6. After your instructor has approved the new circuit design, connect the circuit in the laboratory.

7. Turn on the power and test the circuit for proper operation.

8. **Turn off the power** and disconnect the circuit. Return the components to their proper place.

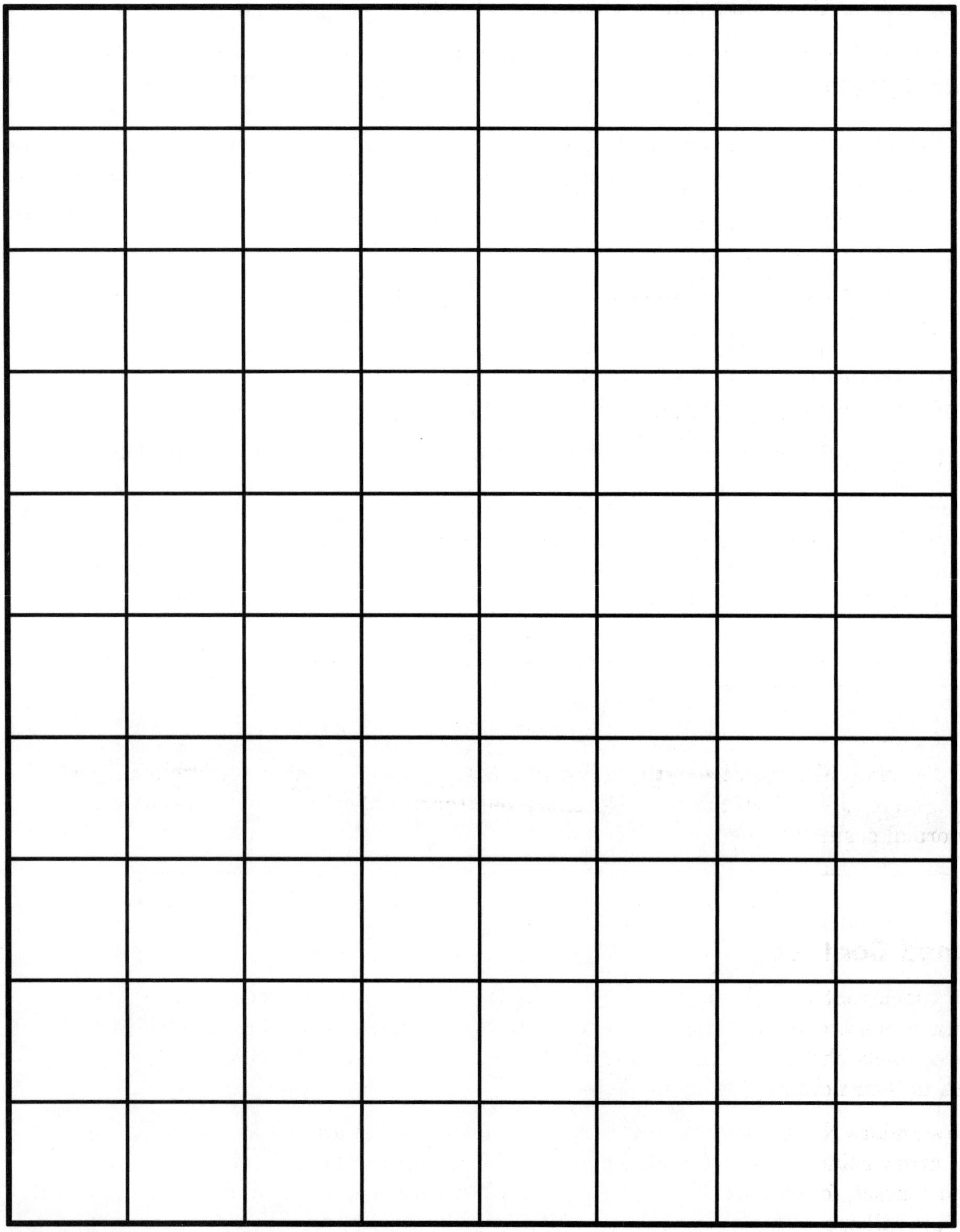

Figure 30-12 Circuit design.

Unit 31 On-Delay Timers

Objectives

After studying this unit, you should be able to:

- Discuss the operation of an on-delay timer.
- Draw the NEMA contact symbols used to represent both normally open and normally closed on-delay contacts.
- Discuss the difference in operation between pneumatic and electronic timers.
- Connect a circuit in the laboratory employing an on-delay timer.

Timers can be divided into two basic types: on-delay and off-delay. Although there are other types such as one shot and interval, they are basically an on- or off-delay timer. In this unit, the operation of on-delay timers is discussed. The operating sequence of an on-delay timer is as follows:

When the coil is energized, the timed contacts will delay changing position for some period of time. When the coil is de-energized, the timed contacts will return to their normal position immediately. In this explanation, the word "timed contacts" is used. The reason is that some timers contain both timed and instantaneous contacts. When using a timer of this type, care must be taken to connect to the proper set of contacts.

> **Helpful Hint**
>
> When the coil is energized, the timed contacts will delay changing position for some period of time. When the coil is de-energized, the timed contacts will return to their normal position immediately.

Timed Contacts

The timed contacts are controlled by the action of the timer, whereas the instantaneous contacts operate like any standard set of contacts on a control relay; when the coil energizes, the contacts change position immediately, and when the coil de-energizes they change back to their normal position immediately.

The standard NEMA symbols used to represent on-delay contacts are shown in Figure 31-1. The arrow points in the direction the contact will move after the delay period. The normally open contact, for example, will close after the time delay period, and the normally closed contact will open after the time delay period.

Instantaneous Contacts

Instantaneous contacts are drawn in the same manner as standard relay contacts. Figure 31-2 illustrates a set of instantaneous contacts controlled by timer TR. The instantaneous contacts are often used as holding or sealing contacts in a control circuit. The control circuit shown in Figure 31-3 illustrates an on-delay timer used to delay the starting of a motor.

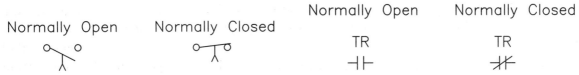

Figure 31-1 NEMA standard symbols for on-delay contacts.

Figure 31-2 Instantaneous contact symbols.

When the start push button is pressed, TR coil energizes and the normally open instantaneous TR contacts close immediately to hold the circuit. After the preset time period, the normally open TR timed contacts will close and energize the coil of M starter, which connects the motor to the line.

When the stop button is pressed and TR coil de-energizes, both TR contacts return to their normal position immediately. This de-energizes M coil and disconnects the motor from the line.

Control Relays Used with Timers

Not all timers contain instantaneous contacts. Most electronic timers, for example, do not. When an instantaneous contact is needed and the timer does not have one available, it is common practice to connect the coil of a control relay in parallel with the coil of the timer (Figure 31-4). In this way the electronic timer will operate with the timer. In the circuit shown in Figure 31-4, both coils TR and CR will energize when the start button is pressed. This causes CR contact to close and seal the circuit.

Time Delay Methods

Although there are two basic types of timers, there are different methods employed to obtain a time delay. One of the oldest methods still in general use is the pneumatic timer. Pneumatic timers use a bellows or diaphragm and operate on the principle of air displacement. Some

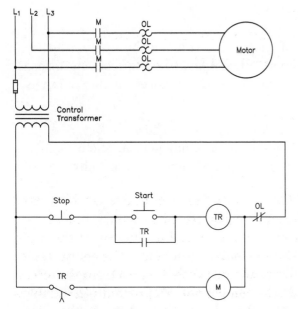

Figure 31-3 An instantaneous timer contact is used as the holding contact.

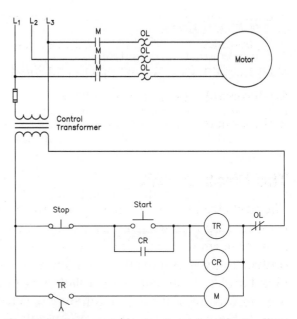

Figure 31-4 control relay furnishes the instantaneous contact.

type of needle valve is generally used to regulate the airflow and thereby regulate the time delay. Pneumatic timers are simple in the way they contain a coil, contacts, and some method of adjusting the amount of time delay. Because of their simplicity of operation, when control circuits are in the design stage, the circuit logic is generally developed with the assumption that pneumatic timers will be used. After the circuit logic has been developed, it may be necessary to make changes that will accommodate a particular type of timer.

Another very common method of providing a time delay is with an electric clock similar to a wall clock. These timers contain a small single-phase synchronous motor. As a general rule, most clock timers can be set for different full-scale values by changing the gear ratio.

Electronic timers are becoming very popular for several reasons:

1. They are much less expensive than pneumatic or clock timers.
2. They have better repeat accuracy than pneumatic or clock timers.
3. Most can be set for 0.1-second delays and many can be set to an accuracy of 0.01 second.
4. Many electronic timers are intended to be plugged into an 8- or 11-pin tube socket. This makes replacing the timer much simpler and takes less time.

LABORATORY EXERCISE

Name _____ Date _____

Materials Required

Three-phase power supply

Control transformer

2 double-acting push buttons (N.O./N.C. on each button)

2 three-phase motor starters with at least one normally open auxiliary contact

Dayton Solid-State Timer—model 6A855 or equivalent and 11-pin socket

8-pin control relay and 8-pin socket

2 three-phase motors or equivalent motor loads

The First Circuit

The first circuit to be connected is shown in Figure 31-4. In this circuit, it will be assumed that an 11-pin timer is being used and that the coil is connected to pins 2 and 10, and a set of normally open timed contacts is connected to pins 1 and 3. The coil of the 8-pin control relay is connected to pins 2 and 7 and a normally open contact is connected to pins 1 and 3. When using control devices that are connected with 8- and 11-pin sockets, it is generally helpful to place pin numbers beside the component. To prevent pin numbers from being confused with wire numbers, a circle will be drawn around the pin numbers (Figure 31-5).

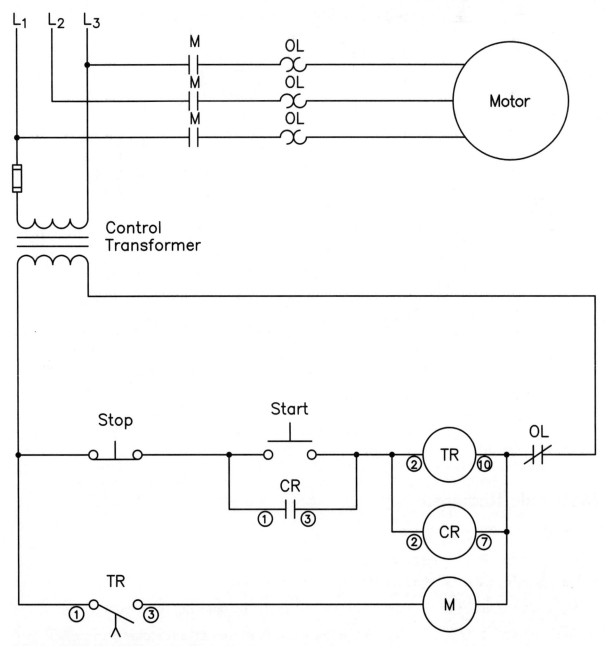

Figure 31-5 Placing pin numbers beside the components.

Connecting Circuit #1

1. Using the circuit shown in Figure 31-5, place wire numbers beside the components.

2. Connect the control part of the circuit by following the wire numbers placed beside the components. Note the pin numbers beside the coils and contacts of the timer and control relay.

3. Plug the timer and control relay into their appropriate sockets. Set the timer to operate as an on-delay timer and set the time period for 5 seconds.

4. After checking with the instructor, turn on the power and test the operation of the circuit.

5. **Turn off the power.**

6. If the control part of the circuit operated correctly, connect the motor or equivalent motor load.

7. Turn on the power and test the total circuit for proper operation.

8. **Turn off the power** and disconnect the circuit.

Discussing Circuit #2

In the next circuit, two motors are to be started with a 5-second time delay between the starting of the first motor and the second motor. In this circuit a normally open auxiliary contact on starter 1M is used as the holding contact, making the use of the control relay unnecessary.

When the start button is pressed, coils 1M and TR energize immediately. This causes motor #1 to start operating and timer TR to begin timing. After 5 seconds, TR contacts close and connect motor #2 to the line. When the stop button is pressed, or if an overload on either motor should occur, all coils will be de-energized and both motors will stop.

Connecting Circuit #2

1. Using the circuit shown in Figure 31-6, place pin numbers beside the timer coil and normally open contact.

2. Place wire numbers on the circuit in Figure 31-6.

3. Connect the control part of the circuit.

4. Turn on the power and test the circuit for proper operation.

5. **Turn off the power.**

6. If the control part of the circuit operated properly, connect the motors or equivalent motor loads.

7. Turn on the power and test the circuit for proper operation.

8. **Turn off the power** and disconnect the circuit.

Review Questions

1. Explain the operation of an on-delay timer.

   _____

   _____

   _____

2. Explain the difference between timed contacts and instantaneous contacts.

   _____

   _____

   _____

3. Refer to the circuit shown in Figure 31-3. If the timer has been set for a delay of 10 seconds, explain the operation of the circuit when the start button is pressed.

   _____

   _____

   _____

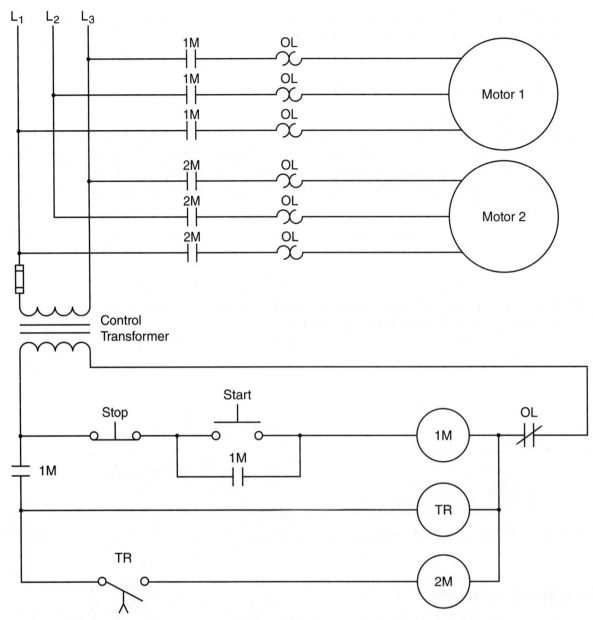

Figure 31-6 Motor 2 starts after motor 1.

4. In the circuit shown in Figure 31-3, is it necessary to hold the start button closed for a period of at least 10 seconds to ensure that the circuit will remain energized? Explain your answer.

_____

_____

_____

5. Assume that the timer in Figure 31-3 is set for a delay of 10 seconds. Now assume that the start button is pressed, and after a delay of 8 seconds the stop button is pressed. Will the motor start 2 seconds after the stop button is pressed?

_____

_____

_____

6. What is generally done to compensate when a set of instantaneous timer contacts is needed and the timer does not contain them?

_____

_____

_____

7. Refer to the circuit shown in Figure 31-6. Assume that it is necessary to stop the operation of both motors after the second motor has been operating for a period of 10 seconds. Using the space provided in Figure 31-7, redraw the circuit to turn off both motors after the second motor has been in operation for 10 seconds. (*Note*: It will be necessary to use a second timer.)

_____

_____

_____

8. After your instructor has approved the design change, connect the new circuit in the laboratory and test it for proper operation.

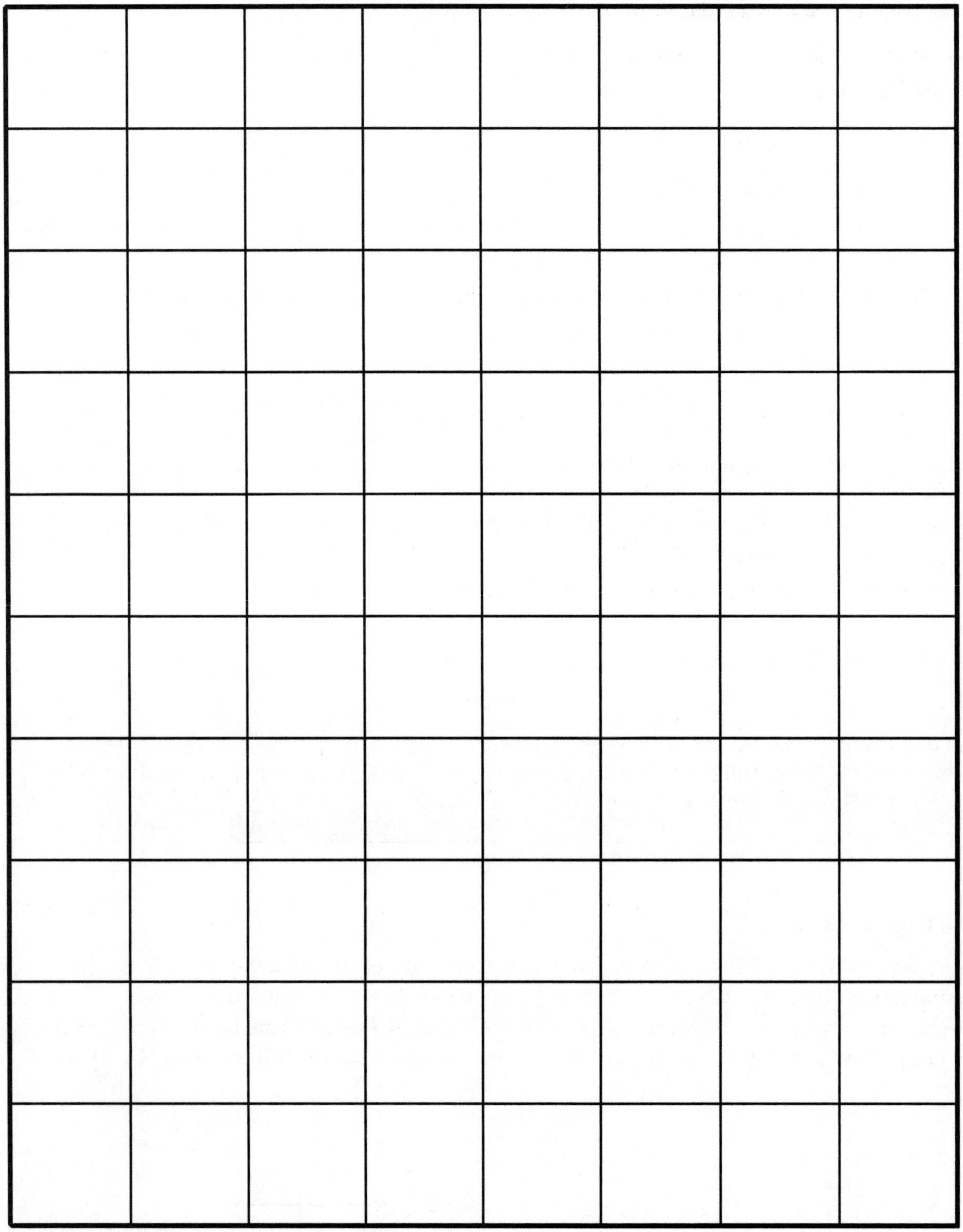

Figure 31-7 Circuit redesign.

Unit 32 Off-Delay Timers

Objectives

After studying this unit, you should be able to:

- Discuss the operation of an off-delay timer.
- Draw the NEMA contact symbols used to represent both normally open and normally closed off-delay contacts.
- Discuss the difference in operation between pneumatic and electronic timers.
- Connect a circuit in the laboratory employing an off-delay timer.

The logic of an off-delay timer is as follows: When the coil is energized, the timed contacts change position immediately. When the coil is de-energized, the timed contacts remain in their energized position for some period of time before changing back to their normal position. Figure 32-1 shows the standard NEMA contact symbols used to represent an off-delay timer. Notice that the arrow points in the direction the contact will move after the time delay period. The arrow indicates that the normally open contact will delay reopening and that the normally closed contact will delay reclosing. Like on-delay timers, some off-delay timers will contain instantaneous contacts as well as timed contacts, and some will not.

Helpful Hint

When the coil is energized, the timed contacts change position immediately. When the coil is de-energized, the timed contacts remain in their energized position for some period of time before changing back to their normal position.

Example Circuit #1

The circuit shown in Figure 32-2 illustrates the logic of an off-delay timer. It will be assumed that the timer has been set for a delay of 5 seconds. When switch S_1 closes, TR coil energizes. This causes the normally open TR contacts to close immediately and turn on the lamp. When switch S_1 opens, TR coil will de-energize, but the TR contacts will remain

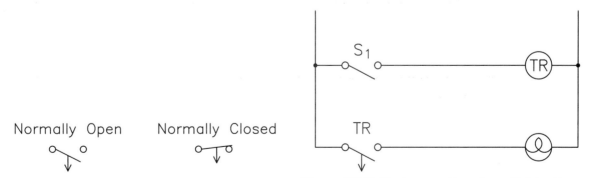

Normally Open   Normally Closed

Figure 32-1 NEMA standard symbols for off-delay contacts.

Figure 32-2 Basic operation of an off-delay timer.

closed for 5 seconds before they reopen. Notice that the time delay period does not start until the coil is de-energized.

Example Circuit #2

In the second example, it is assumed that timer TR has been set for a delay of 10 seconds. Two motors start when the start button is pressed. When the stop button is pressed, motor #1 stops operating immediately, but motor #2 continues to run for 10 seconds (Figure 32-3). In this circuit, the coil of the off-delay timer has been placed in parallel with motor starter 1M, permitting the action of the timer to be controlled by the first motor starter.

Example Circuit #3

Now assume that the logic of the previous circuit is to be changed so that when the start button is pressed both motors still start at the same time, but when the stop button is pressed, motor #2 must stop operating immediately and motor #1 continues to run for 10 seconds. In this circuit the action of the timer must be controlled by the operation of starter 2M instead of starter 1M (Figure 32-4). In the circuit shown in Figure 32-4, a control relay is used to energize both motor starters at the same time. Notice that timer coil TR energizes at the same time as starter 2M, causing the normally open TR contacts to close around the CR contact connected in series with coil 1M.

When the stop button is pressed, coil CR de-energizes and all CR contacts open. Power is maintained to starter 1M, however, by the now closed TR contacts. When the CR contact connected in series with coils 2M and TR opens, these coils de-energize, causing motor #2 to stop operating and starting the time sequence for the off-delay timer. After a delay of 10 seconds, TR contacts reopen and de-energize coil 1M, stopping the operation of motor #1.

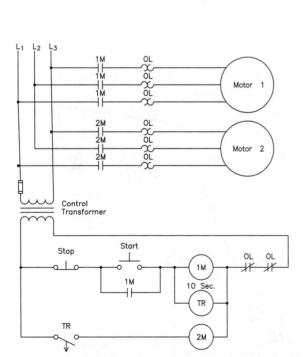

Figure 32-3 Off-delay motor circuit using pneumatic timer.

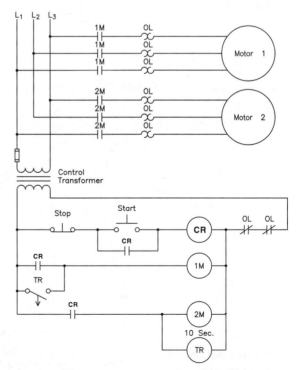

Figure 32-4 Motor 1 stops after motor 2.

Using Electronic Timers

In the circuits shown in Figure 32-3 and Figure 32-4, it was assumed that the off-delay timers were of the pneumatic type. It is common practice to develop circuit logic assuming that the timers are of the pneumatic type. The reason for this is that the action of a pneumatic timer is controlled by the coil being energized or de-energized. The action of the timer is dependent on air pressure, not an electric circuit. This, however, is generally not the case when using solid-state time delay relays. Solid-state timers that can be used as off-delay timers are generally designed to be plugged into an 11-pin tube socket. The pin connection for a Dayton model 6A855 timer is shown in Figure 32-5. Although this is by no means the only type of electronic timer available, it is typical of many.

Notice in Figure 32-5 that power is connected to pins 2 and 10. When this timer is used in the on-delay mode, there is no problem with the application of power because the time sequence starts when the timer is energized. When power is removed, the timer de-energizes and the contacts return to their normal state immediately.

An off-delay timer, however, does not start the timing sequence until the timer is de-energized. Since this timer depends on an electronic circuit to operate the timing mechanism, power must be connected to the timer at all times. Therefore, some means other than disconnecting the power must be used to start the timing circuit. This

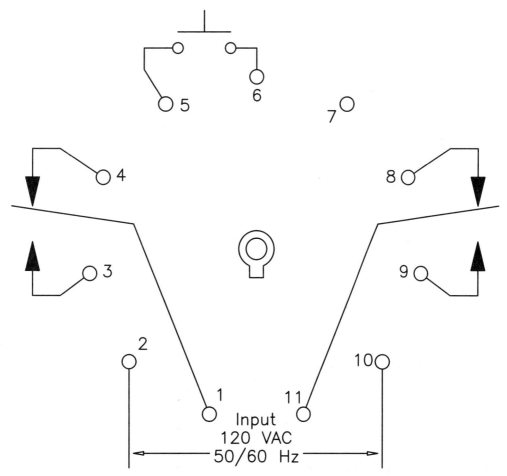

Figure 32-5 Connection diagram for a Dayton model 6A855 timer.

particular timer uses pins 5 and 6 to start the operation. The diagram in Figure 32-5 uses a start switch to illustrate this operation. When pins 5 and 6 are shorted together, it has the effect of energizing the coil of an off-delay timer and all contacts change position immediately. The timer will remain in this state as long as pins 5 and 6 are short circuited together. When the short circuit between pins 5 and 6 is removed, it has the effect of de-energizing the coil of a pneumatic off-delay timer and the timing sequence will start. At the end of the time period, the contacts will return to their normal position.

LABORATORY EXERCISE

Name _____ Date _____

Materials Required:

Three-phase power supply

Control transformer

2 double-acting push buttons (N.O./N.C. on each button)

2 three-phase motor starters with at least one normally open auxiliary contact

Dayton Solid-State Timer—model 6A855 or equivalent

11-pin control relay and two 11-pin sockets

2 three-phase motors or equivalent motor loads

Amending Circuit #1

The circuit in Figure 32-3 has been amended in Figure 32-6 to accommodate the use of an electronic timer. Notice in this circuit that power is connected to pins 2 and 10 of the timer at all times. Since the action of the timer in the original circuit is that the coil of the timer operates at the same time as starter coil 1M, an auxiliary contact on starter 1M will be used to control the action of timer TR. When the start button is pressed, coil 1M energizes and all 1M contacts close. This connects motor #1 to the line, the 1M contact in parallel with the start button seals the circuit, and the normally open 1M contact connected to pins 5 and 6 of the timer closes and starts the operation of the timer. When timer pins 5 and 6 become shorted, the timed contact connected in series with 2M coil closes and energizes starter 2M.

When the stop button is pressed, coil 1M de-energizes and all 1M contacts return to their normal position, stopping the operation of motor #1. When the 1M contacts connected to timer pins 5 and 6 reopen, the timing sequence of the timer begins. After a delay of 10 seconds, timed contact TR reopens and disconnects starter coil 2M from the circuit. This stops the operation of motor #2.

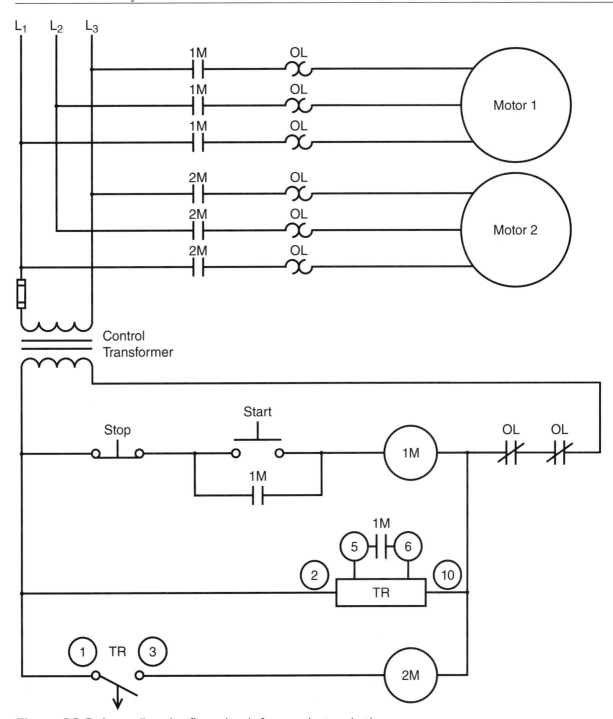

Figure 32-6 Amending the first circuit for an electronic timer.

Amending Circuit #2

Circuit #2 will be amended in much the same way as circuit #1. The timer must have power connected to it at all times (Figure 32-7). Notice in this circuit that the action of the timer is controlled by starter 2M instead of 1M. When coil 2M energizes, a set of normally open 2M contacts closes and shorts pins 5 and 6 of the timer. When coil 2M de-energizes, the 2M auxiliary contacts reopen and start the time sequence of timer TR.

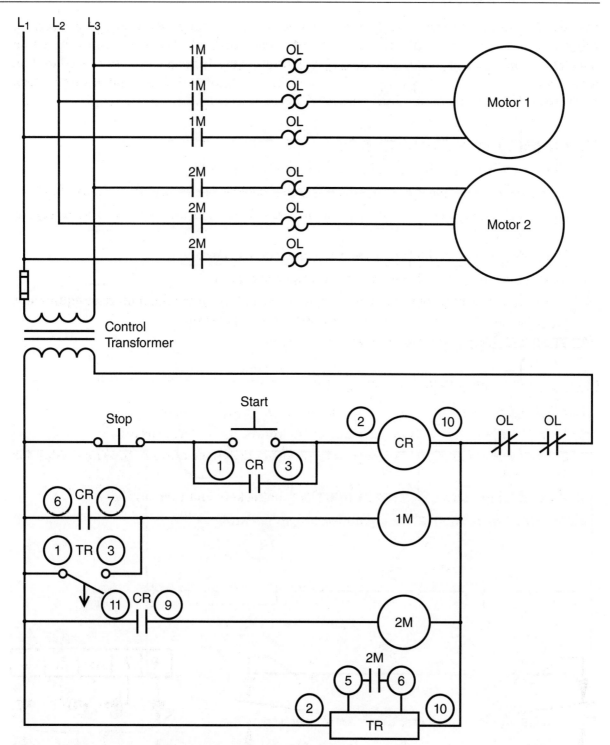

Figure 32-7 Amending circuit 2 for an electronic timer.

Circuit #2 assumes the use of an 11-pin control relay instead of an 8 pin. An 11-pin control relay contains three sets of contacts instead of two. Figure 32-8 shows the connection diagram for most 11-pin control relays. Notice that normally open contacts are located on pins 1 and 3, 6 and 7, and 9 and 11. The coil pins are 2 and 10. Pin numbers have been placed beside the components in Figure 32-7.

Connecting the First Circuit

1. Place wire numbers on the schematic shown in Figure 32-6.
2. Using an 11-pin tube socket, connect the control part of the circuit in Figure 32-6.
3. Set the electronic timer to operate as an off-delay timer and set the time delay for 10 seconds.
4. Plug the timer into the tube socket and turn on the power.
5. Test the control part of the circuit for proper operation.
6. If the control portion of the circuit operated properly, connect the motors or equivalent motor loads and test the entire circuit for proper operation.
7. **Turn off the power** and disconnect the circuit.

Connecting the Second Circuit

1. Place wire numbers on the schematic diagram shown in Figure 32-7.
2. Using two 11-pin tube sockets, connect the control part of the circuit.
3. Set the electronic timer to operate as an off-delay timer and set the time delay for 10 seconds.
4. Plug the timer and control relay into the tube sockets and turn on the power.
5. Test the control part of the circuit for proper operation.

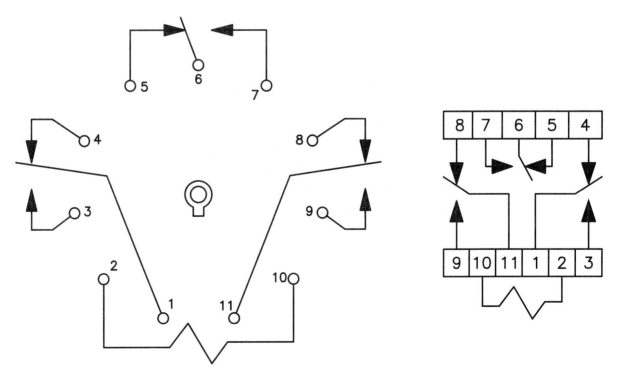

Figure 32-8 Connection diagram for an 11-pin control relay.

6. If the control portion of the circuit operated properly, connect the motors or equivalent motor loads and test the entire circuit for proper operation.
7. **Turn off the power** and disconnect the circuit.
8. Return the components to their proper location.

Review Questions

1. Describe the operation of an off-delay timer.

   _____

   _____

   _____

2. Why is it common practice to develop circuit logic assuming all timers are of the pneumatic type?

   _____

   _____

   _____

3. Refer to the schematic diagram shown in Figure 32-6. Assume that starter coil 2M is open. Describe the action of the circuit when the start button is pressed and when the stop button is pressed.

   _____

   _____

   _____

4. Refer to the circuit shown in Figure 32-7. Assume that when the start button is pressed, motor #1 starts operating immediately, but motor #2 does not start. When the stop button is pressed, motor #1 stops operating immediately. Which of the following could cause this condition?
   a. 1M coil is open.
   b. 2M coil is open.
   c. Timer TR is not operating.
   d. CR coil is open.

5. Refer to the circuit shown in Figure 32-7. When the start button is pressed, both motors #1 and #2 start operating immediately. When the stop button is pressed, motor #2 stops operating immediately, but motor #1 remains running and does not turn off after the time delay period has expired. Which of the following could cause this condition?
   a. CR contacts are shorted together.
   b. 2M auxiliary contacts connected to pins 5 and 6 of the timer did not close.
   c. 2M auxiliary contacts connected to pins 5 and 6 of the timer are shorted.
   d. The stop button is shorted.

6. Refer to the circuit shown in Figure 32-7. Assume that timer TR is set for a delay of 10 seconds. Now assume that timer TR is changed from an off-delay timer to an on-delay timer. Explain the operation of the circuit.

   _____

   _____

   _____

7. Using the space provided in Figure 32-9, modify the circuit in Figure 32-7 to operate as follows:

   a. When the start button is pressed, motor #1 starts running immediately. After a delay of 10 seconds, motor #2 begins running. Both motors remain operating until the stop button is pressed or an overload occurs.

   b. When the stop button is pressed, motor #2 stops operating immediately, but motor #1 continues to operate for a period of 10 seconds before stopping.

   c. An overload on either motor will stop both motors immediately.

   d. Assume the use of electronic timers in final design.

8. After your instructor has approved the modification, connect your circuit in the laboratory.

9. Turn on the power and test the circuit for proper operation.

10. **Turn off the power**, disconnect the circuit, and return the components to their proper location.

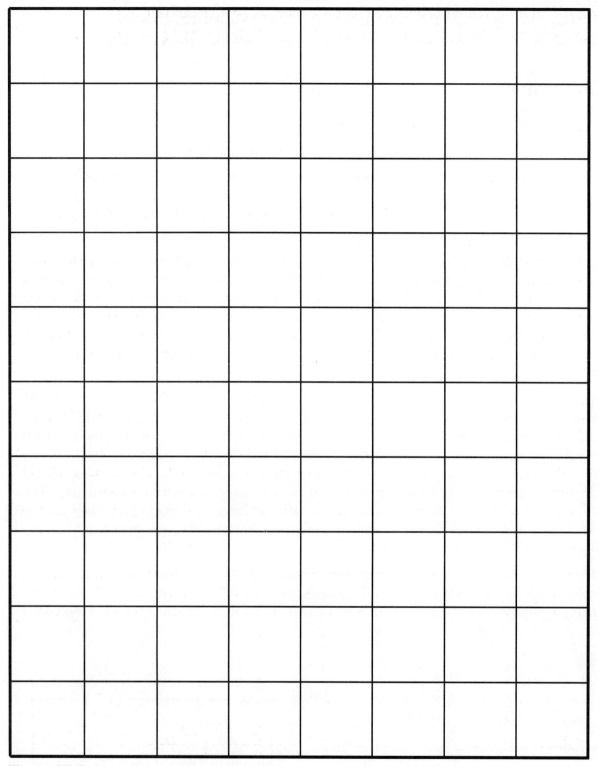

Figure 32-9 Amending the circuit design.

Unit 33 Changing the Logic of an On-Delay Timer to an Off-Delay Timer

Objectives

After studying this unit, you should be able to:

- Discuss the difference in logic between on- and off-delay timers.
- Draw a schematic diagram of a circuit that will change the logic of an on-delay timer into the logic of an off-delay timer.
- Connect an on-delay timer circuit that will operate with the logic of an off-delay timer.

Some manufacturers purchase on-delay timers only. The reason for this is that most timing circuits require the logic of an on-delay timer. If it should become necessary to construct a circuit with the logic of an off-delay timer, it is a relatively simple matter to build a circuit using an on-delay timer that will operate with the same logic as an off-delay timer. A circuit of this type is shown in Figure 33-1. The basic idea is to cause the timer to start operating when a control component is turned off instead of on. Control relay CR is used to perform this function.

In the circuit shown in Figure 33-1, starter 1M is to energize immediately when switch S1 closes. When switch S1 opens, starter 1M should remain energized for some period of time before de-energizing. This is the logic of an off-delay timer. This logic can be accomplished by using an on-delay timer and the circuit shown in Figure 33-1. When switch S1 closes, CR coil energizes and all CR contacts change position (Figure 33-2). The normally closed CR contact connected in series with TR coil opens to prevent the timer energizing. The normally open CR contact connected in series with starter coil 1M closes and energizes the coil. This causes both 1M auxiliary contacts to close. Starter 1M is now energized, but the timer has not started its time sequence.

When switch S1 is reopened, CR coil de-energizes and all CR contacts return to their normal position (Figure 33-3). When the CR contact connected in series with starter coil 1M re-opens, a current path is maintained though the now closed 1M auxiliary contact connected

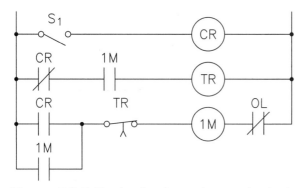

Figure 33-1 Basic circuit to change the logic of an on-delay timer into an off-delay timer.

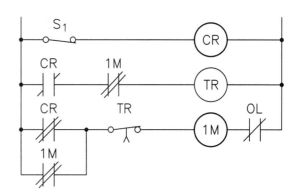

Figure 33-2 Starter 1M energizes immediately, but the timer does not start timing.

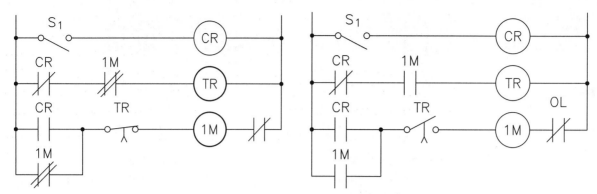

Figure 33-3 Switch S₁ opens and starts the timer.

Figure 33-4 Starter 1M de-energizes when timer contact TR opens.

in parallel with the open CR contact. When the CR contact connected in series with timer coil TR closed, it provided a path to coil TR and the timer began its time sequence.

At the end of the timing sequence, timed contact TR opens and de-energizes coil 1M, causing all 1M contacts to return to their normal position (Figure 33-4). The auxiliary 1M contact connected in series with timer coil TR opens and de-energizes coil TR. This causes contact TR to reclose and the circuit is back to the beginning state, as shown in Figure 33-1.

Changing an Existing Schematic

The circuit shown in Figure 33-5 is an off-delay timer circuit for the control of two motors. It is assumed that the timer used in this circuit is a pneumatic timer. This circuit was discussed in the previous unit. Both motors start when the start button is pressed. When the stop button is pressed, motor #2 stops operating immediately, but motor #1 continues to operate for a period of 10 seconds. Now assume that it is necessary to change the circuit logic to permit an on-delay timer to be used.

Notice in the circuit in Figure 33-5 that timer coil TR is energized or de-energized at the same time as starter coil 2M. In the amended circuit, starter 2M will control the starting of on-delay timer TR (Figure 33-6). A set of 1M auxiliary contacts prevents coil TR from being energized until starter 1M has been energized. To understand the operation of the circuit, trace the logic through each step of operation. Assume that the start button is pushed and coil CR energizes. This causes all CR contacts to close and connect starters 1M and 2M to the line (Figure 33-7). Both 1M auxiliary contacts close, but the normally closed 2M auxiliary contact connected in series with TR coil opens and prevents it from starting its time sequence.

When the stop button is pressed, CR coil de-energizes and all CR contacts return to their normal position (Figure 33-8). Motor starter 1M remains energized because of the closed 1M auxiliary contact connected in parallel with the CR contact. When starter 2M de-energizes, the normally closed auxiliary contact connected in series with timer coil TR recloses and on-delay timer TR begins its timing sequence.

After a delay of 10 seconds, timed contact TR opens and disconnects starter coil 1M from the line (Figure 33-9). This stops the operation of motor #1 and returns all 1M auxiliary contacts to their normal position. When timer TR de-energizes, timed contact TR returns to its normally closed position and the circuit is back to its original state, as shown in Figure 33-6.

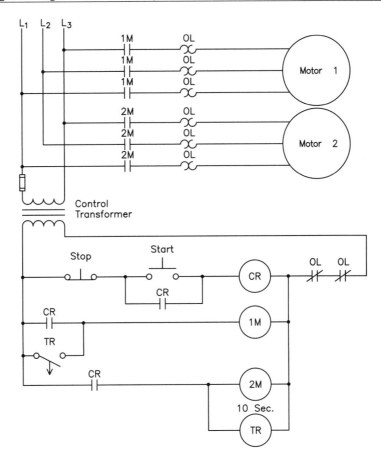

Figure 33-5 Off-delay timer circuit using a pneumatic timer.

LABORATORY EXERCISE

Name _____ Date _____

Materials Required

Three-phase power source

Control transformer

2 three-phase motors or equivalent motor load

2 three-phase motor starters with at least two normally open and one normally closed auxiliary contacts

8-pin or 11-pin on-delay timer with appropriate socket

11-pin control relay with 11-pin socket

Connecting the Circuit

1. Using the circuit shown in Figure 33-6, place pin numbers beside the control compo-nents that mount into tube sockets. These components will probably be the control relay and the timer. Be sure to place pin numbers beside contacts as well as coils. Circle the pin numbers to distinguish them from wire numbers.

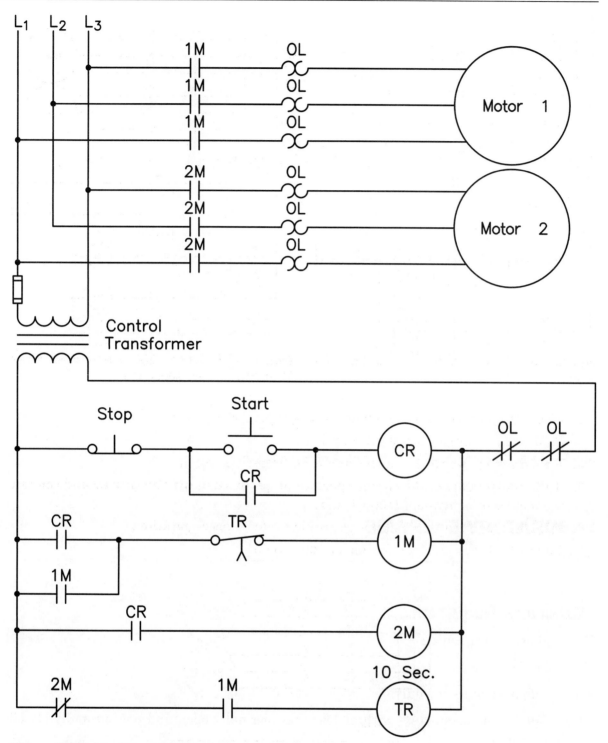

Figure 33-6 Modifying the circuit for an on-delay timer.

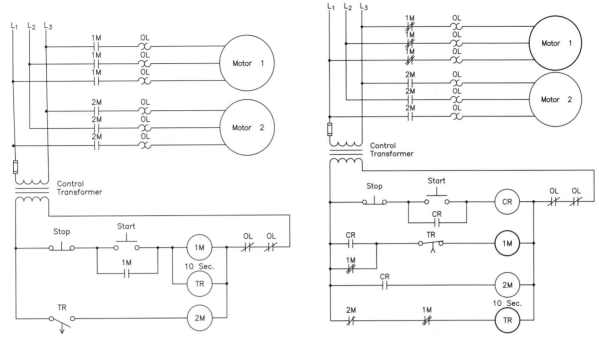

Figure 33-7 Both motors start at the same time.

Figure 33-8 Starter 2M de-energizes; timer TR starts its time sequence.

2. Place wire numbers on the schematic diagram.

3. Connect the control part of the circuit.

4. Turn on the power and test the circuit for proper operation.

5. If the control part of the circuit operates properly, **turn off the power** and connect the motors or equivalent motor loads.

6. Turn on the power and test the entire circuit for proper operation.

7. **Turn off the power** and disconnect the circuit.

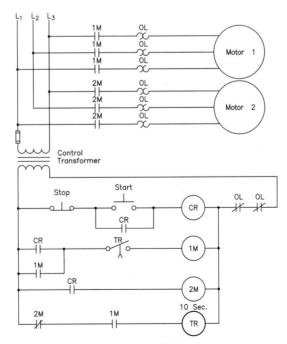

Figure 33-9 Timed contact de-energizes starter 1M.

Review Questions

1. Why do some companies purchase only on-delay timers?

_____

_____

_____

2. Refer to the circuit shown in Figure 33-10. This circuit assumes the use of a pneumatic off-delay timer. It is also assumed that the timer is set for a delay of 10 seconds. Describe the operation of this circuit when the start push button is pressed.

_____

_____

_____

3. Assume that the circuit in Figure 33-10 is in operation. Describe the action of the circuit when the stop button is pressed.

_____

_____

4. The circuit shown in Figure 33-10 employs a pneumatic off-delay timer. Redraw the circuit in the space provided in Figure 33-11 to use an electronic on-delay timer. Make certain that the logic of the circuit is the same.

5. After your instructor has approved the redrawn circuit, connect the circuit in the laboratory.

6. Turn on the power and test the circuit for proper operation.

7. **Turn off the power** and return the components to their proper place.

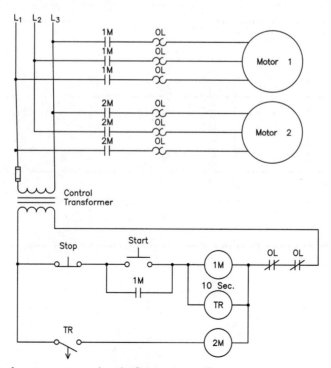

Figure 33-10 Motor 1 stops operating before motor 2.

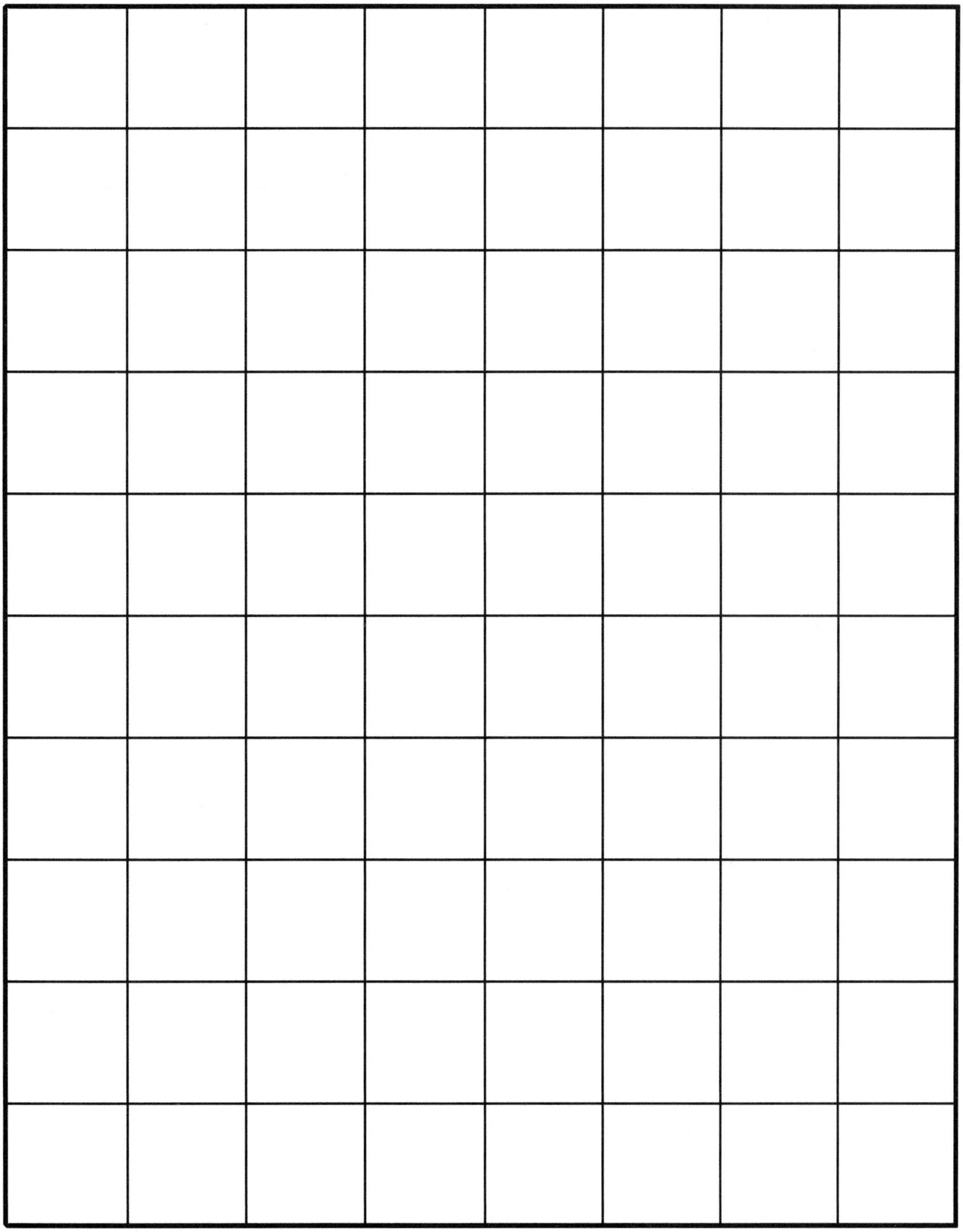

Figure 33-11 Circuit redesign.

Unit 34   Designing a Printing Press Circuit

Objectives

After studying this unit, you should be able to:

- Describe a step-by-step procedure for designing a motor control circuit.
- Design a basic control circuit.
- Connect the completed circuit in the laboratory.

LABORATORY EXERCISE

Name _____ Date _____

Materials Required

Three-phase power supply

Three-phase motor starter

8- or 11-pin on-delay relay with appropriate socket

Three-phase motor or equivalent motor load

Pilot light

Buzzer or simulated load

Control transformer

8-pin control relay and 8-pin socket

In this experiment a circuit for a large printing press will be designed in a step-by-step procedure. The owner of a printing company has the following concern when starting a large printing press:

The printing press is very large and the surrounding noise level is high. There is a danger that when the press starts, a person unseen by the operator may have his or her hands in the press. To prevent an accident, I would like to install a circuit that sounds an alarm and flashes a light for 10 seconds before the press actually starts. This would give the person time to get clear of the machine before it starts.

To begin the design procedure, list the requirements of the circuit. List not only the concerns of the owner but also any electrical or safety requirements that the owner may not be aware of. Understand that the owner is probably not an electrical technician and does not know all the electrical requirements of a motor control circuit.

1. There must be a start and stop push-button control.
2. When the start button is pressed, a warning light and buzzer turn on.

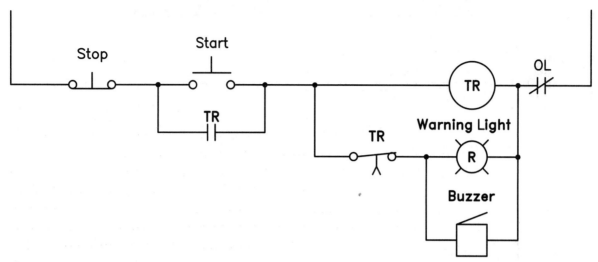

Figure 34-1 First step in the circuit design.

3. After a delay of 10 seconds, the warning light and buzzer turn off and the press motor starts.

4. The press motor should be overload protected.

5. When the stop push button is pressed, the circuit will de-energize even if the motor has not started.

To begin design of the circuit, fulfill the first requirement of the logic: "When the start button is pressed, a warning light and buzzer turn on for a period of 10 seconds." This first part of the circuit can be satisfied with the circuit shown in Figure 34-1. In this example a timer is used because the warning light and buzzer are to remain on for only 10 seconds. Since the warning light and buzzer are to turn on immediately when the start button is pressed, a normally closed timed contact is used. This circuit also assumes that the timer contains an instantaneous contact that is used to hold the circuit in after the start button is released.

The next part of the logic states that after a delay of 10 seconds the warning light and buzzer are to turn off and the press motor is to start. As the present circuit is shown in Figure 34-1, when the start button is pressed, TR coil will energize. This causes the normally open instantaneous TR contacts to close and hold TR coil in the circuit when the start button is released. At the same time, timer TR starts its timing sequence. After a delay of 10 seconds, the normally closed TR timed contact connected in series with the warning light and buzzer will open and disconnect them from the circuit.

The only remaining circuit logic is to start the motor after the warning light and buzzer have turned off. This can be accomplished with a normally open timed contact controlled by timer TR (Figure 34-2). At the end of the timing sequence, the normally closed TR contact will open and disconnect the warning light and buzzer. At the same time, the normally open TR timed contact will close and energize the coil of M starter. The normally closed overload contact connected in series with the rest of the circuit will de-energize the entire circuit in the event of motor overload.

Now that the logic of the control circuit has been completed, the motor load can be added, as shown in Figure 34-3.

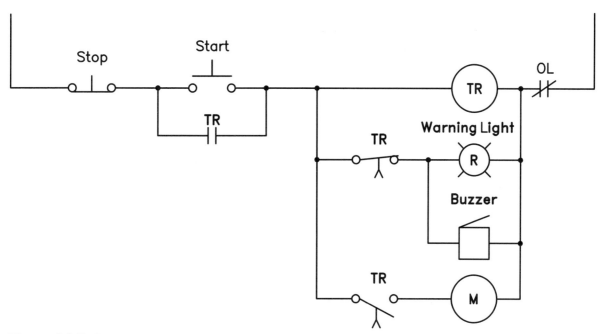

Figure 34-2 Completing the circuit logic.

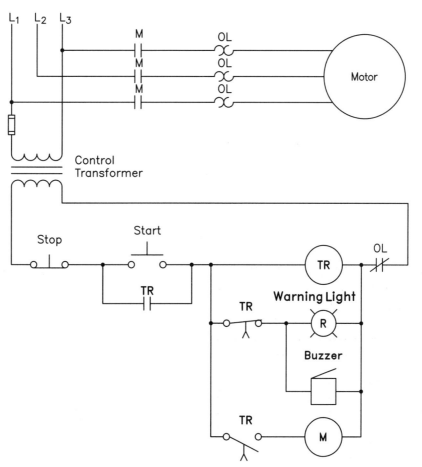

Figure 34-3 The complete circuit.

Addressing a Potential Problem

The completed circuit shown in Figure 34-3 assumes the use of a timer that contains both timed and instantaneous contacts. This contact arrangement is common for certain types of timers such as pneumatic and some clock timers, but most electronic timers do not contain instantaneous contacts. If this is the case, a control relay can be added to supply the needed instantaneous contact by connecting the coil of the control relay in parallel with the coil of TR timer (Figure 34-4).

Connecting the Circuit

1. It will be assumed that the timer in this circuit is the electronic type. Therefore, it will be assumed that a control relay will be used to provide the normally open holding contacts. Assuming the use of an electronic on-delay timer and an 8-pin control relay, place pin numbers beside the components of the timer and control relay shown in Figure 34-4. Circle the numbers to distinguish them from wire numbers.

2. Place wire numbers beside the components in Figure 34-4.

3. Connect the control portion of the circuit. (*Note*: It may be necessary to use a pilot light for the buzzer if one is not available.)

4. Turn on the power and test the control part of the circuit for proper operation.

5. **Turn the power off.**

6. If the control part of the circuit operated properly, connect the motor or simulated motor load to the circuit.

7. Turn on the power and test the entire circuit for proper operation.

8. **Turn off the power** and return the components to their proper location.

Review Questions

1. What should be the first step when beginning the design of a control circuit?

   _____

   _____

   _____

2. Why is it sometimes necessary to connect the coil of a control relay in parallel with the coil of a timer?

   _____

   _____

   _____

3. Refer to the circuit shown in Figure 34-3. Assume that the on-delay timer is replaced with an off-delay timer. Describe the action of the circuit when the start button is pressed.

   _____

   _____

   _____

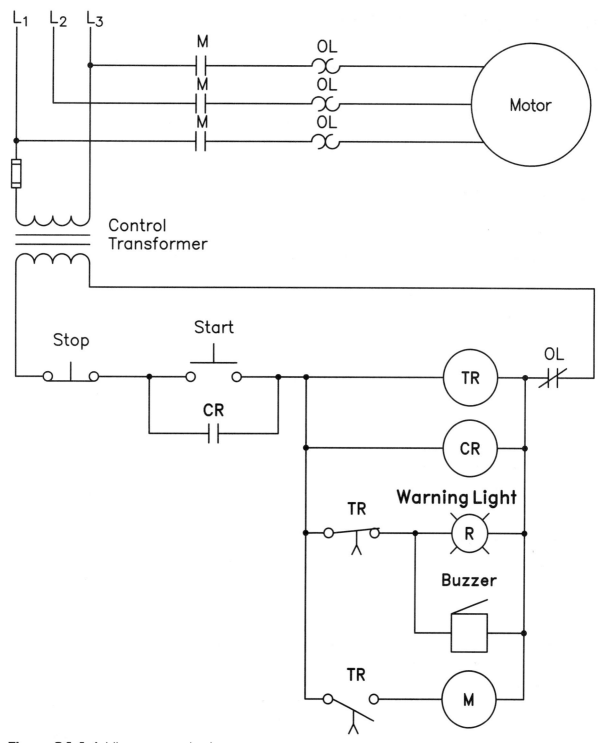

Figure 34-4 Adding a control relay.

4. Describe the operation of the circuit when the stop button is pressed. Assume the circuit is running with an off-delay timer as described in question 3.

_____

_____

_____

5. Refer to the circuit shown in Figure 34-4. Assume the owner decides to change the logic of the circuit as follows:

When the operator presses the start button, a warning light and buzzer turn on for a period of 10 seconds. During this 10 seconds, the operator must continue to hold down the start button. If the start button should be released, the timing sequence will stop and the motor will not start. At the end of 10 seconds, provided the operator continues to hold the start button down, the warning light and buzzer will turn off and the motor will start. When the motor starts, the operator can release the start button and the press will continue to run.

Amend the circuit in Figure 34-4 to meet the requirement.

Unit 35 Sequence Starting and Stopping for Three Motors

Objectives

After studying this unit, you should be able to:

- Discuss the step-by-step procedure for designing a circuit.
- Change a circuit designed with pneumatic timers into a circuit to use electronic timers.
- Connect the circuit in the laboratory.
- Troubleshoot the circuit.

LABORATORY EXERCISE

Name _____ Date _____

Materials Required

Three-phase power supply

Control transformer

2 eight-pin control relays and 8-pin sockets

3 three-phase motor starters

4 electronic timers (Dayton model 6A855 or equivalent) and 11-pin sockets

3 three-phase motors or equivalent motor loads

In this experiment a circuit will be designed and connected. The requirements of the circuit are as follows:

1. Three motors are to start in sequence from motor #1 to motor #3.
2. There is to be a time delay of 3 seconds between the starting of each motor.
3. When the stop button is pressed, the motors are to stop in sequence from motor #3 to motor #1.
4. There is to be a time delay of 3 seconds between the stopping of each motor.
5. An overload on any motor will stop all motors.

When designing a control circuit, satisfy one requirement at a time. This may at times lead to an unforeseen dead end, but don't let these dead ends concern you. When they happen, back up and redesign around them. In this example the first part of the circuit is to start three motors in sequence from motor #1 to motor #3 with a 3-second delay

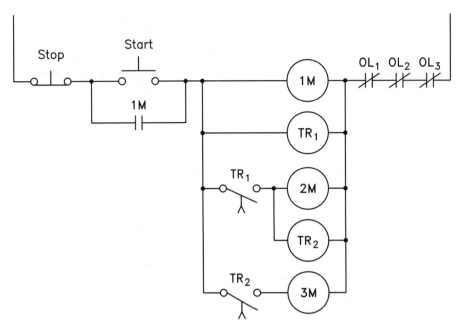

Figure 35-1 The motors start in sequence from 1 to 3.

between the starting of each motor. This is also the time to satisfy the requirement that an overload on any motor will stop all motors. The first part of the circuit can be satisfied by the circuit shown in Figure 35-1. (Note: In this experiment the motor connections will not be shown because of space limitations. It is to be assumed that the motor starters are controlling three-phase motors. It is also assumed that all timers are set for a delay of 3 seconds.)

When the start button is pressed, coils 1M and TR_1 energize. Starter 1M starts motor #1 immediately, and timer TR_1 starts its time sequence of 3 seconds. After a delay of 3 seconds, timed contact TR_1 closes and energizes coils 2M and TR_2. Starter 2M starts motor #2 and timer TR_2 begins its 3-second timing sequence. After a delay of 3 seconds, timed contact TR_2 closes and energizes motor #3. The motors have been started in sequence from #1 to #3 with a delay of 3 seconds between the starting of each motor. This satisfies the first part of the circuit logic.

The next requirement is that the circuit stop in sequence from motor #3 to motor #1. To fulfill this requirement, power must be maintained to starters 2M and 1M after the stop button has been pushed. In the circuit shown in Figure 35-1, this is not possible. Since all coils are connected after the M auxiliary holding contact, power will be disconnected from all coils when the stop button is pressed and the holding contact opens. This circuit has proven to be a dead end. There is no way to fulfill the second requirement with the circuit connected in this manner. Therefore, the circuit must be amended in such a manner that it will not only start in sequence from motor #1 to motor #3 with a 3-second time delay between the starting of each motor but also be able to maintain power after the start button is pressed. This amendment is shown in Figure 35-2.

To modify the circuit so that power can be maintained to coils 2M and 1M, a control relay has been added to the circuit. Contact $1CR_2$ prevents power from being applied to coils 1M and TR_1 until the start button is pressed.

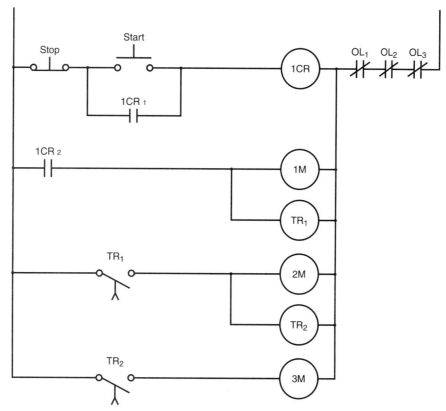

Figure 35-2 A control relay is added to the circuit.

Designing the Second Part of the Circuit

The second part of the circuit states that the motors must stop in sequence from motor #3 to motor #1. Do not try to solve all the logic at once. Solve each problem as it arises. The first problem is to stop motor #3. In the circuit shown in Figure 35-2, when the stop button is pressed, coil 1CR will de-energize. This will cause contact $1CR_2$ to open and de-energize coils 1M and TR_1. Contact TR_1 will open immediately and de-energize coils 2M and TR_2, causing contact TR_2 to open immediately and de-energize coil 3M. Notice that coil 3M does de-energize when the stop button is pressed, but so does everything else. The circuit requirement states that there is to be a 3-second time delay between the stopping of motor #3 and motor #2. Therefore, an off-delay timer will be added to maintain connection to coil 2M after coil 3M has de-energized (Figure 35-3).

The same basic problem exists with motor #1. In the present circuit, motor #1 will turn off immediately when the stop button is pressed. To help satisfy the second part of the problem, another off-delay relay must be added to maintain a circuit to motor #1 for a period of 3 seconds after motor #2 has turned off. This addition is shown in Figure 35-4.

Motors #2 and #1 will now continue to operate after the stop button is pressed, but so will motor #3. In the present design, none of the motors will turn off when the stop button is pressed. To understand this condition, trace the logic step-by-step. When the start button is pressed, coil 1CR energizes and closes all 1CR contacts. When contact $1CR_2$ closes, coils 1M and TR_1 energize. After a period of 3 seconds, timed contact TR_1 closes and energizes coils 2M, TR_2, and TR_4. Timed contact TR_4 closes immediately to bypass contact $1CR_2$. After a delay of 3 seconds, timed contact TR_2 closes and energizes coils 3M and TR_3.

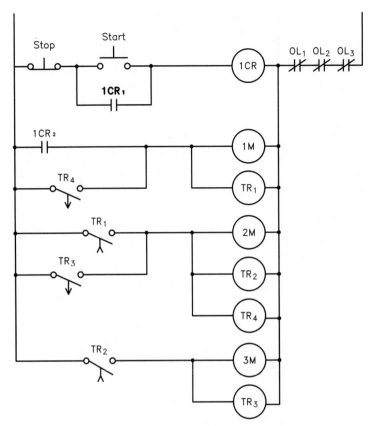

Figure 35-3 Timer TR$_3$ prevents motor 2 from stopping.

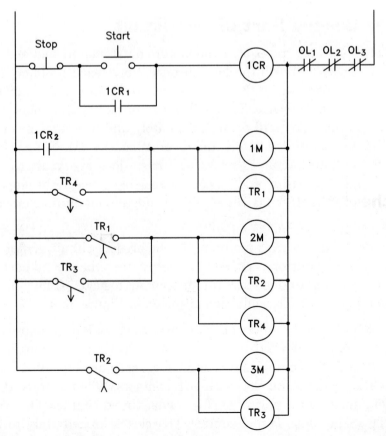

Figure 35-4 Off-delay timer TR$_4$ prevents motor 1 from stopping.

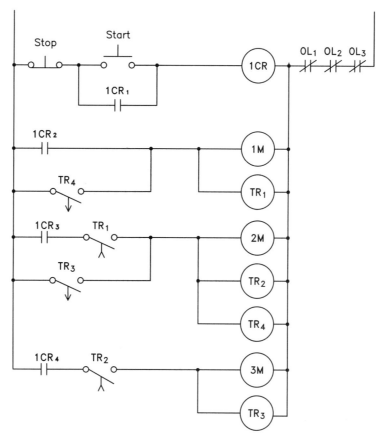

Figure 35-5 Control relay contacts are added to permit the circuit to turn off.

Timed contact TR_3 closes immediately and bypasses contact TR_1. When the stop button is pressed, coil 1CR de-energizes and all 1CR contacts open, but a circuit is maintained to coils 1M and TR_1 by contact TR_4. This prevents timed contact TR_1 from opening to de-energize coils 2M, TR_2, and TR_4, which in turn prevents timed contact TR_2 from opening to de-energize coils 3M and TR_3. To overcome this problem, two more contacts controlled by relay 1CR will be added to the circuit (Figure 35-5). The circuit will now operate in accord with all the stated requirements.

Modifying the Circuit

The circuit in Figure 35-5 was designed with the assumption that all the timers are of the pneumatic type. When this circuit is connected in the laboratory, 8-pin control relays and electronic timers will be used. The circuit will be amended to accommodate these components. The first change to be made concerns the control relays. Notice that the circuit requires the use of four normally open contacts controlled by coil 1CR. Since 8-pin control relays have only two normally open contacts, it will be necessary to add a second control relay, 2CR. The coil of relay 2CR will be connected in parallel with 1CR, which will permit both to operate at the same time (Figure 35-6).

Timers TR_1 and TR_2 are on-delay timers and do not require an adjustment in the circuit logic to operate. Timers TR_3 and TR_4, however, are off-delay timers and do require changing the circuit. The coils must be connected to power at all times. Assuming the use of a Dayton timer model 6A855, power would connect to pins 2 and 10. Starter 3M will be used

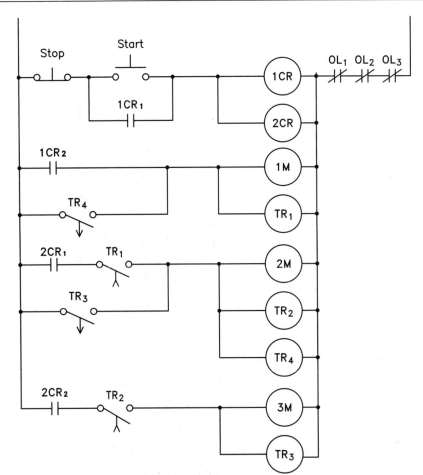

Figure 35-6 Adding a control relay to the circuit.

to control the action of timer TR_3 by connecting a 3M normally open auxiliary contact to pins 5 and 6 of timer TR_3 (Figure 35-7). Starter 2M will control the action of timer TR_4 by connecting a 2M normally open auxiliary contact to pins 5 and 6 of that timer. The circuit is now complete and ready for connection in the laboratory.

Connecting the Circuit

1. Using the circuit shown in Figure 35-7, place pin numbers beside the proper components. Circle the pin numbers to distinguish them from wire numbers.
2. Place wire numbers on the schematic.
3. Connect the control circuit in the laboratory.
4. Turn on the power and test the circuit for proper operation.
5. **Turn off the power** and connect the motor loads to starters 1M, 2M, and 3M.
6. Turn on the power and test the complete circuit.
7. **Turn off the power.**
8. Disconnect the circuit and return the components to their proper places.

Review Questions

Refer to the circuit in Figure 35-7 to answer the following questions. It is assumed that all the timers are set for a delay of 3 seconds.

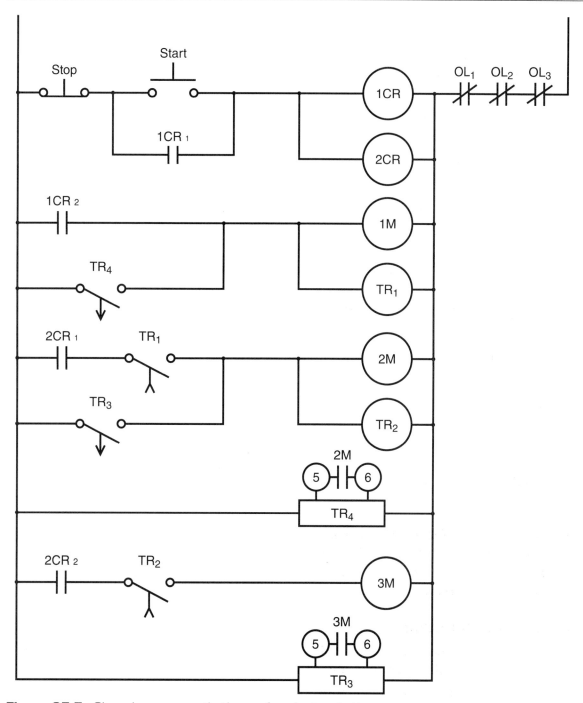

Figure 35-7 Changing pneumatic timers for electronic timers.

1. When the start button is pressed, motor #1 starts operating immediately. Three seconds later motor #2 starts, but motor #3 never starts. When the stop button is pressed, motor #2 stops operating immediately. After a delay of 3 seconds, motor #1 stops running. Which of the following could not cause this condition?

   a. TR_3 coil is open.

   b. 3M coil is open.

   c. TR_2 coil is open.

   d. 2CR coil is open.

2. When the start button is pressed, motor #1 starts operating immediately. Motor #2 does not start operating after 3 seconds, but after a delay of 6 seconds motor #3 starts operating. When the stop button is pushed, motors #3 and #1 stop operating immediately. Which of the following could cause this condition?

   a. 2CR coil is open.

   b. TR_1 coil is open.

   c. TR_3 coil is open.

   d. 2M coil is open.

3. When the start button is pressed, all three motors start normally with a 3-second delay between the starting of each motor. When the stop button is pressed, motor #3 stops operating immediately. After a delay of 3 seconds, both motors #2 and #1 stop operating at the same time. Which of the following could cause this problem?

   a. Timer TR_1 is defective.

   b. Timer TR_2 is defective.

   c. Timer TR_3 is defective.

   d. Timer TR_4 is defective.

4. When the start button is pressed, nothing happens. None of the motors start. Which of the following could *not* cause this problem?

   a. Overload contact OL_1 is open.

   b. 1CR relay coil is open.

   c. 2CR relay coil is open.

   d. The stop button is open.

5. When the start button is pressed, motor #1 does not start, but after a delay of 3 seconds motor #2 starts, and 3 seconds later motor #3 starts. When the stop button is pressed, motor #3 stops running immediately and after a delay of 3 seconds motor #2 stops running. Which of the following could cause this problem?

   a. Starter coil 1M is open.

   b. TR_1 timer coil is open.

   c. Timer TR_4 is defective.

   d. 1CR coil is open.

Unit 36 Hydraulic Press Control Circuit

Objectives

After studying this unit, you should be able to:

- Discuss the operation of this hydraulic press control circuit.
- Connect the circuit in the laboratory.
- Operate the circuit using toggle switches to simulate limit and pressure switches.

LABORATORY EXERCISE

Name _____ Date _____

Materials Required

Three-phase power supply

Control transformer

Three-phase motor starter with at least two normally open auxiliary contacts

5 double-acting push buttons (N.O./N.C. on each button)

Pilot light

3 toggle switches that can be used to simulate two limit switches and one pressure switch

1 three-phase motor or equivalent motor load

2 solenoid coils or lamps to simulate solenoid coils

3 control relays with three sets of contacts (11-pin) and 11-pin sockets

3 control relays with two sets of contacts (8-pin) and 8-pin sockets

The next circuit to be discussed is a control for a large hydraulic press (Figure 36-1). In this circuit, a hydraulic pump must be started before the press can operate. Pressure switch PS closes when there is sufficient hydraulic pressure to operate the press. If switch PS should open, it will stop the operation of the circuit. A green pilot light is used to tell the operator that there is enough pressure to operate the press.

Two run push buttons are located far enough apart so that both of the operator's hands must be used to cause the press to cycle. This is to prevent the operator from getting his hands in the press when it is operating. Limit switches UPLS and DNLS are used to determine when the press is at the bottom of its downstroke and when it is at the top of its upstroke. In the event one or both of the run push buttons are released during the cycle, a reset button can be used to reset the press to its top position. The up solenoid causes the press to travel upward when it is energized, and the down solenoid causes the press to travel downward when it is energized.

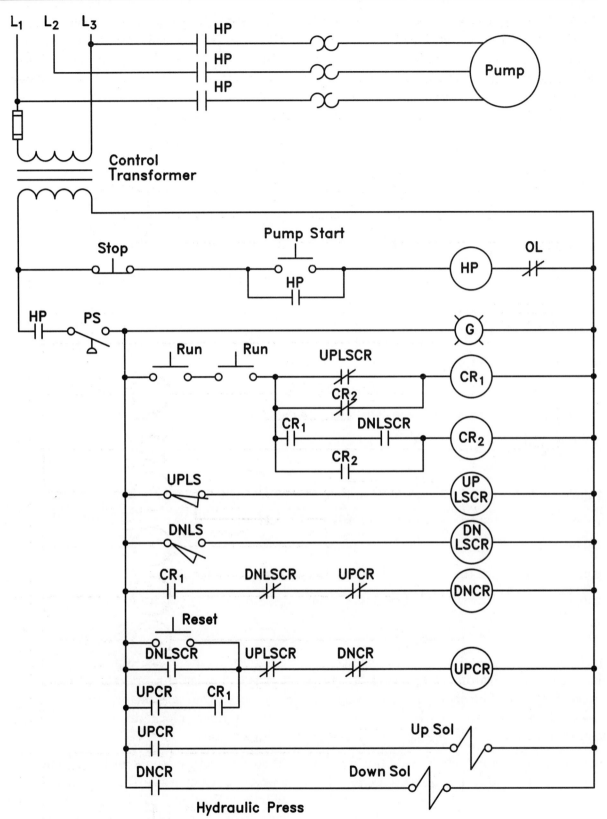

Figure 36-1 Hydraulic press control circuit.

To understand the operation of this circuit, assume that the press is in the up position. Notice that limit switch UPLS is shown normally open held closed. This limit switch is connected normally open, but when the press is in the up position it is being held closed. Now assume that the hydraulic pump is started and that the pressure switch closes. When pressure switch PS closes, the green pilot light turns on and UPLSCR (Up Limit Switch Control Relay) energizes, changing all UPLSCR contacts (Figure 36-2).

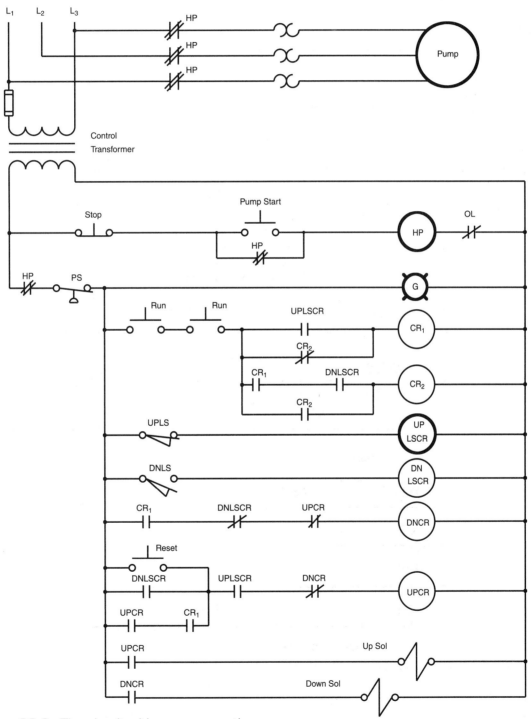

Figure 36-2 The circuit with pump operating.

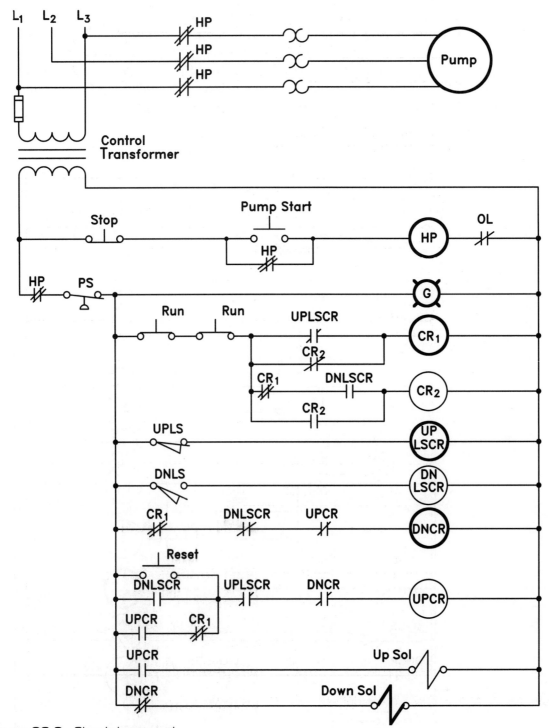

Figure 36-3 Circuit is started.

When both run push buttons are held down, a circuit is completed to CR$_1$ relay, causing all CR$_1$ contacts to change position (Figure 36-3). The CR$_1$ contact connected in series with the coil of DNCR closes and energizes the relay, causing all DNCR contacts to change position. The DNCR contact connected in series with the down solenoid coil closes and energizes the down solenoid.

As the press begins to move downward, limit switch UPLS opens and de-energizes coil UPLSCR, returning all UPLSCR contacts to their normal position (Figure 36-4).

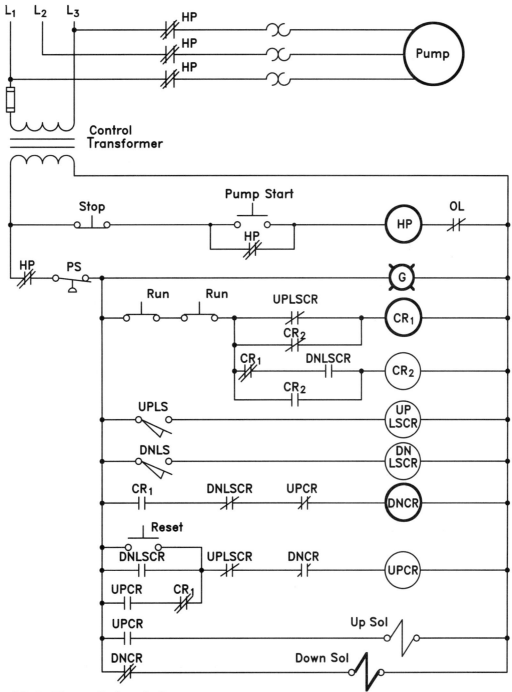

Figure 36-4 The up limit switch opens.

When the press reaches the bottom of its stroke, it closes down limit switch DNLS. This energizes the coil of the down limit switch control relay, DNLSCR, causing all DNLSCR contacts to change position (Figure 36-5). The normally open DNLSCR contact connected in series with the coil of CR$_2$ closes and energizes that relay, causing all CR$_2$ contacts to change position. The normally closed DNLSCR contact connected in series with DNCR coil opens and de-energizes that relay. All DNCR contacts return to their normal positions. The normally open contact connected in series with the down solenoid coil opens and de-energizes the solenoid. The normally closed DNCR contact connected in series with UPCR coil recloses and provides a current path to that relay.

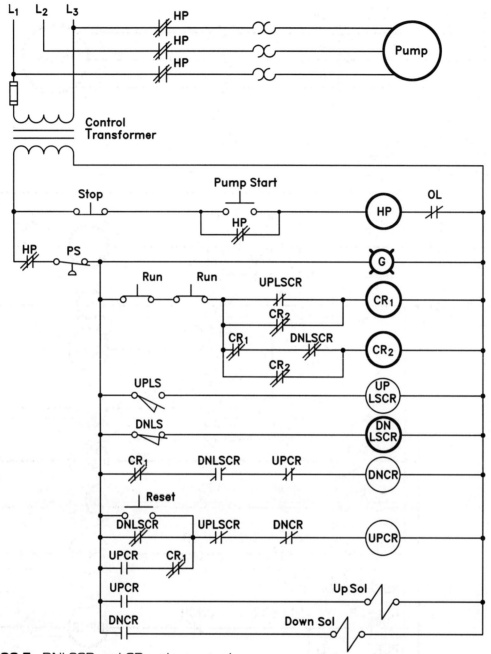

Figure 36-5 DNLSCR and CR$_2$ relays energize.

The UPCR contact connected in series with coil DNCR opens and prevents coil DNCR from re-energizing when coil DNLSCR de-energizes. The normally open UPCR contact connected in series with the up solenoid closes and provides a current path to the up solenoid. When the press starts upward, limit switch DNLS reopens and de-energizes coil DNLSCR. A circuit is maintained to UPCR coil by the now closed UPCR contact connected in series with the CR$_1$ contact (Figure 36-6).

The press will continue to travel upward until it reaches its upper limit and closes limit switch UPLS, energizing coil UPLSCR (Figure 36-7). This causes both UPLSCR contacts to change position. The UPLSCR contact connected in series with coil UPCR opens and de-energizes the up solenoid. Notice that control relays CR$_1$ and CR$_2$ are still energized.

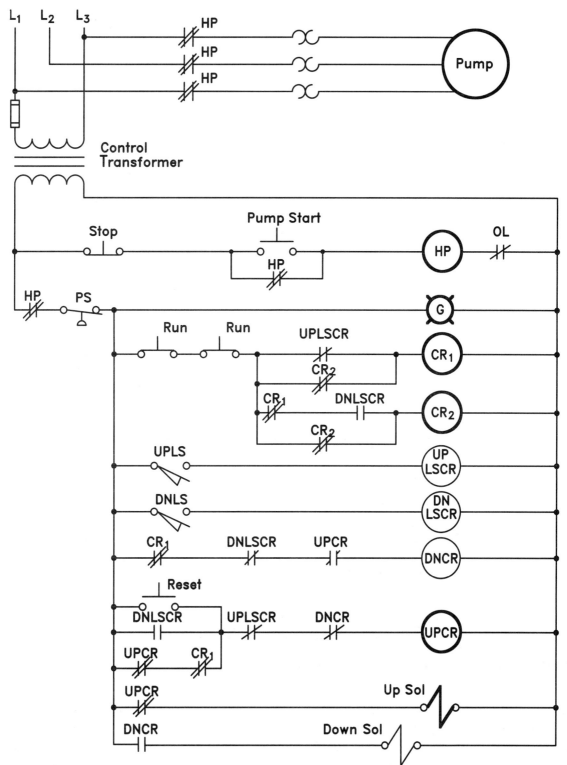

Figure 36-6 Limit switch DNLS reopens.

Before the press can be re-cycled, one or both of the run buttons must be released to break the circuit to the control relays. This will permit the circuit to reset to the state shown in Figure 36-2. If for some reason the press should be stopped during a cycle, the reset button can be used to return the press to the starting position.

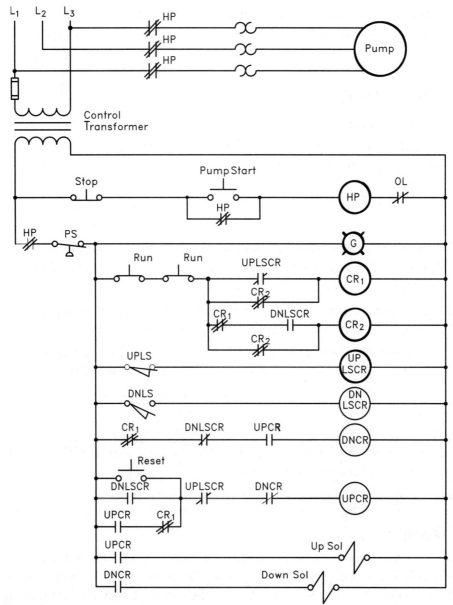

Figure 36-7 The press completes the cycle.

Connecting the Circuit

In this exercise toggle switches will be used to simulate the action of the pressure switch and the two limit switches. Lights may also be substituted for the up and down solenoid coils.

1. Refer to the circuit shown in Figure 36-1. Count the number of contacts controlled by each of the control relays to determine which should be 11-pin and which should be 8-pin. Relays that need three contacts will have to be 11-pin, and relays that need two contacts may be 8-pin.

2. After determining whether a relay is to be 11-pin or 8-pin, identify the relay with some type of marker that can be removed later. Identifying the relays as CR_1, CR_2, and so on can make connection much simpler.

3. Place the pin numbers on the schematic in Figure 36-1 to correspond with the contacts and coils of the control relays. Circle the numbers to distinguish them from wire numbers.

4. Place wire numbers beside each component on the schematic.

5. Connect the circuit. (*Note:* When connecting the two run push buttons, connect them close enough together to permit both to be held closed with one hand.)

Testing the Circuit

To test the circuit for proper operation:

1. Set the toggle switches used to simulate the pressure and down limit switch in the open (off) position. Set the toggle used to simulate the up limit switch in the closed (on) position.

2. Press the "pump start" button and the motor or simulated motor load should start operating.

3. Close the pressure switch. The pilot light and UPLSCR relay should energize.

4. Press and hold down both of the run push buttons. Relays CR_1 and DNCR should energize. The down solenoid should also turn on.

5. The press is now traveling in the down direction. Open the up limit switch. This should cause UPLSCR to de-energize. The down solenoid should remain turned on.

6. Close the down limit switch to simulate the press reaching the bottom of its stroke. DNLSCR, CR_2, and UPCR should energize. The press is now starting to travel upward.

7. Open the down limit switch. DNLSCR should de-energize, but the UPCR should remain energized.

8. Close the up limit switch to simulate the press reaching the top of its stroke. The up solenoid should turn off. Control relays CR_1 and CR_2 should both remain on as long as the two run buttons are held closed.

9. To restart the cycle, release the run buttons and reclose them.

Review Questions

1. Assume that the hydraulic pump is running and the pilot light is turned on indicating that there is sufficient pressure to operate the press. Now assume that the up limit switch is not closed. What will be action of the circuit if both run buttons are pressed?

_____

_____

_____

2. Assume that the press is in the middle of its downstroke when the operator releases the two run push buttons. Explain the action of the circuit.

_____

_____

_____

3. Referring to the condition of the circuit as stated in question 2, what would happen if the two run push buttons are pressed and held closed? Explain your answer.

_____

_____

_____

4. Referring to the condition of the circuit as stated in question 2, what would happen if the reset button is pressed and held closed? Explain your answer.

_____

_____

_____

5. Assume that the press traveled to the bottom of its stroke and then started back up. When it reached the middle of its stroke, the power was interrupted. After the power has been restored, if the two run buttons are pressed, will the press continue to travel upward to complete its stroke, or will it start moving downward?

_____

_____

_____

Unit 37 Design of Two Flashing Lights

Objectives

After studying this unit, you should be able to:

- Design a circuit from a written statement of requirements.
- Connect the circuit in the laboratory after the design has been approved.

LABORATORY EXERCISE

Name _____ Date _____

Materials Required

Materials depend on the circuit design

In the space provided in Figure 37-1, draw a schematic diagram of a circuit that will fulfill the following requirements. Use two separate timers. Do not use an electronic timer set in the repeat mode. Remember that there is generally more than one way to design any circuit. Try to keep the design as simple as possible. The fewer components a circuit has, the less it is likely to fail.

1. An on-off toggle switch is used to connect power to the circuit.
2. When the switch is turned on, two lights will alternately flash on and off. Light #1 will be turned on when light #2 is turned off. When light #1 turns off, light #2 will turn on.
3. The lights are to flash at a rate of on for 1 second and off for 1 second.

When completed, have your instructor approve the design. After the design has been approved, connect it in the laboratory.

Review Questions

1. When designing a control circuit that requires the use of a timing relay, what type of timer is generally used during the design?

   _____

   _____

   _____

2. Should schematic diagrams be drawn to assume that the circuit is energized or de-energized?

   _____

   _____

   _____

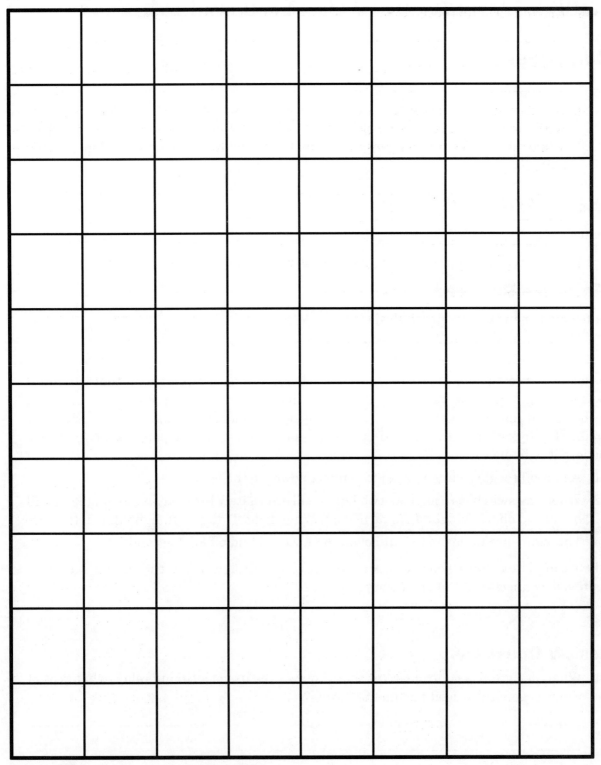

Figure 37-1 Design of two flashing lights.

3. Explain the difference between a schematic and a wiring diagram.

   _____

   _____

   _____

4. In a forward-reverse control circuit, a normally closed F contact is connected in series with the R starter coil, and a normally closed R contact is connected in series with the F starter coil. What is the purpose of doing this and what is this contact arrangement called?

   _____

   _____

   _____

5. What type of overload relay is not sensitive to changes in ambient temperature?

   _____

   _____

   _____

Unit 38 Design of Three Flashing Lights

Objectives

After studying this unit, you should be able to:

- Design a motor control circuit using timers.
- Discuss the operation of this circuit.
- Connect this circuit in the laboratory.

LABORATORY EXERCISE

Name _____ Date _____

Materials Required

Materials depend on the design of the circuit

The design of this circuit will be somewhat similar to the circuit in Unit 37. This circuit, however, contains three lights that turn on and off in sequence. Use the space provided in Figure 38-1 to design this circuit. The requirements of the circuit are as follows:

1. A toggle switch is used to connect power to the circuit. When the power is turned on, light #1 will turn on.
2. After a delay of 1 second, light #1 will turn off and light #2 will turn on.
3. After a delay of 1 second, light #2 will turn off and light #3 will turn on.
4. After a delay of 1 second, light #3 will turn off and light #1 will turn back on.
5. The lights will repeat this action until the toggle switch is opened.

Procedure

1. After the design of your circuit has been approved by your instructor, connect the circuit in the laboratory.
2. Test the circuit for proper operation.
3. Disconnect the circuit and return the components to their proper location.

Review Questions

1. A 60 hp, three-phase squirrel cage induction motor is to be connected to a 480-volt line. What size NEMA starter should be used to make this connection?

   _____

   _____

   _____

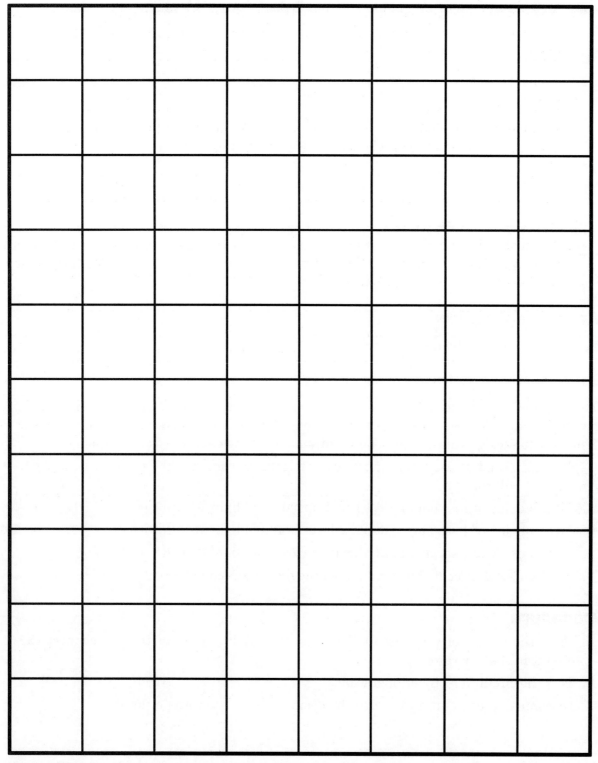

Figure 38-1 Design of three lights that turn on and off in sequence.

2. An electrician is given a NEMA size 2 starter to connect a 30 hp, three-phase squirrel cage motor to a 575 volt line. Should this starter be used to operate this motor?

_____

_____

_____

3. Assume that the motor in question 2 has a design code B. What standard size inverse time circuit breaker should be used to connect the motor?

_____

_____

_____

4. The motor described in questions 2 and 3 is to be connected with copper conductors with type THHN insulation. What size conductors should be used? The termination temperature rating is not known.

_____

_____

_____

5. Assume that the motor in question 2 has a nameplate current rating of 28 amperes and a marked service factor of 1. What size overload heater should be used for this motor?

_____

_____

_____

Unit 39 Control for Three Pumps

Objectives

After studying this unit, you should be able to:

- Analyze a motor control circuit.
- List the steps of operation in a control circuit.
- Connect this circuit in the laboratory.

LABORATORY EXERCISE

Name _____ Date _____

Materials Required

Three-phase power supply

Control transformer

3 motor starters with normally open auxiliary contacts

6 toggle switches to simulate auto-man switches and float switches

8-pin control relay and 8-pin socket

3 three-phase motors or equivalent motor loads

1 normally open and 1 normally closed push button

One of the primary duties of an industrial electrician is to troubleshoot existing control circuits. To troubleshoot a circuit, the electrician must understand what the circuit is designed to do and how it accomplishes it. To analyze a control circuit, start by listing the major components. Next, determine the basic function of each component. Finally, determine what occurs during the circuit operation.

To illustrate this procedure, the circuit previously discussed in Unit 36 will be analyzed. The hydraulic press circuit is shown in Figure 39-1. In order to facilitate circuit analysis, wire numbers have been placed beside the components. The first step will be to list the major components in the circuit.

1. Normally closed stop push button
2. Normally open push button used to start the hydraulic pump
3. Two normally open push buttons used as run buttons
4. Normally open push button used for the reset button
5. Normally open pressure switch
6. Two normally open limit switches

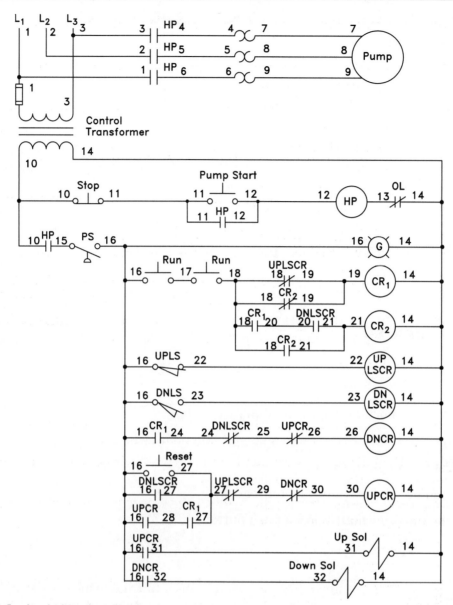

Figure 39-1 Analyzing the circuit.

7. Two solenoid valves

8. Three 8-pin control relays (CR₂, UPLSCR, and DNCR)

9. Three 11-pin control relays (CR₁, DNLSCR, and UPCR)

10. Control transformer

11. Green pilot light

The next step in the process is to give a brief description of the function of each listed component:

1. (Normally closed stop push button)—Used to stop the operation of the hydraulic pump motor.

2. (Normally open push button used to start the hydraulic pump)—Starts the hydraulic pump.

3. (Two normally open push buttons used as run buttons)—Both push buttons must be held down to start the action of the press.

4. (Normally open push button used for the reset button)—Resets the press to the topmost position.

5. (Normally open pressure switch)—Determines whether or not there is enough hydraulic pressure to operate the press.

6. (Two normally open limit switches)—Determine when the press is at the top of its stroke and when it is at the bottom of its stroke.

7. (Two solenoid valves)—The up solenoid valve opens on energize to permit hydraulic fluid to move the press upward. The down solenoid valve opens on energize to permit hydraulic fluid to move the press downward.

8. (Three 8-pin control relays [CR_2, UPLSCR, and DNCR])—Part of the control circuit.

9. (Three 11-pin control relays [CR_1, DNLSCR, and UPCR])—Part of the control circuit.

10. (Control transformer)—Reduces the value of the line voltage to the voltage needed to operate the control circuit.

11. (Green pilot light)—Indicates there is enough hydraulic pressure to operate the pump.

The final step is to analyze the operation of the circuit. To analyze circuit operation, trace the current paths each time a change is made in the circuit. Start by pressing the pump start button.

1. When the pump start button is pressed, a circuit is completed to the coil of starter HP.

2. When coil HP energizes, all HP contacts change position. The three load contacts close to connect the pump motor to the line. The HP auxiliary contact located between wire points 11 and 12 closes to maintain the circuit after the pump start button is released, and the HP auxiliary contact located between wire numbers 10 and 15 closes to provide power to the rest of the circuit.

3. After the hydraulic pump starts, the hydraulic pressure in the system increases and closes the pressure switch.

4. When the pressure switch closes, a current path is provided to the green pilot light to indicate that there is sufficient hydraulic pressure to operate the press. A current path also exists through the normally open held closed up limit switch to control relay (UPLSCR) coil.

5. When UPLSCR relay energizes, both UPLSCR contacts open. The UPLSCR contact located between wire numbers 18 and 19 opens to break a current path to CR_1 coil. UPLSCR contact located between wire numbers 27 and 29 opens to break the current path to coil UPCR.

6. Both run push buttons must be held down to provide a current path through the normally closed CR_2 contact located between wire numbers 18 and 19 to the coil of CR_1 relay.

7. When CR_1 relay coil energizes, the CR_1 contact located between wire numbers 18 and 20 closes to provide a path to CR_2 coil in the event that the DNLSCR contact should close. The CR_1 contact located between wire numbers 16 and 24 closes to provide a current path to the down control relay (DNCR). The CR_1 contact located between wire numbers 28 and 27 closes to provide an eventual current path to the up control relay (UPCR).

8. When DNCR coil energizes, the DNCR contact located between wire numbers 29 and 30 opens to provide interlock with the up control relay. The DNCR contact between wire numbers 16 and 32 closes and provides a current path to the down solenoid valve.

9. When the down solenoid valve energizes, the press begins its downward stroke. This causes the normally open held closed up limit switch to open and de-energize the UPLSCR coil.

10. Both UPLSCR contacts reclose.

11. When the press reaches the bottom of its stroke, the down limit switch located between wire numbers 16 and 23 closes to provide a current path to the coil of the down limit switch control relay (DNLSCR).

12. All DNLSCR contacts change position. The DNLSCR contact located between wire numbers 20 and 21 closes to provide a current path through the now closed CR_1 contact to the coil of CR_2 relay. The DNLSCR contact located between wire numbers 24 and 25 opens and breaks the current path to DNCR relay. The DNLSCR contact located between wire numbers 16 and 27 closes to provide a current path to UPCR relay when the DNCR contact located between 29 and 30 recloses.

13. When CR_2 coil energizes, the normally closed CR_2 contact located between wires 18 and 19 opens to prevent a maintained current path to CR_1 when the UPLSCR contact reopens. The normally open CR_2 contact located between 18 and 21 closes to maintain a current path to the coil of CR_2 in the event that CR_1 or DNLSCR contacts should open.

14. When the DNCR relay coil de-energizes, the DNCR contact located between wires 29 and 30 recloses to permit coil UPCR to be energized. The DNCR contact located between 16 and 32 reopens to break the current path to the down solenoid valve.

15. When the UPCR coil energizes, the normally closed UPCR contact located between wires 25 and 26 opens to provide interlock with the DNCR relay coil. The UPCR contact located between 16 and 28 closes to maintain a circuit through the now closed CR_1 contact to the coil of UPCR. The UPCR contact located between 16 and 31 closes and provides a current path to the up solenoid valve.

16. When the up solenoid valve opens, hydraulic fluid causes the press to begin its upward stroke.

17. When the press starts upward, the down limit switch reopens and de-energizes the coil of DNLSCR relay.

18. When coil DNLSCR de-energizes, the DNLSCR contact located between wires 20 and 21 reopens, but a current path is maintained by the now closed CR_2 contact. The DNLSCR contact located between 24 and 25 recloses, but the current path to DNCR coil remains broken by the UPCR contact located between 25 and 26. The DNLSCR contact located between wires 16 and 27 reopens, but a current path is maintained by the now closed UPCR and CR_1 contacts.

19. When the press reaches the top of its stroke, the up limit switch again closes and provides a current path to the coil of UPLSCR relay.

20. The UPLSCR contact located between wires 18 and 19 opens to break the current path to CR_1 coil. The UPLSCR contact located between wires 27 and 29 opens to break the current path to the coil of UPCR.

21. When CR_1 coil de-energizes, all CR_1 contacts return to their normal position. The CR_1 contact between wires 18 and 20 reopens, CR_1 contact between wires 16 and 24 reopens to prevent a current path from being established to the DNCR relay coil, and CR_1 contact between wires 27 and 28 reopens.

22. When coil UPCR de-energizes, its contacts return to their normal position. The UPCR contacts located between wires 16 and 28 reopen, and the UPCR contact located between wires 16 and 31 reopens to break the circuit to the up solenoid.

23. Before the circuit can be restarted, the current path to relay CR_2 must be broken by releasing one or both of the run push buttons. This will return all contacts back to their original state.

24. In the event the press should be stopped in the middle of its stroke, the up limit switch will be open and coil UPLSCR will be de-energized. The DNCR coil will also be de-energized. If the reset button is pressed and held, a circuit will be completed through the normally closed DNLSCR and DNCR contacts to the coil of UPCR. This will cause the up solenoid valve to energize and return the press to its up position.

Determining What the Circuit Does

The circuit in this experiment is intended to operate three pumps. The pumps are used to pump water from a sump to a roof storage tank. The water in the storage tank is used for cooling throughout the plant. After the water has been used for cooling, it returns to the sump to be recooled. Three float switches are used to detect the water level in the storage tank. As the water is drained out of the tank, the level drops and the float switches turn on the pumps to pump water from the sump back to the storage tank (Figure 39-2).

List the Components

In the space provided, list the major components in the control circuit shown in Figure 39-3.

1. _____

2. _____

3. _____

4. _____

5. _____

6. _____

7. _____

8. _____

9. _____

10. _____

Describe the Components

In the space provided, give a brief description of the function of the components in this circuit.

1. _____

_____

_____

2. _____

_____

_____

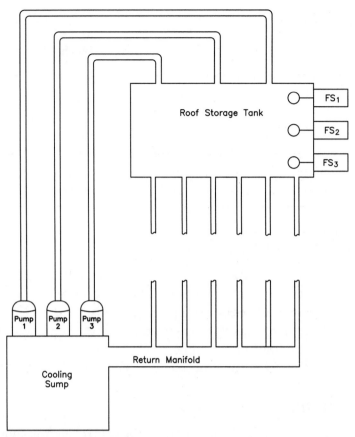

Figure 39-2   Roof-mounted tank for plant cooling system.

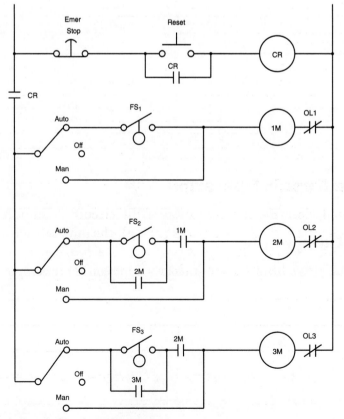

Figure 39-3   Control circuit for three pumps.

3. _____

_____

_____

4. _____

_____

_____

5. _____

_____

_____

6. _____

_____

_____

7. _____

_____

_____

8. _____

_____

_____

9. _____

_____

_____

10. _____

_____

_____

Describing the Circuit Operation

In the space provided, describe the operation of the circuit. Assume that in the normal state the roof storage tank is filled with water, and all the auto-off-man switches are set in the auto position. Also, assume that the three motor starters control the operation of the three pumps, although the pumps are not shown on the schematic.

1. _____

_____

_____

2. _____

_____

3. _____

_____

_____

4. _____

_____

_____

5. _____

_____

_____

6. _____

_____

_____

7. _____

_____

_____

8. _____

_____

_____

9. _____

_____

_____

10. _____

_____

_____

11. _____

_____

_____

12. _____

_____

_____

13. _____

_____

_____

14. _____

_____

_____

15. _____

_____

_____

Review Questions

To answer the following questions, refer to the circuit shown in Figure 39-3.

1. Assume that all three pumps are operating. What would be the action of the circuit if the auto-off-man switch of pump #2 were to be switched to the off position?

_____

_____

_____

2. Assume that the auto-off-man switch of pump #3 is set in the manual position. What will be the operation of the circuit if float switch FS_1 closes?

_____

_____

_____

3. Assume that the roof storage tank empties completely, but none of the pumps have started. Which of the following could not cause this condition?

a. The emergency stop button has been pushed and the control relay is de-energized.

b. The auto-off-man switch of pump #1 has been set in the off position.

c. The auto-off-man switch of pump #1 has been set in the manual position.

d. 1M coil is open.

4. Assume that all three pumps are in operation and OL_3 contact opens. Will this affect the operation of the other two pumps?

_____

_____

_____

5. Assume that FS_2 float switch is defective. If the water level drops enough to close float switch FS_3, will pump #3 start running?

_____

_____

_____

Unit 40 Oil Pressure Pump Circuit for a Compressor

Objectives

After studying this unit, you should be able to:

- Analyze a motor control circuit.
- List the steps of operation in a control circuit.
- Connect this circuit in the laboratory.

LABORATORY EXERCISE

Name _____ Date _____

Materials Required

Three-phase power supply

2 motor starters

Control transformer

2 electronic timers (Dayton model 6A855) and 11-pin tube sockets

2 pilot lights

2 double-acting push buttons

In the circuit shown in Figure 40-1, the oil pump must start for some time before the compressor is started. When the start button is pressed, the oil pump should continue to run for some time after the compressor stops operating.

List the Components

In the space provided, list the circuit components.

1. _____
2. _____
3. _____
4. _____
5. _____
6. _____
7. _____
8. _____
9. _____
10. _____

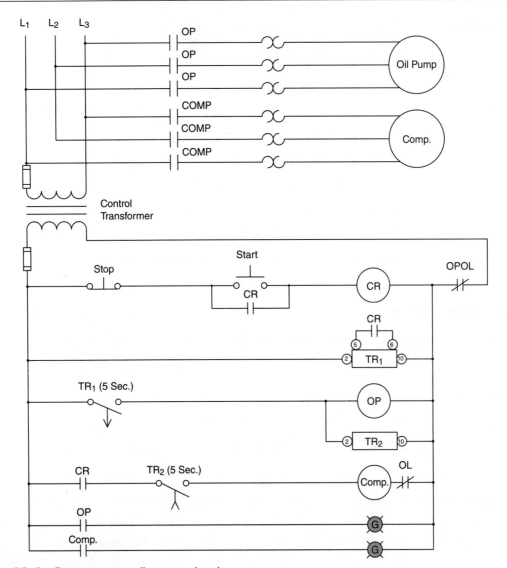

Figure 40-1 Compressor oil pump circuit.

Describe the Components

In the space provided, give a brief description of what function is performed by each component.

1. _____

_____

2. _____

_____

3. _____

_____

4. _____

_____

5. _____

_____

6. _____

_____

7. _____

_____

8. _____

_____

9. _____

_____

10. _____

_____

Circuit Operation

In the space provided, describe the operation of the circuit in a step-by-step sequence.

1. _____

_____

2. _____

_____

3. _____

_____

4. _____

_____

5. _____

_____

6. _____

_____

7. _____

_____

8. _____

_____

9. _____

_____

10. _____

_____

11. _____

_____

12. _____

_____

13. _____

_____

14. _____

_____

15. _____

_____

Connecting the Circuit

1. Connect the circuit shown in Figure 40-1.

2. After checking with the instructor, turn on the power and test the circuit for proper operation.

3. **Turn off the power** and disconnect the circuit. Return the components to their proper location.

Review Questions

To answer the following questions, refer to the circuit shown in Figure 40-1.

1. Assume that the start button is pressed and the oil pump starts operating. After a delay of 5 seconds, the COMP pilot light turns on, but the compressor motor does not start. Which of the following could cause this condition?

   a. TR_2 timer is defective.

   b. COMP starter coil is defective.

   c. The compressor motor is defective.

   d. All of the above.

2. Assume that the circuit is in operation. When the stop button is pressed, both the compressor and oil pump stop operating immediately. Which of the following could cause this condition?

   a. CR relay is defective.

   b. TR_1 timer is defective.

   c. OP starter is defective.

   d. Timer TR_2 is defective.

3. When the start button is pressed, the oil pump starts operating immediately. After a delay of 5 seconds, the oil pump motor turns off. An electrician finds that the control transformer fuse is blown. Which of the following could cause this condition?

   a. TR_1 coil is shorted.

   b. OP coil is shorted.

   c. TR_2 coil is shorted.

   d. COMP coil is shorted.

4. When the start button is pressed, the oil pump motor starts operating immediately. After a long time delay, it is determined that the compressor motor will not start. Which of the following could not cause this condition?

   a. OP coil is defective.

   b. TR_2 coil is defective.

   c. COMP coil is defective.

   d. The compressor overload contact is open.

5. When the start button is pressed, the oil pump motor starts operating immediately. When the start button is released, however, the oil pump motor turns off. The operator then presses the start button and holds it down for a period of 10 seconds. This time the oil pump motor starts operating immediately, but the compressor motor never starts. When the start button is released, the oil pump motor again immediately turns off. Which of the following could cause this condition?

   a. CR coil is defective.

   b. TR_1 coil is defective.

   c. TR_2 coil is defective.

   d. COMP coil is defective.

Unit 41 Autotransformer Starter

Objectives

After studying this unit, you should be able to:

- Discuss the operation of an autotransformer starter.
- Explain the operation of an autotransformer starter.
- Connect an autotransformer starter in the laboratory.

Autotransformer starters are used to reduce the amount of inrush current when starting a large motor. The autotransformer starter accomplishes this by reducing the voltage applied to the motor during the starting period. If the voltage is reduced by one-half, the current will be reduced by one-half, and the torque will be reduced to one-fourth of normal.

There are several different ways to construct an autotransformer starter. Some use three transformers, and others use two transformers. In this experiment, two transformers connected as an open delta will be used. Two 0.5 kVA control transformers will be employed. Since these transformers are to be used as autotransformers, only the high-voltage windings will be connected. The low-voltage windings (X_1 and X_2) will not be used in this experiment. The high-voltage windings can be identified by the markings on the terminal leads of H_1 through H_4. These high-voltage windings are to be connected in series by connecting a jumper between terminals H_2 and H_3. This jumpered point provides a center tap for the entire winding.

Obtaining Enough Contacts

A schematic diagram of this connection is shown in Figure 41-1. Notice that there are a total of five starting contactor (SC) load contacts needed during the starting period. Contactors that contain five load contacts can be purchased, but they are difficult to obtain and they are expensive. For this reason, two three-phase contactors will be used to provide the needed load contacts. This can be accomplished by connecting the coil of SC_1 and SC_2 contactors in parallel with each other.

Circuit Operation

When the start button is pressed, coils CR, TR, SC_1, and SC_2 energize. When the SC_1 and SC_2 load contacts close, the motor is connected to the center tap of the open delta autotransformer. Since the transformers have been center tapped, the motor is connected to half of the line voltage. A basic schematic diagram of this connection is shown in Figure 41-2. The normally closed SC_1 and SC_2 auxiliary contacts connected in series with the R coil open to provide interlock and prevent the R contactor from energizing as long as SC_1 or SC_2 is energized.

After some time, TR timer reaches the end of its timing sequence and the two timed TR contacts change position. The normally closed TR contact connected in series with coils SC_1 and SC_2 opens and de-energizes these contactors. This causes all SC_1 and SC_2 load contacts to open and disconnect the autotransformer from the line. The normally closed SC_1 and SC_2 auxiliary contacts connected in series with R coil reclose.

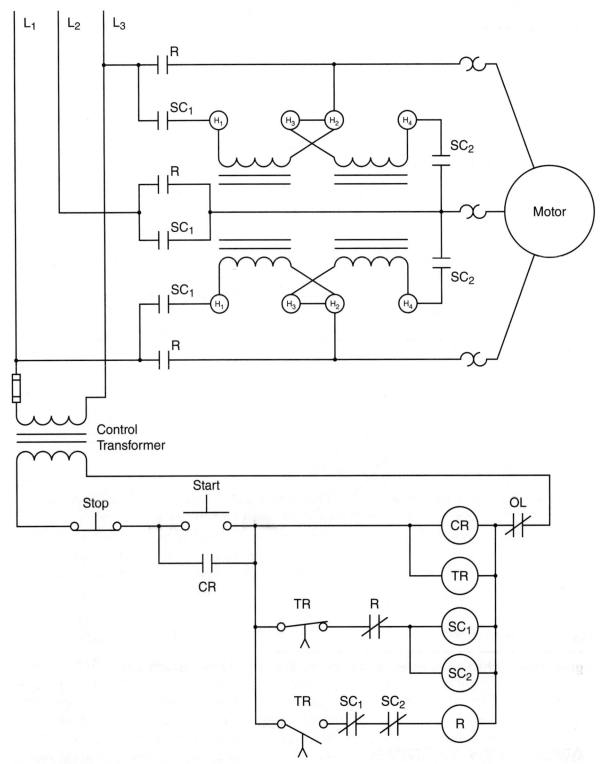

Figure 41-1 Autotransformer starter.

When the normally open TR contact connected in series with R coil closes, the R contactor energizes and closes all R load contacts. This connects the motor directly to the power line. The normally closed R auxiliary contact connected in series with coils SC_1 and SC_2 opens to provide interlock. The motor will continue to run until the stop button is pressed or an overload occurs.

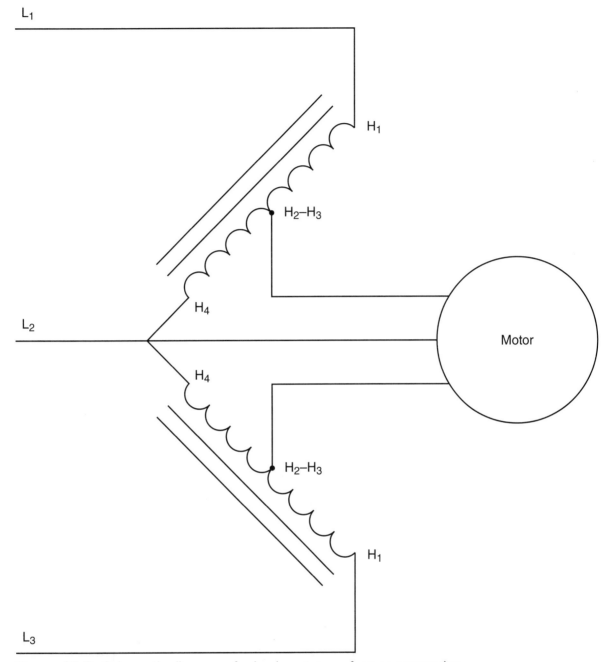

Figure 41-2 Schematic diagram of a basic autotransformer connection.

LABORATORY EXERCISE

Materials Required

Three-phase power supply

Control transformer

3 three-phase contactors with at least one normally open and one normally closed auxiliary contact

2 0.5-kVA control transformers (480/240-120)

Three-phase motor or equivalent motor load

On-delay timer (Dayton model 6A855 or equivalent) and 11-pin tube socket

8-pin control relay and 8-pin tube socket

2 double-acting push buttons (N.O./N.C. on each button)

Three-phase overload relay or three single-phase overload relays with the overload contacts connected in series

(In this circuit it is possible to replace the two SC contactors with a single contactor that contains five load contacts, if one is available. Also, if true contactors are not available, it is permissible to use motor starters for the two SC contactors.)

1. Assuming that relay CR is an 8-pin control relay, and that timer TR is a Dayton model 6A855, place pin numbers beside the components of CR and TR in Figure 41-1. Circle the pin numbers to distinguish them from wire numbers.

2. Place wire numbers beside all circuit components in Figure 41-1.

3. Place corresponding wire numbers beside the components, as shown in Figure 41-3. Make certain to make the connection between H_2 and H_3 on the high-voltage side of the control transformers.

4. Connect the control section of the circuit, as shown in Figure 41-1.

5. Set the timing relay for a delay of 5 seconds.

6. Turn on the power and test the control section of the circuit for proper operation.

7. **Turn off the power.**

8. Connect the load section of the circuit.

9. Turn on the power and test the circuit for proper operation. (*Note*: Connect a voltmeter across the motor or equivalent motor load terminals and monitor the voltage. When the circuit is first energized, the voltage applied to the motor should be one-half the full-line value. After a delay of 5 seconds, the voltage should increase to full value.)

10. **Turn off the power** and disconnect the circuit. Return the components to their proper places.

Review Questions

1. How does the autotransformer reduce the amount of starting current to a motor?

_____

_____

_____

2. Is the autotransformer used in this experiment connected as a wye, delta, or open delta?

_____

_____

_____

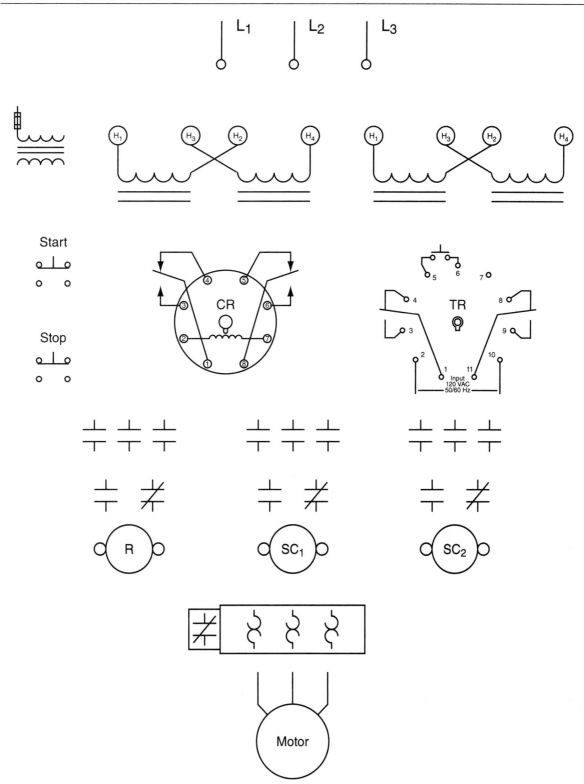

Figure 41-3 Developing a wiring diagram.

3. What is the advantage, if any, of using an open delta connection as opposed to a closed delta or wye?

_____

_____

_____

4. Assume that the line-to-line voltage in Figure 41-1 is 480 volts. Also, assume that when the start button is pressed, the motor starts with 240 volts applied to the motor. When the start button is released, however, the motor stops running. Which of the following could cause this problem?

a. SC_1 coil is open.

b. CR coil is open.

c. TR coil is open.

d. The stop push button is open.

5. Refer to the circuit shown in Figure 41-1. When the start button is pressed, nothing happens for 5 seconds. After 5 seconds, the motor suddenly starts with full voltage connected to it. Which of the following could cause this problem?

a. CR coil is open.

b. TR coil is open.

c. R coil is open.

d. R normally closed auxiliary contact is open.

Index